76-132

Digital Logic
and
Switching Circuits

Digital Logic and Switching Circuits:

Operation and Analysis

JEFFERSON C. BOYCE
Allan Hancock College

PRENTICE-HALL, INC. Englewood Cliffs, New Jersey

Library of Congress Cataloging in Publication

BOYCE, JEFFERSON C.
  Digital logic & switching circuits: Operation
and analysis.

  1. Switching theory. 2. Electronic digital
computers--Circuits. I. Title

        74–16763

    ISBN 0–13–214478–6

© 1975
by PRENTICE-HALL, INC.
Englewood Cliffs, New Jersey

10 9 8 7 6 5 4 3 2 1

Printed in the United States of America

PRENTICE-HALL INTERNATIONAL, INC., London
PRENTICE-HALL OF AUSTRALIA, PTY., LTD., Sydney
PRENTICE-HALL OF CANADA, LTD., Toronto
PRENTICE-HALL OF INDIA PRIVATE LIMITED, New Delhi
PRENTICE-HALL OF JAPAN, INC., Tokyo

To
Jay, Dan, Margi, and Kathi (who gave up her room);
they made this effort necessary:
but especially to Betty,
whose love and understanding
made it possible.

Preface

The digital age has arrived. Techniques that were confined to the highly complex digital computer just a few short years ago are now applied in common home entertainment and commercial devices. Until recently, engineering and technical personnel have been forced to gain their knowledge of digital techniques either from highly technical digital computer orientated texts, or from "trial and error" on the job. This book bridges the gap by exploring the *general* field of digital techniques. It prepares the reader to recognize, understand, analyze, and troubleshoot digital logic equipments *without* the need to relate to a more complex type of digital device such as a computer. The digital computer has been purposely de-emphasized to illustrate the non-computer applications of digital techniques.

The title of this book, DIGITAL LOGIC AND SWITCHING CIRCUITS: OPERATION AND ANALYSIS, appropriately describes its contents. As will be observed as this book is used, the terms *operation* and *analysis* are closely allied. Following development of the basic concepts of digital logic in the first few chapters, it becomes apparent that separation of the two terms is almost impossible. No interconnection of the hardware elements of digital logic can be made that does not require *analysis* to determine the *operation* of the resulting circuit. *Analysis in this book is used to describe application of the relatively few rules of digital logic to circuits to determine the operation of the circuits.* The circuits may be as simple as two or three interconnected logic elements, or as complex as a digital computer. The same techniques apply.

As a result of reading and applying the principles and exercises provided in this book, the reader should be able to follow logic diagram data flow from input to output. He should be able to use simple Boolean Algebra, truth tables, logic maps, and waveforms to analyze and understand the operation of digital logic equipment. Since the approach to logic tends to be hardware oriented rather than theoretical, minimal electronic and mathematical background is required.

Chapters 1 and 2 are introductory in nature, discussing the basic concepts of digital logic, numbers, and counting. The fundamentals of combinational logic (gates) are presented in Chapters 3 and 4, while Chapters 5 and 6 discuss algebraic and graphical simplification of logic expressions. Combinational logic is summarized in Chapter 7, where procedures for analysis and troubleshooting of this type of logic circuit are developed by introducing the use of troubleshooting flow charts.

The knowledge gained during study of combinational logic is applied in Chapters 8 and 9, where sequential (time-dependent) logic is investigated. Chapter 10 integrates combinational and sequential logic by developing analysis and troubleshooting procedures (once again using flow charts) applicable to circuits containing both types of logic.

If any division of this book is to take place, it should be after Chapter 10. All previous chapters are somewhat theoretical in nature, although approached from a practical viewpoint. The remaining chapters discuss specific types of logic circuits. For example, Chapter 11 is devoted entirely to the subject of counters. Implementation of the counting function is discussed from both the individual logic element packaging method to the integrated circuit (IC) packages currently in use.

Registers are discussed in the same manner in Chapter 12, while miscellaneous logic functions such as adders, comparators, multiplexers, etc. are covered in Chapter 13. Chapters 14 and 15 are somewhat related, since they present the number codes used in digital equipment, the converter devices used with such codes, and the devices used to display information.

Chapter 16 summarizes the complete book by showing the analysis of a simple digital subsystem. An overview of the complete digital field is presented in Chapter 17 by discussing a number of common applications at the block diagram level.

It is impossible to list all of the individuals whose influence and suggestions have resulted in the organization of, and the material in, this book. Mr. Edward Francis of Prentice-Hall and Professor Kosow of Staten Island Community College deserve recognition for their untiring efforts during the evolution of this book. My colleagues in The Boeing Company and at Allan Hancock College, whose encouragement and advice has been invaluable, must receive much of the credit for this effort. The many students who suffered through early versions of portions of this book also should not be overlooked. And, most of all, the patience of an understanding family, all of whom gave up so much so that this book could be developed, must be acknowledged.

JEFFERSON BOYCE

Contents

Digital Logic
and
Switching Circuits

1

An Introduction to Digital Logic

Revolution is the mainstay of mankind's survival — not violent revolution, but a steadily progressive change in man's way of life. We shall devote this chapter to a discussion of such basic revolutions as the transportation revolution, the revolution in energy production, and, most importantly, the revolution in machines designed to aid mans' intellectual pursuits.

1–1   AN HISTORICAL BACKGROUND

The Transportation Revolution

Man has progressed rapidly in this century. In the early 1900s, one could travel cross-country only by rail at the convenience of the railroad. Today the same trip is routinely made by auto, at the individual's convenience — a revolutionary accomplishment!

Still more fantastic has been the transportation revolution brought about by the airplane. The first powered flight by the Wright brothers in 1903 led to the beginning (accelerated by World War I) of the first passenger service in 1919 between London and Paris. Today hundreds of millions of passengers and a large volume of freight traffic are carried yearly.

Even "the sky is the limit no longer." Man now travels outside the earth's atmosphere. It is almost commonplace for astronauts (or cosmonauts) to orbit the earth. A few individuals have even travelled to the moon (¼ million miles), landed, explored its surface, and safely returned with samples of another planetary body. For many of us, perhaps, travel within our planetary system may become a reality in this century. Can anyone deny the "transportation revolution?"

The Energy Revolution

Similar progress has been achieved in man's application of energy. The 1900 census showed that the largest percentage of power for industrial use was furnished by steam engines. Today electrical power has taken over almost completely.

The big energy breakthrough occurred when man developed the ability to harness the atom. Although a direct result of wartime effort, peaceful applications of nuclear power will soon revolutionize our lives. The Atomic Energy Commission predicts that by 1980 approximately 150 million kilowatts of electrical power will be produced by U.S. nuclear power plants. In addition, nuclear explosions used for excavation, and nuclear energy that powers seagoing vehicles, desalinates water, and acts as a tracer in medical applications are only some of the hundreds of other practical applications. Only man's imagination can limit new and varied uses for nuclear power.

Intelligent Machines

Last – but not least – in our revolutionary progress is the growing partnership between man and machine. The eighteenth century Industrial Revolution may soon take second billing to an "intellectual revolution." The concept of applying "intelligent machines" to many of man's fields of endeavor has existed for centuries, but only the evolution of electronic techniques made this concept a reality.

One of the earliest such devices that man built was the *abacus*. This digital machine appears in history prior to the birth of Christ. The example shown in Fig. 1–1 is a modified form that represents the alterations made by the Japanese and the Chinese civilizations. A skilled abacus operator can successfully compete with modern, mechanical desktop calculators.

Little progress was made in the development of machine aids until the appearance in 1642 of Blaise Pascal's desk calculator. Pascal's device, using simple gears, could add and subtract. Due to a lack of mechanical precision, other mathematicians could not improve Pascal's machine beyond achieving multiplication.

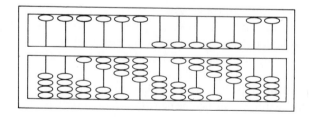

Figure 1–1   Japanese abacus, or Soroban

The next major milestone associated with the development of digital machines occurred in the early 1800s. Charles Babbage envisioned a mechanical device that incorporated many principles of the modern digital computer. His "Difference Engine" was developed to calculate and print mathematical tables. Again, imperfect materials, a shortage of precision tools, withdrawal of government support (after 1843), and a lack of understanding among his associates resulted in the abandonment of his project after several incomplete models had been constructed.

As scientists developed techniques that applied electrical and electronic principles to record storing, account handling, and bookkeeping, new, more versatile machines appeared. Once again, the impetus of wartime activities (World War II) spurred development; in 1940 a computer using relays was devised, and pulse techniques developed for radar were incorporated into applied mathematics. Thus automation requirements, arising from production needs, resulted in machines that could peform routine tasks without human intervention. Soon these production machines were capable of making decisions concerning quality, quantity, and so forth.

The general development of electronic computers in World War II has been followed by an almost endless activity in the design and application of these "intelligent machines." Textbooks, papers, and magazines abound on both technical and non-technical aspects of the "care and feeding" of computers. *The computer merely uses techniques that have existed for years.*

Our primary concern in this text is to explore these basic techniques and their diverse applications; however, we shall pay particular attention to the study of that broad field of digital techniques that can be applied to data communication and to data processing. Since the term *digital techniques* may be unfamiliar, we shall begin by comparing them with common analog techniques.

1–2 ANALOG (LINEAR) VS. DIGITAL TECHNIQUES

Analog (Measuring) Devices

An *analog* (linear) *device* is basically a measuring instrument. The measured analog signal can take on an infinite number of possible values. For example, consider temperature. The liquid thermometer (an analog device) measures and displays an approximate value. With sufficient precision, a thermometer reading of $+70°$F. might actually appear as $+69.92987°$F. There are an infinite number of possible temperatures between $+69°$F. and $+70°$F. The only limitation on resolution is the accuracy of the measuring device.

With a liquid thermometer, however, the *actual* temperature is not really being measured. Instead, the height of an enclosed column of liquid (mercury or alcohol) is made *analogous* to the temperature. The thermometer is a *transducer* that converts temperature into the expansion of a liquid column. Analog devices can also transduce physical phenomena into electrically measurable quantities using magnitudes or phases of voltages and currents, electrical resistances, shaft rotation, and the like. In such cases, the primary factor in the analog device* is the *magnitude* of the electrical signal: that is, how large or how small is it?

Digital (Counting) Devices

Whereas the analog signal is usually a *smooth, continuously varying signal*, the digital signal has a *discrete, discontinuous character.* A major criterion for any digital device is the presence (or absence) of a signal — not its actual value. We want to know not *how much* something has changed but only *if* it has changed. Numerical values are represented directly as *digits* (not as continuously varying quantities). *The digital device counts, while the analog device measures.*

In a digital device, the presence, absence, and repetition rate of the signals are the important factors — not the signal amplitude. The digital system transmits only discrete (step) changes and is discontinuous. (The analog system transmits physical quantity in a smoothly and *continuously* varying manner.)

An ordinary, hand-operated tally counter is a simple example of a digital device (Fig. 1–2). The counting lever must be pushed all the way down to register a count. There is no middle position of the lever that can register a half count. *It either counts one digit, or it does not count.*

Today all digital systems are based on whether something is actually there, whether it is true or false, whether there is a hole in a punched card, and so forth. Even mammoth, multimillion dollar computers are based on this simple yes/no decision. Complexity results when the equipment is instructed to make such decisions millions of times a second.

Figure 1–2   Tally counter (courtesy of Veeder-Root, Hartford, Conn.)

---

* Electronic components used with analog devices or transducers are mentioned in this text only where signals from such devices are to be converted to digital operation.

Digital Applications

Digital techniques are persistently invading fields that formerly used analog techniques. For example, remote measuring equipment, which once depended on analog methods, is now being replaced by digital equipment. Analog measurements are converted at their source to digital signals, then transmitted, processed, stored, and acted upon at some remote point prior to readout and action (Fig. 1—3).

Thus telephone conversations are being broken up by digital equipment, combined into a composite signal, transmitted, and decoded at some distant terminal. Some approaches to telephone transmission actually convert conversations to digital values instead of just "breaking-up" information (Fig. 1—4).

Figure 1—3   Data logging system (courtesy Analog Digital Data Systems, Inc.)

Figure 1—4   Digital telephone system (courtesy of Lynch
Communications Systems, San Francisco, CA.)

Inaccuracies in reading conventional, moving-pointer, measuring instruments are being overcome by applying digital techniques. Digital voltmeters are common today, and so are the number of digital readout instruments available to the technical and scientific disciplines. Figure 1—5 shows a typical digital multimeter design for general testing and measurement.

Many functions formerly performed by analog devices in missile and aircraft guidance and control now use digital techniques (Fig. 1—6). Greater accuracy results, and a tremendous reduction in weight and volume is effected.

Digital techniques are also making inroads in medical electronics, the control of industrial processes and railroad trains, material handling, the monitoring and operating of ocean-going ships, the supervision of industrial plant functions, the centralized control of electrical power transmission and distribution, automobile servicing, machine control (Fig. 1—7), supermarkets (Fig. 1—8), the stock exchange (Fig. 1—9), and other areas. Naturally, many of these functions require the use of a digital computer, but again, the digital computer is just another device that performs tasks with digital assistance.

Since considerable time has been spent discussing digital computers versus digital data handling and processing systems, the following definitions should help to differentiate the two.

Figure 1—5 Digital voltmeter (courtesy of Hickok Electrical
Instrument Co., Cleveland, Ohio)

Figure 1—6 F-111 navigation display unit (left) and computer
control (right) (courtesy of Singer-General Precision, Inc.,
Kearfott Div., Little Falls, N.J.)

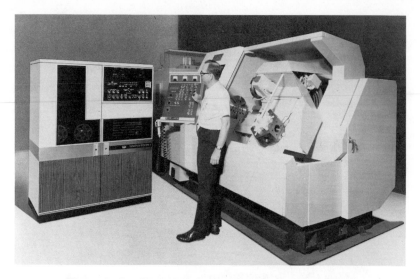

Figure 1–7   Dynapath System 4 numerical controller (courtesy of Bendix Industrial Controls Division, Detroit, Michigan).

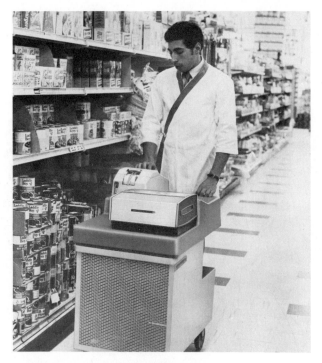

Figure 1–8   Digital supermarket stock control (courtesy of Digitronics Corp., Albertson, L.I., N.Y.)

Figure 1–9　　Stock Exchange securities terminal (courtesy of SYCOR, Inc., Ann Arbor, Michigan).

The digital computer is a device for performing mathematical calculations on numbers represented digitally. Hence any device for manipulating number symbols according to a detailed procedure would qualify.

A data handling and processing system transforms information given to it into a suitably organized presentation according to some criterion and some orderly preplanned procedure.

In science and engineering, for example, data processing systems are used in data reduction, preparation of mathematical and statistical tables, recording and maintaining experimental data, and revisions to libraries of routines.

1–3 ANALOG (LINEAR) CIRCUIT ANALYSIS

Electronic circuits used in digital data handling and processing systems are subject to many of the same analysis techniques as those used for conventional analog (linear) circuits. To establish the portions of the techniques that are similar, therefore, we shall use the analysis of a simple radio receiver as an example.

Block Diagram Analysis

A common way to begin the analysis of an electronic device is to develop an overall understanding of the complete device. This is usually accomplished by using a *block diagram.* Such a diagram indicates what is functionally necessary to meet the specifications for a particular piece of equipment. For example, assume the following specifications.

FM Tuning Range: 152 to 174 mHz
Bandwidth: 100 kHz at 6 dB down
Sensitivity: 6 μV at 5 kHz deviation
Squelch Sensitivity: Opens for 5 μV signal
Audio Output: 1 watt to 8 ohm speaker
Antenna Impedance: 50 to 75 ohms
Power Requirements: 110 to 130 Volts, 60 Hz

The block diagram shown in Fig. 1–10* performs the specified functions.

Once the functions of individual blocks are determined and all of the block functions are integrated, the operation of the radio receiver can be fully explained. Although most readers may be familiar with the operation of this device, such familiarity is not a prerequisite. The block diagram level of analysis purposely avoids using electronic circuit details. Only the *functions* of the blocks and the *flow of data* from block to block are important.

The incoming signal from the antenna is amplified in the RF amplifier, which is tuned so that only the desired signal is amplified. In the mixer, the amplified signal is combined with a local oscillator signal to form the sum and difference frequencies. The RF amplifier, mixer, and oscillator are tuned simultaneously, thus maintaining constant sum and difference frequencies at the

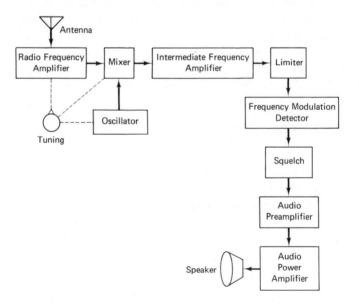

Figure 1–10   Block diagram, typical FM receiver

---

* Power connections are not shown, but are assumed to be present and properly connected.

output of the mixer. The IF amplifiers are tuned to select only the sum or difference frequency, which is amplified in these stages. Since this device is a frequency modulation (FM) receiver, limiter stages are incorporated to remove all amplitude modulation. The FM detector develops an audio output that is a close replica of the original modulation applied to the transmitter. To prevent constant background noise (always present without an incoming signal) from reaching the speaker, a squelch circuit is included. This circuit effectively blocks any input to the audio preamplifier until a signal of a predetermined and preset level is reached. The squelch circuit then opens and allows the desired audio signal to be applied to the preamplifier for amplification. Sufficient audio power to operate the speaker is developed by the audio output stage. Finally, the speaker converts the electrical variations from the output stage into audible sounds.

Schematic Diagram Analysis

If further detail is required, the next level of explanation is usually accompanied by a *schematic diagram.* The schematic diagram (Fig. 1–11) shows such individual discrete components as transistors, capacitors, and resistors – all interconnected to perform the functions of individual blocks in the block diagram.* Although complete schematic diagrams without any correlation to block diagrams are commonly provided, schematic diagrams of individual blocks (circuit design permitting) should be furnished. The schematic diagram in Fig. 1–11 shows a typical mixer stage.

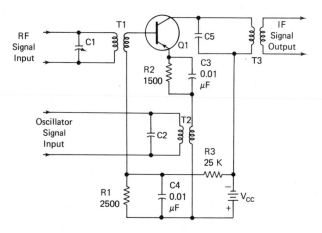

Figure 1–11   Schematic diagram, mixer stage

----

* Discrete component circuit analysis is beyond the scope of this text, but readers with sufficient electronic background should experience no difficulty with this circuit.

1–4 DIGITAL CIRCUIT ANALYSIS

A block diagram may also be used with digital equipment to functionally describe the *system operation.* Due to the complexity of most digital systems, however, it is impractical to provide a complete circuit schematic diagram. Besides, the use of only a few basic circuits is well suited to an intermediate level of analysis. The complexity of digital systems is due not to "exotic" circuits but to multiple use of a few simple circuits.

Block Diagram Analysis

Figure 1–12 shows a simple digital measuring system in block diagram form. Although greatly simplified, this digital measuring/recording system is a typical application of present digital techniques to industrial processes. Transducers, described in Sec. 1–2, provide inputs to this digital system. The electrical outputs of transducers vary considerably, depending on the conversion method used. Therefore a "linear" amplifier, called a *signal conditioner,* is commonly included in series with the transducer output to either amplify or limit to a predetermined range the input to the digital measuring/recording system.

The system shown here contains enough transducers and signal conditioners to measure all required parameters in the monitored process. Upon command from the *control unit,* the *multiplexer* selects one of the many input signals and feeds it to the *analog/digital converter.* Here the analog signal is converted by one of several methods to a pure digital signal and then routed to either the *data display* or the *data recorder,* or both, depending on the requirement. As soon as the conversion of one channel is complete, the multiplexer is stepped to the next channel, and the conversion process is repeated. An internal timing device, known as a *clock,* establishes the basic multiplexing cycle for the complete system. Many variations of sampling times, display times, and recording media exist, and provision is usually made for simple adjustment and selection of both input and internal parameters.

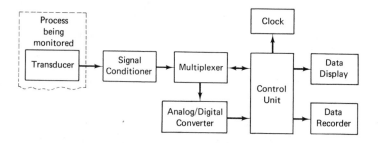

Figure 1–12   Digital measuring/recording system

Schematic Diagram Analysis

Like the radio receiver, the functions of individual blocks are determined and these block functions are then integrated before a full explanation of the operation of the digital system is achieved. In digital systems, however, circuit schematics result in endless repetition of the same circuits. For example, Fig. 1–13, a partial schematic diagram, represents only a small portion of the clock in Fig. 1–12. For simplicity, therefore, only those discrete components that enter directly into the explanation of the operation are defined.

Figure 1–13 shows a commonly employed circuit in digital clocks that divides by 6. For each 6 "input signals" received, 1 "output signal" is generated. A complete electronic circuit would obviously be quite lengthy. However investigation of Fig. 1–13 reveals some interesting aspects. The $Q_1 - Q_2$ circuitry is almost identical to that of $Q_3 - Q_4$ and $Q_5 - Q_6$. The only apparent differences concern $CR_{13} - CR_{14}$ and $CR_{15} - CR_{18}$ (plus some additional diodes, resistors, and capacitors). $CR_{13} - CR_{14}$ and $CR_{15} - CR_{18}$ perform identical functions; so do the $Q_1 - Q_2$, $Q_3 - Q_4$, and $Q_5 - Q_6$ circuits.

Logic Diagram Analysis

If the operation of these circuits can be described by sets of rules and represented by special symbols, then the complete circuit or system can be described by its own set of symbols. This new level of diagram is called a *logic diagram*, and its symbols are called *logic symbols*.

The logic diagram for the circuit diagram in Fig. 1–13 is shown in Fig. 1–14. Note that Fig. 1–14 shows a considerable decrease in complexity compared with Fig. 1–13. Only two symbols appear in the logic diagram. Both the AND GATE symbol (Fig. 1–15a) and the FLIP-FLOP symbol (Fig. 1–15b) represent *functions** – not the specific arrangement of discrete components shown in Fig. 1–13.

Referring to Figs. 1–13 and 1–14, AND GATE 1 corresponds to $CR_{13} - CR_{14}$ and associated components; AND GATE 2 corresponds to $CR_{15} - CR_{18}$ and associated components; and the $Q_1 - Q_2$, $Q_3 - Q_4$, and $Q_5 - Q_6$ circuitry correspond, respectively, to FF_1, FF_2, and FF_3. But since we need to understand only the functions in logic diagrams, it is sufficient to realize that the AND GATE *combines signals* according to predetermined rules and that the FLIP-FLOP *stores information*. Actually, the "divide-by-6" device *counts* the number of times the input signal changes and also provides a change in output signal for each sixth input-signal change. Counting is accomplished through the binary number system (see Ch. 2), and the condition of FLIP-FLOPS 1, 2, and 3 at any specific time is determined by how many input-signal changes have taken place. Thus the FLIP-FLOPS store numbers (at least temporarily)

---

* Numerous examples of electronic implementation of logic functions appear in Appendix D.

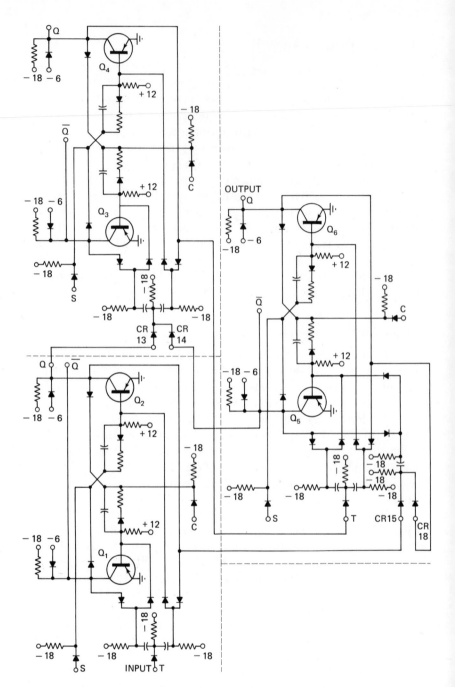

Figure 1—13   Partial schematic diagram — clock

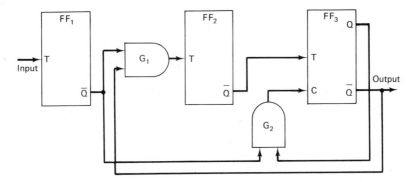

Figure 1—14   Logic diagram, divide-by-six circuit

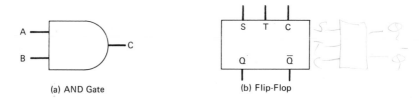

(a) AND Gate   (b) Flip-Flop

Figure 1—15   Logic symbols

in the binary number system. The AND GATES determine what numbers are stored and route control signals so that the device will recycle back to the 0 count after detection of the sixth input change and generation of the output signal.

The digital circuit analyses performed throughout this text use the MIL-STD-806B logic symbols,* a portion of which appears in Appendix B. As these symbols appear in later chapters, the complete rules associated with their operation will be developed.

1—5 LOGIC ANALYSIS TECHNIQUES

HIGH and LOW

In addition to the new symbology and a new level of diagram, we must develop a new approach to signal analysis. As we mentioned earlier, a radio receiver functions by selecting incoming signals of a particular frequency and

_____
* MIL-STD-806B (Graphic Symbols for Logic Diagrams) is the established standard of notation. However, minor changes may occur in the future, and the latest revision may therefore be obtained from the Superintendent of Documents, U.S. Government Printing Office, Washington, D.C. 20402.

amplitude, but a digital measuring device is activated by the *presence* (or *absence*) *of signals* and the *rate* at which they arrive.

Most digital devices operate on the principle of *bistable circuits*. This means that the devices are stable when they are in *either* of two states: *one state is represented by the presence of a signal, and the other state is represented by the absence of a signal.* In logic diagram analysis, *the relative level of the signal is the criterion.* A signal may be either *HIGH* or *LOW*. The actual value of the voltage or current that will represent *HIGH* is determined by the device designers. Likewise, the actual value of voltage or current that represents *LOW* does not really affect the analysis of a logic diagram. *The important consideration is that* HIGH *represents one of the stable states and* LOW *represents the other.*

Binary Number System

As we mentioned briefly, counting is not performed by the familiar *decimal system.* Ten separate stable states would be needed to represent all the decimal digits (0 through 9, inclusively); electronically, this becomes an almost insurmountable task. A number system using only 2 stable states does exist. This system is called the *binary number system* and has only *two values*, 0 and 1. The binary number system is vitally important to any logic circuit analysis, so it will be discussed in detail in later chapters.

Boolean Algebra

Mathematics is a useful tool in digital circuit analysis. Algebra, trigonometry, and calculus are commonly used in analyzing *analog circuits.* However, since all logic operations in *digital circuits* depend on either the existence or absence of a signal, a variable may have only *one* of two possible values. Reduced to a two-valued system, both mathematical operations and the analysis of digital circuits are greatly simplified. *The mathematics of two-valued logic is called Boolean algebra.* The basic postulates and theorems of Boolean algebra appear in Appendix A, but their applications are explained in detail as we encounter them throughout the text.

Maps and Truth Tables

The use of graphic methods for linear and nonlinear systems analysis is widespread. Characteristic curves of active devices such as semiconductors and vacuum tubes are applied to the design of electronic circuit operations. Experimental data are derived, tables manufactured, and graphs of responses of both

active and passive devices developed. In the digital field, graphic techniques*
are not as complicated, but they still are quite useful. Such techniques include
truth tables, Venn diagrams, Veitch diagrams, Karnaugh maps, and so forth.

At this point, it might be useful to examine the application of a truth
table to the analysis of the AND GATE. *A truth table is a tabular listing of all*
possible combinations of input conditions to a logic device, and a resultant out-
put for each case. As such, a truth table is a useful method of condensing all
information concerning the functional characteristics of a logic device.
Figure 1−16 is a simplified truth table for any 2-input AND GATE. Logic
devices and combinations thereof may be analyzed either by truth tables or
by the other techniques we mentioned earlier.

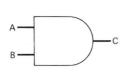

Input		Output
A	B	C
Low	Low	Low
Low	High	Low
High	Low	Low
High	High	High

Figure 1−16   Truth table, 2-input AND gate

Waveforms

Many of the most common measuring instruments used in analog and
linear applications are not useful in digital system analysis, since the information most needed relates to the absence or presence of a signal and the rate at
which it occurs. The *cathode-ray oscilloscope* (CRO) is, however, a useful
measurement device. Within all digital devices, signals are constantly changing
from HIGH to LOW, and in general, digital devices are *time-sequenced.* By
using the CRO the time relationships between internal and external signals may
be obtained. The commonly encountered waveforms of linear and analog
signals seldom appear in digital equipment. Instead, constantly changing HIGH
and LOW signals are routed throughout the circuitry and may be observed on
the CRO or other special logic-oriented test devices. These waveforms are unique
to each type of circuit being investigated. Figure 1−17 shows the waveforms
that might be encountered in the divide-by-6 circuit.

Waveforms are used in conjunction with logic diagrams, Boolean algebra,
and graphic methods to integrate the analysis of digital equipment. All of these
methods of analysis will be discussed in detail later.

---

* The most common graphic techniques are examined in subsequent chapters.

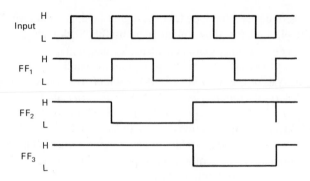

Figure 1–17   Waveforms, divide-by-six circuit

PROBLEM APPLICATIONS

1–1. Trace the evolution of digital computers from ancient times to the present day.

1–2. Define the term *analog*.

1–3. Define the term *digital*.

1–4. List as many applications as possible where digital equipment is replacing or supplementing analog equipment.

1–5. Compare analog and digital signals and their characteristics.

1–6. What is the purpose of a block diagram?

1–7. What is the purpose of a schematic diagram?

1–8. What is the purpose of a logic diagram?

1–9. Why are logic diagrams used in conjunction with digital equipment?

1–10. Why is the binary number system used in digital equipment?

1–11. What is a truth table?

1–12. Why are waveforms used in troubleshooting digital circuits?

1–13. What is the most practical digital-equipment troubleshooting instrument?

2

Numbers
and Counting

2–1   THE DECIMAL NUMBER SYSTEM

Our constant, almost casual use of the decimal number system may prevent us from utilizing some of its more important features. A quick review of
certain properties and characteristics of the decimal system would therefore be
helpful.

The *radix* or *base* of a number system is the total number of *unique*
(different) *symbols* available in that system. The largest-valued symbol always
has a magnitude of *one less than the radix.* For example, in the decimal number
system there are *ten* different symbols: 0, 1, 2, 3, 4, 5, 6, 7, 8, and 9, where the
largest, 9, is one less than 10 (the base).

Counting

A table of progression of decimal number system symbols is shown in
Fig. 2–1. Note especially the treatment of numbers greater in value than the
basic symbols of the number system. The 0 symbol follows the highest-valued
symbol each time all of the allowable symbols have been used in the counting
sequence. When the 0 symbol appears, the results of accumulating the counts
from 1 through 9 must be "shifted" into the column (position) immediately to
the left of the 0 symbol, and the basic counting sequence must be resumed in
the present column. The 0 symbol is called a *shift marker.* This action can be
seen at numbers 10, 20, and 100 in Fig. 2–1.

The number 006 (commonly written as 6) represents 6 units, while the
number 060 (commonly written as 60) represents the quantity sixty. This
method of using Arabic numeral symbols to indicate a magnitude or quantity is
actually a type of shorthand. Each position in a decimal number has a value

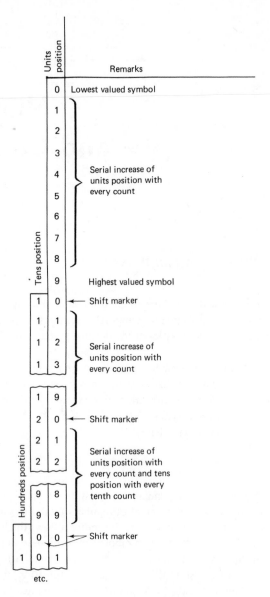

Figure 2–1   Decimal counting concept

that is ten times the value of the next position to the right. *In other words, every positional value is a multiple of ten and can be expressed as ten raised to some exponent.*

> *Example 2–1.* Express 123 in exponential form.
> *Solution:* $123 = 1 \times 100 + 2 \times 10 + 3 \times 1$
> $\qquad\qquad = 1 \times 10^2 + 2 \times 10^1 + 3 \times 10^0$

Note: Any number raised to the exponent 0 equals 1.

This progression of increasing exponents can be continued as far to the left of the decimal point as desired. The same progression can also be extended to the right of the decimal point, but then the exponents are *negative*. For example, the first position to the right of the decimal point is the "tenth's" position; it has a value of 10^{-1} or $1/10^1$.

Any number in the decimal number system can be represented like those in Fig. 2–2. The symbol placed in any position indicates how many *multiples*

Decimal Number	10^n	10^3 Thousands	10^2 Hundreds	10^1 Tens	10^0 Units	. Decimal Point	10^{-1} Tenths	10^{-2} Hundredths	10^{-3} Thousandths	10^{-m}	
	10^n	10^3	10^2	10^1	10^0	.	10^{-1}	10^{-2}	10^{-3}	10^{-m}	Power
	n	3	2	1	0	.	-1	-2	-3	$-m$	Position
		1000	100	10	1	.	0.1	0.01	0.001		Base 10 Equiv. value
1					1	.					
10				1	0	.					
123			1	2	3	.					
1234.56		1	2	3	4	.	5	6			
0.789						.	7	8	9		
10.10				1	0	.	1	0			
53				5	3	.					
45.312				4	5	.	3	1	2		
146.64			1	4	6	.	6	4			
2927.975		2	9	2	7	.	9	7	5		

Figure 2–2   Representation of decimal numbers

of that power of 10 are in the total quantity represented by the number. Study the examples in Fig. 2–2 carefully.*

The general equation for any decimal number is

$$N = \sum_{i=-m}^{i=n} (A_i)(10^i) \qquad (2-1)$$

$$N = A_n 10^n + A_{n-1} 10^{n-1} + \cdots + A_1 10^1 + A_0 10^0$$

$$+ A_{-1} 10^{-1} + \cdots A_{-m} 10^{-m}$$

where A = admissible mark or symbol $(0, 1, 2, 3, 4, 5, 6, 7, 8, 9)$, n = upper limit, and m = lower limit.

> *Example 2–2.* Express decimal number 1234.56 in the form of Eq. (2–1).
> *Solution:* $n = 3$; $m = 2$
> $N = 1 \times 10^3 + 2 \times 10^2 + 3 \times 10^1 + 4 \times 10^0 + 5 \times 10^{-1} + 6 \times 10^{-2}$
> $N = 1000 + 200 + 30 + 4 + 0.5 + 0.06$
> $N = \mathbf{1234.56}$

2–2 THE BINARY NUMBER SYSTEM

Radix and Symbols

Now that some of the concepts and characteristics of the decimal number system have been established, we can ask the next logical question. Is it possible to have a number system with a radix less than 10? Logical analysis of the requirements for any number system should answer this question. To begin: a *position marker* is needed to indicate positional shift; hence the zero is required. In addition to zero, at least one other symbol with a different value is required, thus establishing the need for the one.

Now comes the next question. Are two symbols sufficient to completely define a number system? Perhaps this question can best be answered by attempting to count in the decimal-system manner.

Counting

With a radix of 2 (just two symbols), the maximum number that can be shown with a single digit is the number 1 (one). If we want to show the decimal number two, the 1 must be moved one place to the left and the 0 must be used

---

* Numerous exercises are presented at the end of this chapter to assist in completely reviewing the principles of the decimal number system.

Decimal	Binary	Comments
0	0000	Identical
1	0001	Identical
2	0010	Shift to the left required
3	0011	A 2 and a 1 makes 3
4	0100	Another shift to left required
5	0101	A 4 and a 1 makes 5
6	0110	A 4 and a 2 makes 6
7	0111	A 4 and a 2 and a 1 makes 7
8	1000	Another shift to left required
9	1001	An 8 and a 1 makes 9
10	1010	An 8 and a 2 makes 10
11	1011	An 8 and a 2 and a 1 makes 11
12	1100	An 8 and a 4 makes 12
13	1101	An 8 and a 4 and a 1 makes 13
14	1110	An 8 and a 4 and a 2 makes 14
15	1111	An 8 and a 4 and a 2 and a 1 makes 15

Figure 2–3   Binary counting table

to indicate that a shift has taken place. So the decimal number two (2 in binary) is 10 (read as "one–zero" not "ten"). See Fig. 2–3 for the development of the *binary* (base 2) *counting table.*

Obviously, the ability to count exists in number systems other than base 10. Figure 2–3 has not been extended beyond count 15, but it is entirely possible to count as *high* in a binary (base 2) number system as in a decimal (base 10) number system.

Figure 2–4 shows how any number in the binary system may be represented. Note that this figure appears similar to Fig. 2–2, which represented the decimal system. The only change is that *multiples of powers of* 2* (instead of powers of 10) are now being used. This is only reasonable, since the base of the binary system is 2! Each of the examples† in Fig. 2–4 should be studied carefully.

An equation similar to Eq. (2–1) can be used to express any binary number:

$$N = \sum_{i=-m}^{i=n} A_i 2^i \tag{2-2}$$

$$N = A_n 2^n + A_{n-1} 2^{n-1} + \cdots + A_1 2^1 + A_0 2^0 + \cdots A_{-m} 2^{-m}$$

where A = admissible mark or symbol $(0, 1)$, n = upper limit, and m = lower limit.

---

* A table of powers of 2 is found in Appendix E.

† As with the decimal number system, numerous practice exercises are provided at the end of this chapter.

Decimal Number	2^n	Eights 2^3	Fours 2^2	Twos 2^1	Ones 2^0	Binary Point	Halves 2^{-1}	Fourths 2^{-2}	Eighths 2^{-3}	2^{-m}	
	n	3	2	1	0	.	-1	-2	-3	$-m$	Position
		8	4	2	1	.	$\frac{1}{2}$	$\frac{1}{4}$	$\frac{1}{8}$		Base 10 equiv. value
1					1	.					
2				1	0	.					
5			1	0	1	.					
15		1	1	1	1	.					
10.5		1	0	1	0	.	1	0			
0.875					0	.	1	1	1		
7.125		0	1	1	1	.	0	0	1		
8.625		1	0	0	0	.	1	0	1		
1.25		0	0	0	1	.	0	1	0		
5.375		0	1	0	1	.	0	1	1		

Figure 2—4   Representation of binary numbers

Example 2–3. Express binary number 1010.10 in the form of Eq. (2–2).
Solution: $n = 3, \quad m = 2$
$$N = 1 \times 2^3 + 0 \times 2^2 + 1 \times 2^1 + 0 \times 2^0 + 1 \times 2^{-1} + 0 \times 2^{-2}$$
$$N = 8 + 0 + 2 + 0 + 0.5$$
$$N = \mathbf{10.5}$$

Any Number System

Actually, *any number* may be used as the base for a number system.
Figure 2—5 shows the skeleton of a general number system with a radix r. Note
that the same format is used as had been used for the base 2 and base 10
systems. A number with any base can be expressed as a summation of products:

$$N = \sum_{i=-m}^{i=n} A_i r^i \tag{2–3}$$

$$N = A_n r^n + A_{n-1} r^{n-1} + \cdots + A_1 r^1 + A_0 r^0 + A_{-1} r^{-1} + \cdots + A_{-m} r^{-m}$$

Figure 2—5   General number system

where A = admissible mark or symbol, r = radix or base of the number system, n = upper limit, and m = lower limit.

This general equation allows investigation of other number systems used in digital equipments; e.g., the *octal* (base 8), *hexadecimal* (base 16), and other *coded binary systems.* However, in one way or another, the great majority of the operations within modern digital equipment can be tied directly into the binary number system.

Need for the Binary Number System

The binary system eliminates the extreme difficulty of *accurately* representing decimal digits in electronic equipment. Ten discrete levels are required to allow use of all the symbols in the decimal system. And these ten discrete levels must be sufficiently distinguishable from each other to assure that no mistakes in symbol recognition occur. Such criteria are overly restrictive in most electronic data-handling devices.

In the binary system, however, only *two discrete levels* are needed. Fortunately, decimal values can be coded into equivalent binary values so that the binary system may be used. Two arbitrary levels of voltage or current can be established, thus leaving no doubt about which binary value is represented. The value 0 is assigned to one of the arbitrary levels and the value 1 to the other level.

Nearly all of the data handling and processing equipment discussed in this text uses the binary system. Nevertheless one of the basic requirements in working with and understanding digital equipment is the ability to convert from decimal to binary and vice versa.

2—3 NUMBER SYSTEM CONVERSIONS

It is very easy to confuse a number in the binary system with a number in the decimal system. For example, binary 1010 is equivalent to decimal 10. To minimize confusion, therefore, we have devised a *system of subscripts.* Thus the *base* of the number system appears as a subscript — binary 1010 becomes 1010_2 and decimal 10 becomes 10_{10}. In most cases, however, the number system being used is obvious, so it is unnecessary to employ subscripts.

Decimal to Binary (D/B) Conversion

Decimal-to-binary (D/B) *integer conversion* is initially approached on a "reasoning" basis. After establishing that such conversions are easily performed, a general solution can then be presented. For example, consider D/B conversion of 14_{10} (decimal 14). The only "tools" needed are a table of powers of 2 (Appendix E) and a knowledge of the principles of positional notation, described in Eqs. (2–1), (2–2), and (2–3). Just as 14_{10} is really $1 \times 10^1 + 4 \times 10^0$, the equivalent binary numbers may be made up of sums of products of admissible binary symbols and powers of 2. Examination of the table of powers of 2 shows that 14_{10} is less than 1×2^4 (16_{10}), so 1×2^4 cannot be used. However 1×2^3 (8_{10}) is less than 14_{10}, so 14_{10} must include 1×2^3. A remainder of 6_{10} is left, a value that must be examined in the same manner as was 14_{10}. 1×2^3 has already been used, so 1×2^2 (4_{10}) must be the next value. Note that 6_{10} includes 1×2^2 with a remainder of 2_{10}. Continuing the process, we discover that 2_{10} includes 1×2^1 with a remainder of 0. Therefore

$$14_{10} = 1 \times 2^3 + 1 \times 2^2 + 1 \times 2^1 + 0 \times 2^0 \quad \text{or} \quad \mathbf{1110_2}$$

This operation is called the *direct method of* D/B *conversion.* Its formal, step-by-step procedure is listed below.

1. Successively subtract the largest possible power of 2 from the decimal number.
2. Continue performing step 1 until a zero remainder occurs. (Negative remainders are not allowed.)
3. Write a 1 for each power of 2 subtracted.
4. Write a 0 for each power of 2 not used.
5. Write the 1s and 0s in the order in which the subtractions are performed. All powers of 2 used (from the largest through and including 2^0) must be accounted for by using either 1 or 0.

Sample conversions are shown in Examples 2–4 and 2–5.

Example 2–4. Convert decimal number 14 to a binary number using the direct method.

Solution:
$$
\begin{array}{ll}
14 & \\
-\underline{8} \;\; (2^3) & \quad 1 \;\; \text{Most Significant Digit (MSD)} \\
6 & \\
-\underline{4} \;\; (2^2) & \quad 1 \\
2 & \\
-\underline{2} \;\; (2^1) & \quad 1 \\
0 & \\
-\underline{0} \;\; (2^0) & \quad 0 \;\; \text{Least Significant Digit (LSD)} \\
0 &
\end{array}
$$

$$1 \times 2^3 + 1 \times 2^2 + 1 \times 2^1 + 0 \times 2^0 = 14_{10}$$
$$\mathbf{1110_2} = 14_{10}$$

Example 2–5. Convert 25_{10} to a binary number using the direct method.

Solution: 25
$\quad\quad \underline{-16}$ (2^4) 1 (MSD)
$\quad\quad\quad 9$
$\quad\quad \underline{-\ 8}$ (2^3) 1
$\quad\quad\quad 1$
$\quad\quad \underline{-\ 1}$ (2^0) 1 (LSD)
$\quad\quad\quad 0$

$\left\{ \begin{array}{l} \text{Note that 4 } (2^2) \text{ and} \\ 2 \ (2^1) \text{ cannot be subtracted from 1;} \\ \text{they must be included (Rule 5).} \end{array} \right.$

$$1 \times 2^4 + 1 \times 2^3 + 0 \times 2^2 + 0 \times 2^1 + 1 \times 2^0 = 25_{10}$$
$$11001_2 = 25_{10}$$

It is possible (and sometimes quite practical) to perform decimal binary conversion by the *successive division method.* A table of powers of 2 is not required — all we have to know is how to divide by 2! Here is the detailed step-by-step procedure and some typical examples.

1. Successively divide the decimal number by 2.
2. Place the quotients directly beneath the dividend.
3. Place the remainders opposite the quotients.
4. The equivalent binary number becomes the remainders — the *final* remainder being the *most significant digit* (MSD) and the *first* remainder being the *least significant digit* (LSD).

Example 2–6. Convert 14_{10} to a binary number using the successive division method.

Solution: 2 $\lfloor$14
$\quad\quad$ 2 $\lfloor\ \underline{7}$ 0 (LSD)
$\quad\quad$ 2 $\lfloor\ \underline{3}$ 1
$\quad\quad$ 2 $\lfloor\ \underline{1}$ 1 $1110 = 14_{10}$
$\quad\quad\quad\ \ 0$ 1 (MSD)

Example 2–7. Convert decimal number 25 to a binary number using the successive division method.

Solution: 2 $\lfloor$25
$\quad\quad$ 2 $\lfloor$12   1 (LSD)
$\quad\quad$ 2 $\lfloor\ \underline{6}$ 0
$\quad\quad$ 2 $\lfloor\ \underline{3}$ 0
$\quad\quad$ 2 $\lfloor\ \underline{1}$ 1 $11001_2 = 25_{10}$
$\quad\quad\quad\ \ 0$ 1 (MSD)

Decimal-to-binary conversion of fractional numbers can be performed by applying the following steps.

1. Successively multiply the decimal fraction by 2.
2. Ignore any integer results when performing the multiplication.
3. Place the products directly below the multiplicand.
4. Work toward zero or until the number of digits required for degree of precision is reached.
5. The first integer part obtained is the MSD.

Some typical examples of this method of conversion are shown below.

Example 2–8. Convert 0.8125_{10} to a binary number.

Solution: $\boxed{0}\cdot$ (8125) x 2

(MSD) $\boxed{1}\cdot$ (6250) x 2

   $\boxed{1}\cdot$ (2500) x 2 $0.8125_{10} = \mathbf{0.1101_2}$

   $\boxed{0}\cdot$ (5000) x 2

(LSD) $\boxed{1}\cdot$ (0000)

Example 2–9. Convert 0.7893 to a binary number.

Solution: $\boxed{0}\cdot$ (7893) x 2

(MSD) $\boxed{1}\cdot$ (5786) x 2

   $\boxed{1}\cdot$ (1572) x 2

   $\boxed{0}\cdot$ (3144) x 2

   $\boxed{0}\cdot$ (6288) x 2 $0.7893_{10} = \mathbf{0.1100101_2}$

   $\boxed{1}\cdot$ (2576) x 2

   $\boxed{0}\cdot$ (5152) x 2

(LSD) $\boxed{1}\cdot$ (0304)

Binary to Decimal (B/D) Conversion

Conversion from decimal to binary provides the most efficient method of conditioning inputs to digital data processing and computing equipment. Once the required operations are performed, however, a problem arises. Since the binary system is not employed by everyone, the results of the performed operations may not be easily readable. Therefore it may be desirable to reconvert binary numbers into decimal numbers. Fortunately, decimal-binary (D/B) conversion processes have binary-decimal (B/D) inverses.

The *direct method of* B/D *conversion* is accomplished by using either a B/D conversion table (Appendix E) or a powers of 2 table (Appendix E). Two basic rules apply.

1. A 1 in any digit position indicates that the corresponding power of 2 is to be used.
2. A 0 indicates that the power of 2 is not required.

Some examples are presented below.

Example 2–10. Convert 11011_2 to a decimal number using the direct method.

Solution: $11011_2 = 1 \times 2^4 + 1 \times 2^3 + 0 \times 2^2 + 1 \times 2^1 + 1 \times 2^0$
$$= 1 \times 16 + 1 \times 8 + 0 \times 4 + 1 \times 2 + 1 \times 1$$
$$= 16 + 8 + 2 + 1$$
$$11011_2 = 27_{10}$$

Example 2–11. Convert 0.11011_2 to a decimal number using the direct method.

Solution: $0.1101_2 = 1 \times 2^{-1} + 1 \times 2^{-2} + 0 \times 2^{-3} + 1 \times 2^{-4}$
$$= \tfrac{1}{2} + \tfrac{1}{4} + 0 + \tfrac{1}{16}$$
$$= 0.5 + 0.25 + 0.0625$$
$$0.1101_2 = 0.8125_{10}$$

Although it is not absolutely necessary, B/D conversion of mixed numbers is usually performed separately for the integer and for the fractional portions. The individual parts of the number are then combined, as Example 2–12 illustrates.

Example 2–12. Convert 101101.0101_2 to a decimal number using the direct method.

Solution: $101101 = 1 \times 2^5 + 0 \times 2^4 + 1 \times 2^3 + 1 \times 2^2 + 0 \times 2^1 + 1 \times 2^0$
$$= 1 \times 32 + 0 \times 16 + 1 \times 8 + 1 \times 4 + 0 \times 2 + 1 \times 1$$
$$= 32 + 0 + 8 + 4 + 0 + 1$$
$$= 45_{10}$$
$0.0101 = 0 \times 2^{-1} + 1 \times 2^{-2} + 0 \times 2^{-3} + 1 \times 2^{-4}$
$$= 0 + \tfrac{1}{4} + 0 + \tfrac{1}{16}$$
$$= 0.3125$$

Therefore $101101.0101_2 = 45.3125_{10}$

B/D conversion of integers may be approached by the *successive multiplication method*, sometimes referred to as the *double-dabble method*, which is the inverse of the successive division method used in D/B conversion. This method of B/D conversion can be performed by applying the following steps.

1. Double the highest order binary digit.
2. Add this doubled value to the next lower order binary digit and record the sum.
3. Double the sum obtained.
4. Add this doubled value to the next lower order binary digit and record the new sum.
5. Continue Steps 3 and 4 until the last (lowest-order) binary digit has been added to the previously doubled sum and a final sum has been obtained. The final sum is the decimal equivalent of the binary number.

Examples 2–13 and 2–14 demonstrate the double-dabble method.

Example 2–13. Convert 10100_2 to a decimal number using the double-dabble method.

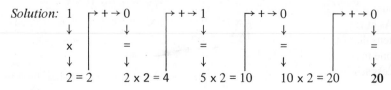

Example 2–14. Convert 11100_2 to a decimal number using the double-dabble method.

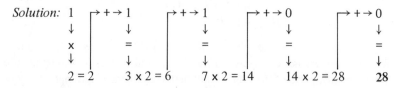

2–4 THE COUNTING PROCESS

Both the ability to convert from one number system to another and the ability to count are useless without a means to mechanize the counting process. Paper and pencil sufficed for centuries, but now high-speed counting requirements demand new methods. Once again, since the decimal number system is relatively well understood, the mechanics of decimal counting have been used to develop a method of mechanization.

Decimal Counting

The concepts of decimal counting were explained in Section 2–1. Now consider, as an example, the odometer on an automobile (Fig. 2–6). As the miles go by, the units dial increases. Then, as the units dial rolls past 9 to 0, the

Figure 2–6 Odomoter mechanism (courtesy Veeder-Root, Hartford, Conn.)

tens dial increases by 1, as shown in the decimal counting table in Fig. 2—1. The process will continue, following the procedure described in Sec. 2—1.

Another example is the simple adding machine. When adding 5 to 6, the sum produced exceeds the highest-valued symbol available in the number system. As in the odometer example, passing from 9 to 0 on the units dial causes the tens dial to increase by 1. Thus increasing the units dial on the adding machine one more time results in a sum of 11. The drawing in Fig. 2—7 makes this process self-evident.

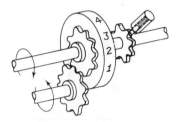

Figure 2—7   "Carry" mechanism

Binary Counting

Binary counting is easier to mechanize than decimal counting, but examples of binary counters are not common. The device* shown in Fig. 2—8, however, does verify the Binary Counting Table in Fig. 2—3. Continuous reference should be made to Fig. 2—8 while reviewing the binary counting operation.

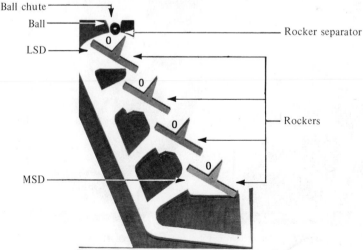

Figure 2—8   Mechanical binary counter

* All major parts of the mechanical binary counter are labeled to conform with nomenclature used in our explanation.

All "rockers" are initially positioned so that the 1s are covered and the 0s are uncovered. This presets the counter to 0000. The first ball dropping through the chute is deflected to the *left* by the Least Significant-Digit (LSD) rocker separator. The weight of the ball causes the rocker to tip to the 1 position, and the ball is then routed to the storage area via the return chute. One ball has been used (Fig. 2–9a), so the binary counter now displays 0001 (1_{10}).

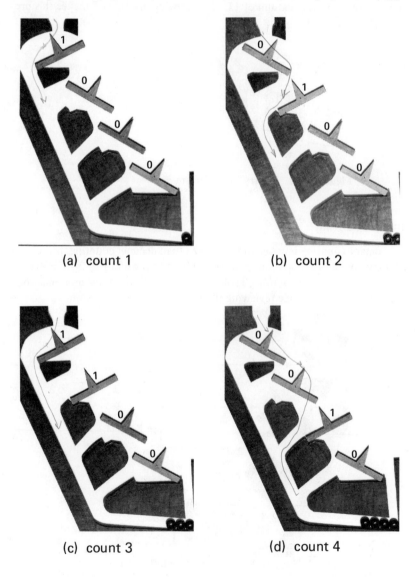

(a) count 1   (b) count 2

(c) count 3   (d) count 4

Figure 2–9   Mechanical binary counter, counts 1, 2, 3, and 4

The second ball through the chute strikes the *right* side of the LSD rocker, causing it to tip to the right which covers the 1 and uncovers the 0. The ball rolls from the right side of the LSD rocker to the left side of the next rocker. Ball-weight causes this rocker to tip to the left, thus covering the 0 and uncovering the 1. The second ball is now routed to the storage area via the return chute. Two balls have been used (Fig. 2–9b), so the binary counter now displays 0010 (2_{10}).

The third ball through the chute strikes the left side of the LSD rocker, tipping it to the 1 position. The ball is routed to the storage area via the return chute. The binary counter display (Fig. 2–9c) is now 0011 (3_{10}).

The fourth ball through the chute strikes the right side of the LSD rocker, tipping it to the 0 position. The ball rolls from the LSD rocker to the right side of the next rocker, tipping it to the 0 position. It then rolls from this rocker to the left side of the next rocker. This tips to the 1 position, and the fourth ball returns to the storage area. The binary counter (Fig. 2–9d) now displays 0100 (4_{10}).

As balls continue to drop through the chute, the counter is actually reproducing the Binary Counter Table in Fig. 2–3. We recommend that the reader continue this exercise until decimal count 15 is reached — then proceed to one additional count (16_{10}) to gain a comprehensive understanding of the mechanics of binary counting.

2–5 THE FLIP-FLOP (FF)

The mechanical binary counter performs its required functions adequately but slowly. If operational speed is needed, a *bistable electronic circuit*, called a *flip-flop* (FF), should be used. This bistable FF duplicates the action of each of the bistable rockers on the mechanical counter. (p. 16)

Physical Form

Flip-flops (FFs) are available in different physical forms. In early digital computers, electromechanical relays performed the FF function. Vacuum-tube circuit FFs are still found in some older equipment. But in recent years, semiconductor and integrated-circuit FFs have replaced practically all vacuum tubes and relays.

Discrete component assemblies, consisting of individual transistors, diodes, capacitors, and resistors, have been a feature of digital devices for a number of years. They are found in many different sizes and shapes. Now semiconductor technology has developed subminiature assemblies that contain *all* the active and passive functions of a FF on a *single chip of silicon* less than 0.05 inch square and 0.006 inch thick. These assemblies, called *integrated circuits* (ICs), package complex logic functions into a space no larger than a discrete transistor.

In the future, scientists and engineers, instead of merely providing multiple FFs in a subminiature package, will furnish complete operations in the same package. Devices containing numerous FFs, gates, and so forth, which are already interconnected in a single module, are already commonly available. Complex ICs, capable of performing high-level mathematical operations, have made their appearance. Figure 2–10 shows this evolution in digital logic packaging from vacuum-tube circuitry to the complex IC.

(a)

(b)

Figure 2–10   Digital logic packaging: (a) vacuum tube to IC; (b) integrated circuit closeup (courtesy IBM, White Plains, NY)

Symbology

Because we are not concerned with the individual electronic parts and their interconnections, a single symbol is sufficient to represent the FF. Although many different symbols are used, this text will adhere to the United States military standard containing graphic symbols for use with logic diagrams (MIL-STD-806B). As each symbol* is introduced, its use will be explained.

The FF symbol that we have selected was first shown in Fig. 1−15b. Its shape† (and that of most logic symbols) denotes the logical operation performed by the device it represents. The standard FF has *three inputs*, labeled S, C, and T. The S is the SET *input*; the C is the CLEAR or RESET *input*; and the T is the TRIGGER or TOGGLE *input*. In many cases, only the inputs required to perform the FFs operation in the logic circuit are shown. Two outputs are provided, named the Q *output* and the $\bar{Q}$ *output*. ($\bar{Q}$ is spoken and read as "NOT Q.")

Rules of Operation

The reaction of a FF to the original application of power is both unimportant and indeterminate. Most analog devices assume a specific input/output (I/O) relationship when power is initially applied. But since the FF is bistable, it may take on *either* of its two states upon application of power. Many different terms are used to describe these states: ON and OFF, SET and RESET, SET and CLEAR, ONE and ZERO, 1 and 0, and Q and $\bar{Q}$. The basic rules of FF operations found within the framework of MIL-STD-806B provide a means of determining the state of a FF.

> (1)
> If the flip-flop is in the set state, the Q output is HIGH
> and the $\bar{Q}$ output is LOW.

Now it becomes necessary to define what we mean by HIGH and LOW. Since there are many ways of implementing the FF, different combinations of circuit voltages or currents can be used to indicate HIGH and LOW. So, we must again paraphrase MIL-STD-806B.

> A HIGH output is relatively high in respect to the LOW
> output.

Imagine a physical analogy: if you are on the "up" side of a teeter-totter, you are *relatively high* with respect to the "down" side; or if you are on the third floor of a building, you are relatively high compared with the ground floor. For the electronically oriented reader, 0 volts might be LOW, and +10 volts might be HIGH; while −10 volts might be LOW and 0 volts might be HIGH. It is the

---

* Appendices B and C contain all symbols used in this text and their relationship to other commonly used symbols.

† It is important in drawing symbols to use the proper size and shape. Templates conforming to MIL-STD-806B are readily available and should be used when drawing logic diagrams.

separation between two discrete values of some parameter that determines HIGH and LOW. In any case, HIGH is *always* higher than LOW! Figure 2–11 presents a visual representation of this concept.

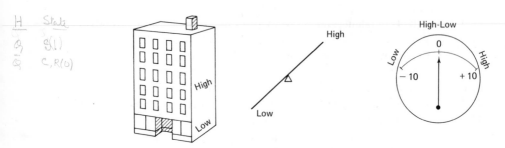

Figure 2–11   HIGH and LOW analogies

Flip-Flop Operation

A common method of illustrating FF operation is to draw diagrams representing HIGHs and LOWs. (*H* normally denotes HIGH, and *L* denotes LOW.) These diagrams are called *waveforms* and are nothing more than the "pictures" that would appear on a cathode-ray oscilloscope (CRO) when the FF is operating. All waveforms are labeled to firmly define which level is HIGH and which is LOW.

As we stated earlier, if the Q output is HIGH, the FF is in the SET or the 1 state. If, however, the $\overline{Q}$ output is HIGH, the FF is in the CLEAR, RESET, or 0 state. A very interesting aspect of the FF is that the Q output and the $\overline{Q}$ output are always *opposite* to each other. When the Q output is HIGH, the $\overline{Q}$ output is LOW and vice-versa. This is equivalent to having available *both* a binary number and its ones complement. That availability is quite advantageous in performing arithmetic operations in the binary number system.

No matter what state the FF is in, it can be put into the SET state by applying a HIGH-going signal to the SET input; similarly a HIGH-going transition on the CLEAR input places the FF into the CLEAR state. A HIGH-going transition on the TRIGGER input also causes the FF to change state; if it is in the CLEAR state it will go SET, and if it is in the SET state it will go CLEAR. All of the operations of the FF may be seen in Fig. 2–12, which is a waveform diagram of FF states based on specific input signals.

This diagram represents the conventional presentations encountered in instruction handbooks for digital data processing equipment. Such diagrams are *time-oriented*, where the time starts on the *left* side of the page and progresses to the right. Time intervals are represented by subscripted letters (t_0, t_1, etc.) with t_0 usually representing the start of an operation, as it does in Fig. 2–12. The FFs response to various inputs (keyed to time intervals) is listed in Table 2–1.

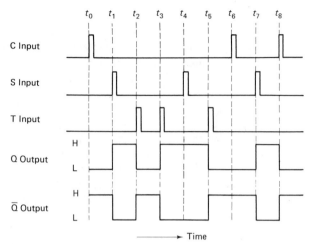

Figure 2—12   Flip-flop operation

Table 2—1   FF response to various inputs at selected time intervals

Time	Input Terminal	Response
t_0	CLEAR	FF goes to CLEAR state
t_1	SET	FF goes to SET state
t_2	TOGGLE	FF changes state (CLEAR)
t_3	TOGGLE	FF changes state (SET)
t_4	SET	FF stays in SET state
t_5	TOGGLE	FF changes state (CLEAR)
t_6	CLEAR	FF stays in CLEAR state
t_7	SET	FF goes to SET state
t_8	CLEAR	FF goes to CLEAR state

One specific combination of input signals in Fig. 2—12 is conspicuous by its absence. When actuating signals are applied to 2 or more inputs *simultaneously*, the results are *indeterminate* — the FF *may* or *may not* change state! Such a situation is unacceptable, and precautions must be taken to prevent such *multiple-input* signals. Methods of preventing indeterminate operation will be discussed in later chapters.

Often it is inconvenient to use either CROs or other test instruments capable of displaying waveforms. An alternate method of determining the state of a FF employs *indicator lamps*. For example, by using electronic circuitry called *lamp drivers* in conjunction with indicator lamps, it is possible to cause the lamp to be turned on fully when the input is HIGH and turned off fully when the input is LOW. Connection of the lamp driver to the Q output of the FF enables visual indication of the state of the FF. Thus when the FF is in the SET state and the Q output is HIGH, the indicator lamp is ON; when the FF is in the CLEAR state and the Q output is LOW, the indicator lamp is OFF.

The waveforms in Fig. 2—12 can be modified, as they were in Fig. 2—13, to show visual indications correlated with waveforms and FF states. The use of visual indicators in place of waveform display is commonly encountered not only in such digital devices as computers but also in training devices (Fig. 2—14).

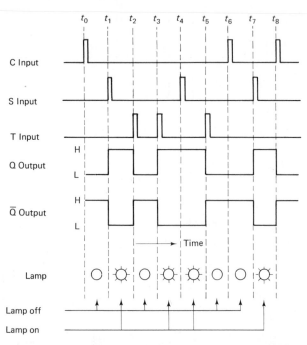

Figure 2—13   Flip-flop operation with visual indication

Figure 2—14   Digital techniques training device (courtesy Digital Equipment Corp., Maynard, Mass.)

2–6 A BINARY COUNTER

Logic Diagram

An examination of our explanations of FF operation and binary counting should clearly show that it is possible to count only 0 and 1 with a single FF. To count to numbers greater than 1 in the binary system, we must devise a way to count the second, third, and fourth orders (and so on). If properly connected, *one* FF may be used to represent *each* binary order. A logic diagram to accomplish this counting function is shown in Fig. 2–15.

Note that the logic diagram and waveforms in Fig. 2–15 do not agree with conventional drawing practices. However, to simplify the explanation of circuit operation, a certain amount of "artistic license" was exercised. Each FF was given a name that corresponds to the binary order that it represents. The right-

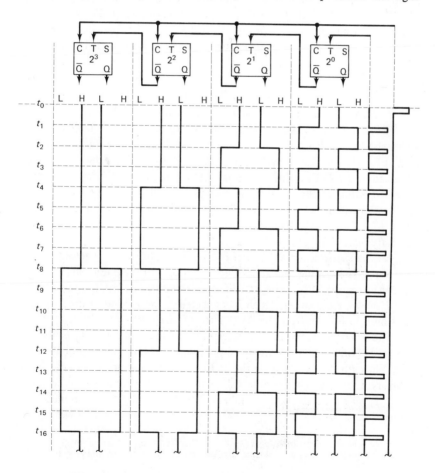

Figure 2–15   Binary counter and waveforms

most FF is labeled 2^0; the FF immediately to the left is labeled 2^1. Note the correspondence between FF positions, names, and order when compared with conventional presentation of binary numbers, i.e., LSD on the right and MSD on the left.

Also note that the accompanying waveforms are "sideways," occurring from top to bottom instead of from left to right. Required power to each FF is assumed to be applied, and it is further assumed that FF states prior to t_0 either are unimportant or are to be disregarded,

To assure that counting begins with 0 at t_0, an actuating signal is applied simultaneously to the C (CLEAR) input of all FFs. Note that all C inputs are connected together in Fig. 2–15 and that the CLEAR pulse occurs on the time scale at t_0. Thus all FFs start out in the 0 state, and the binary number represented is 0000_2. The T (TRIGGER) input* of the LSD (2^0) FF is connected directly to the source of actuating signals to be counted. The $\overline{Q}$ output of the LSD FF connects directly to the T input of the next-most-significant digit FF (2^1). In a similar manner, $\overline{Q}$ outputs of the other FFs connect to the T inputs of next-most-significant digits, until the MSD FF is reached.

Analysis

At t_1 the first actuating signal to be counted is applied to the T input of the 2^0 FF, causing it to change state. The Q output (which was LOW) goes HIGH, and the $\overline{Q}$ output (which was HIGH) goes LOW. Since the $\overline{Q}$ output of the 2^0 FF is directly connected to the T input of the 2^1 FF, a changing signal is coupled to the 2^1 FF. However, it is a LOW-going signal and causes no change. If the 2^1 FF doesn't change state, no change is coupled to the T input of the 2^2 FF and it will not change state. (Similarly, no change occurs in the 2^3 FF.) With the 2^3, 2^2, and 2^1 FFs in the RESET state and the 2^0 FF in the SET state, the binary number represented is 0001.

No change occurs in any of the FFs until the arrival of the actuating signal at t_2. The leading edge (HIGH-going portion) of the pulse causes the 2^0 FF to change from the SET to the RESET state, resulting in a HIGH-going $\overline{Q}$ output. This HIGH-going transition, which is connected to the T input of the 2^1 FF, causes the 2^1 FF to go from the RESET to the SET state.

The $\overline{Q}$ output of the 2^1 FF, which is connected to the T input of the 2^2 FF, has no effect on the 2^2 FF state, since the change is LOW-going – not HIGH-going. Since no change occurs in the 2^2 FF, no change occurs in the 2^3 FF and the binary number 0010 is established. Note that this corresponds to decimal number 2, which represents the number of input-actuating signals applied to the counter.

---

* Methods of generating trigger inputs are discussed later. At this point, it is assumed that such methods do exist and that the trigger inputs are arriving at specific times, as pointed out in the diagram.

The arrival of the third actuating trigger pulse, t_3, will once again cause the 2^0 FF to change state. In going from RESET to SET state, the $\overline{Q}$ output goes from HIGH to LOW. Again, according to the rules for operation of the FF, a LOW-going signal transition will not actuate the next FF; therefore the 2^1 FF will not change state. Consequently, both the 2^2 and the 2^3 FFs remain unchanged. The binary number now stored is 0011_2, which corresponds to 3_{10}.

Application of the t_4 trigger pulse causes the 2^0 FF to RESET, and the resulting HIGH-going transition of the $\overline{Q}$ output triggers the 2^1 FF from SET to RESET state. Since the $\overline{Q}$ output of the 2^1 FF then changes from LOW to HIGH, it also activates the T input of the 2^2 FF, causing it to be SET. An evaluation of the FF states now shows the following results.

> 2^0 RESET, storing a binary 0
> 2^1 RESET, storing a binary 0
> 2^2 SET, storing a binary 1
> 2^3 RESET, storing a binary 0

The counter stores the number 4_{10}, represented as 0100_2.

The reader should continue our logic analysis of the counter in order to gain valuable experience in this method of analysis. In particular, the reaction of the counter at t_{16} should be critically analyzed — a number of interesting conclusions may be reached. For example, in this particular logic circuit, a higher order FF can change state only when the next-lowest-order FF changes from RESET to SET. Examination of the waveforms for the 2^0 FF shows that it changes state *every time* an input trigger pulse is received. The 2^1 FF changes state for *every two* input triggers; the 2^2 FF changes state for every *four* input triggers; and the 2^3 FF changes state for every *eight* input triggers. Note the correspondence between FF names and the number of times they change state. Obviously, the name of the FF (which represents the positional order in the binary system) not only defines the decimal value of that position but also defines how many inputs may be counted prior to change of state. This implies a division capability, which will be discussed in detail later.

The State Table

A table of FF state correlated with input triggers may be derived using a complete logic analysis of the counter and the resulting waveforms (Fig. 2–16). The table starts with all FFs in the RESET state (all 0s). Then, when input triggers are applied to the T input of the 2^0 FF, the individual FFs change state, as described earlier. The *state table* therefore describes counter operations in a manner as meaningful as the waveforms. The reader should carefully compare Fig. 2–16 with Fig. 2–15 to establish this correlation.

2^3	2^2	2^1	2^0
0	0	0	0
0	0	0	1
0	0	1	0
0	0	1	1
0	1	0	0
0	1	0	1
0	1	1	0
0	1	1	1
1	0	0	0
1	0	0	1
1	0	1	0
1	0	1	1
1	1	0	0
1	1	0	1
1	1	1	0
1	1	1	1

Figure 2—16   Binary counter state table

EXERCISES

Convert the following decimal numbers to exponential form.

2— 1.   7
2— 2.   206
2— 3. 5280
2— 4.   0.6836
2— 5.   914.254

Convert the following exponential numbers to decimal numbers.

2— 6. 5×10^0
2— 7. $1 \times 10^2 + 3 \times 10^1 + 7 \times 10^0$
2— 8. $3 \times 10^3 + 6 \times 10^2 + 4 \times 10^1 + 2 \times 10^0$
2— 9. $1 \times 10^{-1} + 2 \times 10^{-2} + 4 \times 10^{-3} + 3 \times 10^{-4}$
2—10. $4 \times 10^2 + 7 \times 10^1 + 6 \times 10^0 + 8 \times 10^{-1} + 3 \times 10^{-2} + 9 \times 10^{-3}$

Convert the following binary numbers to exponential form.

2—11.   1
2—12.   101
2—13.   11001
2—14.   0.1101
2—15. 1100111.1011

Convert the following exponential numbers to binary numbers.

2—16. $0 \times 2^1 + 1 \times 2^0$
2—17. $1 \times 2^2 + 1 \times 2^1 + 0 \times 2^0$
2—18. $1 \times 2^5 + 1 \times 2^4 + 0 \times 2^3 + 1 \times 2^2 + 0 \times 2^1 + 1 \times 2^0$
2—19. $0 \times 2^{-1} + 1 \times 2^{-2} + 1 \times 2^{-3} + 0 \times 2^{-4} + 1 \times 2^{-5}$
2—20. $1 \times 2^3 + 1 \times 2^2 + 0 \times 2^1 + 1 \times 2^0 + 0 \times 2^{-1} + 1 \times 2^{-2}$

Convert the following decimal numbers to their binary equivalents.

2–21.	3	2–31.	126
2–22.	7	2–32.	400
2–23.	8	2–33.	2048
2–24.	12	2–34.	3000
2–25.	13	2–35.	4000
2–26.	16	2–36.	0.5
2–27.	24	2–37.	0.125
2–28.	38	2–38.	0.03125
2–29.	79	2–39.	0.42
2–30.	84	2–40.	3.1416

Convert the following binary numbers to their decimal equivalents.

2–41.	101	2–51.	110101
2–42.	111	2–52.	0.0011
2–43.	1101	2–53.	0.0101
2–44.	01101	2–54.	1.01
2–45.	10111	2–55.	1.011101
2–46.	11010	2–56.	11.101
2–47.	0011101	2–57.	0101.1001
2–48.	100000	2–58.	1101.0001
2–49.	101101	2–59.	11011.1101
2–50.	110001	2–60.	10000100000.010001

PROBLEM APPLICATIONS

2– 1. Define the term *radix*.

2– 2. Why is the binary number system used in digital logic?

2– 3. What is the decimal value of the largest number that can be represented by a 10-digit binary number?

2– 4. How many binary digits would it take to represent the decimal number 1000?

2– 5. Explain the general concept of decimal counting, i.e., the use of symbols and shift-markers.

2– 6. Explain the general concept of binary counting, i.e., the use of symbols and shift-markers.

2– 7. What is a *flip-flop* (FF)?

2– 8. Draw the standard symbol for a FF and label all inputs and outputs.

2– 9. Define HIGH and LOW as used in digital logic.

2–10. How can the state of a FF be determined?

2–11. Why are waveform drawings used to show FF operation?

2–12. Define the activating signal that causes FF transition.

2–13. What is the function of the CLEAR input on a FF?

2–14. What is the function of the SET input on a FF?

2–15. What is the function of the TRIGGER input on a FF?

2–16. Completely describe the rules of operation for a standard FF.

2–17. Draw a timing diagram showing the Q output terminal of a FF with 5 sequential pulses applied to the TRIGGER input. Assume that the FF was RESET at t_0.

2—18.   Assume that a FF is in the RESET state. What state will the FF assume
        after a HIGH-to-LOW transition is applied to the
        (a) CLEAR input?
        (b) TRIGGER input?
        (c) SET input?

2—19.   A FF used as part of a binary counter is labeled as "2^4." Explain what
        this name means and what can be determined about the FF from its label.

2—20.   Draw a timing diagram showing the Q output of a 3-stage binary counter,
        assuming the following sequence.
        (a) The counter is initially cleared.
        (b) Seven pulses are applied to the T input of the first stage.
        (c) A LOW-to-HIGH transition is applied to the CLEAR input.

2—21.   Draw a logic diagram of a 5-stage binary counter and explain its opera-
        tion for each input pulse to be counted.

2—22.   What is the maximum count attainable from a 5-stage binary counter?

2—23.   How many FFs are required to store decimal number 10 in the binary
        scale?

2—24.   Using your imagination, list at least five potential applications for a FF.

3

Basic Combinational Logic

3–1 THE FUNDAMENTALS OF COMBINATIONAL LOGIC

Most digital equipment consists of two general types of logic circuits: *combinational* and *sequential*. In a *combinational logic circuit*, the output depends on the *input combination* present when the output is examined. But a *sequential logic circuit* contains memory elements, so the output depends not only on the state of the input at any given instant but also on the *entire input history* of the circuit. A complete logic system can contain both types of circuits.

Combinational circuits can make *decisions*, which range from the very simple to the extremely complex. This capability is one of the primary reasons for the application of digital techniques to many fields formerly dominated by analog techniques. Understanding how these decisions are made requires some knowledge of the mathematics of logic (Boolean algebra). It also requires the application of such "tools" as *truth tables*, *symbology*, and *waveforms*. This chapter will explore the fundamentals of combinational (*decision-making*) elements.

Basic Logic Operations

Logic operations are quite similar to algebraic operations. We will therefore begin by reviewing one of the simplest algebraic operations – *addition*. Practically everyone uses addition daily, but how often is the structure of this operation considered?

Addition is frequently considered to be a *primitive concept* because it is defined only by the properties assumed for it. The addition operation combines two or more numbers to obtain a new number, called the sum. Figure 3–1 illustrates this concept.

Algebraic operators are similar to grammatical verbs, since they indicate "doing something." Thus the arithmetic statement 3 + 5 = 8 may be translated to 5 *must be added* to 3 to obtain 8. Note the *action meaning* of the + symbol.

In algebra letter symbols represent variables whose numerical values range from plus infinity to minus infinity and whose number form can be real, imaginary, complex, rational, irrational, integral, or fractional. The addition operation (+) in algebra can therefore be expressed in the following manner.

$$a + b = c$$

1	1	2
1	2	3
2	2	4
3	5	8
5	4	9

The other algebraic operations, subtraction, multiplication, division, exponentiation, and root extraction, may be approached in a similar manner — definitions* are developed from whatever is to be done to a specified group of numbers. Hence the crucial facts for the reader to remember are that an algebraic operator indicates *what is to be done* and that its meaning is *strictly a matter of definition.*

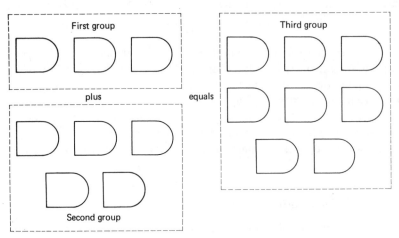

3 AND gates (First Group) added to 5 AND gates (Second Group)
results in a new grouping (Third Group) of 8 AND gates.

Figure 3–1   Concept of addition

* Once the definition of subtraction is developed, the other four operations may be derived from the addition and subtraction definitions.

This system of mathematics, using these six fundamental operations, can provide a solution for most of our everyday problems. For example, both the electronics technician and the engineer employ the system as a tool in understanding and designing complex equipment. In the specialized field of digital equipment, however, a simplified system of mathematics is applied. It is called *Boolean algebra* and utilizes the principles of *symbolic logic.*

The basic operations in symbolic logic are CONJUNCTION, DISJUNCTION, and NEGATION. Each of these operations, like those in algebra, can be described in terms of the effect it produces when it is used with specified variables and constants. In symbolic logic, letter symbols represent variables that are always *two-valued* (binary);* these may be expressed as one or zero, true or false, pulse or no pulse, plus voltage or minus voltage, and so forth.

The first logic operation, CONJUNCTION, can be defined as follows.

c is true only when a and b are simultaneously true.

The symbol for CONJUNCTION** is ($\cdot$), so the logic function just defined is written as $a \cdot b = c$ or $ab = c$. $ab = c$ is read as "a AND b is equal to c." Thus the logic operation AND has a mathematical name, CONJUNCTION, and a symbol ($\cdot$), just as the algebraic operation "times" uses the mathematical name multiplication and the symbol ($\cdot$). Table 3–1 shows all possible conjunctive combinations of the variables a and b having a resultant value of c.

The second logic operation, DISJUNCTION, can be defined in the following manner.

c is true if either a is true or b is true, or if both a and b are true at the same time.

The symbol for DISJUNCTION† is (+), so the logic function just defined is written as $a + b = c$. $a + b = c$ is read as "a OR b is equal to c." Thus the logic

Table 3–1   CONJUNCTION

$$a \cdot b = c$$

a	b	c
0(F)	0(F)	0(F)
0(F)	1(T)	0(F)
1(T)	0(F)	0(F)
1(T)	1(T)	1(T)

No other possibilities exist

---

*For the purpose of our discussion, 1 and true are used synonymously, as are 0 and false; i.e., **true = 1** and **false = 0**.

** Due to its similarity to conventional algebraic multiplication, this operation is sometimes called *logical multiplication.*

† Due to its similarity to conventional algebraic addition, this logic operation is sometimes called *logical addition.*

operation OR has a mathematical name, DISJUNCTION, and a symbol (+), just as the algebraic operation "plus" uses the mathematical name addition and the symbol (+). Table 3–2 shows all possible disjunctive combinations of the variables a and b having a resultant value of c.

The last of the fundamental logic operations is NEGATION and can be defined as follows.

> If b is the negation of a, then a is true if b is false and a is false if b is true.

The symbol for NEGATION* is an overbar placed above a variable $(\overline{b})$, so the logic operation just defined is written as $a = \overline{b}, \overline{a} = b$. $a = \overline{b}$ is read as "a is equal to NOT b," conversely, $\overline{a} = b$ is read as "NOT a is equal to b." Table 3–3 shows all combinations of variables a and b under the operation of NEGATION.

Other types of logic operations exist. However, these can always be derived from the three fundamental operations that we have just discussed.

Table 3–2   DISJUNCTION

$a + b = c$

a	b	c
0(F)	0(F)	0(F)
0(F)	1(T)	1(T)
1(T)	0(F)	1(T)
1(T)	1(T)	1(T)

No other possibilities exist

Table 3–3   NEGATION

a	b
0(F)	1(T)
1(T)	0(F)

Logic Variables and Constants

In Chapter 1 Boolean algebra was defined as the mathematics of logic. Thus the basic logic operations of Boolean algebra link existing techniques and knowledge with the new field of digital logic. Just as in other branches of mathematics, certain postulates and theorems† are used to define the characteristics of Boolean algebra. We will employ only a few of these rules here, but as additional rules are required, they will be developed and explained.

Since the basic operations of digital logic (CONJUNCTION, DISJUNC-TION, and NEGATION) have been defined, we can now turn to an examination of the elements to which they will be applied — Boolean *constants* and *variables*.

---

* No *direct* correlation to this operation exists in conventional algebra.

† Appendix A is a composite listing of these postulates and theorems.

In ordinary algebra all integers and fractions may be used as constants. Since a *constant* is something (a value, quantity, etc.) that has a *fixed meaning*, conventional algebra contains a large number of constants. Fortunately, Boolean algebra contains only *two* possible constants, 0 and 1.

A *variable*, on the other hand, is a quantity that can assume the value of *any constant* at *any time*. Since there are only two constants in Boolean algebra, Boolean variables may assume only one of the two values (i.e., 0 or 1). Variables are conventionally denoted by capital letters — A, B, Q, X, and so forth.

The NEGATION operation is also commonly called INVERSION. Table 3–3, which defined NEGATION, showed that when the variable a is 0, the NEGATION of a is 1; also when a is 1, the NEGATION of a is 0. Since variables may assume the values of constants, this NEGATION statement could also be written as

$$a = 0, \quad \bar{a} = 1 \quad \text{and} \quad \bar{a} = 0, \ a = 1$$

therefore $\quad\quad\quad\quad 0 = \bar{1} \quad \text{and} \quad \bar{0} = 1$

When a Boolean constant is *inverted*,* it is made equal to the other constant.

The definition of the AND operation was included in our discussion of the CONJUNCTION operation. So, from Table 3–1, it becomes apparent that if two or more constants are ANDed together, all of the constants must be 1 for the result to be 1. Since 1 has been previously equated with the "true" form of a variable and since variables can assume the values of constants, if two or more constants are ANDed together, all must be true for the result to be true. For example, if $A \cdot B = 1$, both A and B must be equal to 1. If either of the variables were 0, then a 0 would be ANDed with a 1, and the result would have to be 0.

Truth Tables

As indicated in Ch. 1, a *truth table* for a logic device is a tabular listing of all possible combinations of input, and the resultant output for each combination. It represents a useful method of condensing all information concerning the functional characteristics of a logic device and provides a visualization that makes these characteristics more readily perceivable. The name "truth table" comes from formal *symbolic logic*,† where the truth or falsity of statements is investigated.

In symbolic logic, English-language phrases are combined into sentences, and the truth value of these sentences is then investigated. Of course, many rules govern the types and forms of phrases allowed. But one or two examples should be sufficient to help us bridge the gap between symbolic logic and the digital applications of logic.

---

* In many texts, a variable without an overbar (e.g., A) is considered to be in its "true" form, while the "inverted" form of the variable has an overbar (e.g., $\bar{A}$).

† Many texts have been written about formal symbolic logic, so the reader who desires to delve into its intricacies will have no difficulty in locating the appropriate material.

Phrases used in symbolic logic are called propositions and are usually in the *declarative* form. Since these phrases are assertive in nature, they contain an inherent truth value; i.e., *they are either true or false.* A phrase such as "the FF is in the SET state" is a good example. If the FF is SET, the phrase is true; if it is not SET (CLEAR), the phrase is false. "The TRIGGER pulse is present" represents another phrase with inherent truth value. If the TRIGGER pulse is present, the phrase is true; if it is absent, the phrase is false.

Phrases may be combined by using *logical connectives* (previously called logical operators) to form more complex phrases. When two phrases are joined by the logical operator AND, however, new meanings may appear. Thus the preceding phrases become "the FF is in the SET state AND the TRIGGER pulse is present." The truth value of this new phrase now depends not only on the truth value of each of the individual phrases but also on the definition of the logical operator AND. An analysis, using English-language statements, might appear as follows.

> If the FF is not SET (CLEAR), the truth value of the first
> phrase is false. If the TRIGGER pulse is not present,
> the truth value of the second phrase is false. Therefore,
> if the FF is CLEAR and the TRIGGER pulse is absent, the
> the truth value of the complex statement is false.

All such possible combinations in symbolic logic may be so examined — at the obvious expense of a great many words!

The application of a form (mathematical) of symbolic logic in which a special set of symbols replaces the language statements would greatly simplify the presentation. Thus letting the variable A represent "the FF is in the SET state" and the variable B represent "the TRIGGER pulse is present," the complete statement can be written as "$A \cdot B$." All possible combinations of A and B can now be placed in matrix or table form, and the truth value of $A \cdot B$ examined quite easily. Table 3–4 defines the original logical statement.

This discussion may be carried one step further — to investigate the relationships of dependent and independent variables. A small addition to the previous statement will demonstrate these relationships.

> "The FF is CLEARed if the FF is SET AND the
> TRIGGER pulse is present."

A has already been defined as "the FF is SET" and B as "the TRIGGER pulse is present." The additional phrase "the FF will be CLEARed" is *dependent* on the condition of A and B, just as it would be in conventional algebra. By letting X represent this dependent variable, the complete statement becomes $X = A \cdot B$. Modifying the truth table to reflect this new statement results in Table 3–5.

Note that each row of the last column gives the truth value of the function that is determined by the truth values entered in the preceding columns of that row.

By applying techniques borrowed from its "parent" — symbolic logic — digital logic becomes a more understandable and therefore more useful tool.

Table 3—4 Truth table defining the logical statement $A \cdot B$

A	B	$A \cdot B$
F	F	F
F	T	F
T	F	F
T	T	T

Table 3—5 Truth table defining the logical equation $X = A \cdot B$

A	B	X
F	F	F
F	T	F
T	F	F
T	T	T

An Introduction to Boolean Algebra

Having discussed constants, variables, and basic operations, we shall now investigate some simple *theorems* (laws) of Boolean algebra. The tables used to define CONJUNCTION, DISJUNCTION, and NEGATION provide some insight into a method of proof for certain theorems. So the basic operation tables have been reproduced in Fig. 3—2, along with a reference designation for each row. The information contained in each of these rows constitutes the *basic postulates* of Boolean algebra. A list of the postulates (keyed to Fig. 3—2) follows:

①	$0 \cdot 0 = 0$	(P—1)
②	$0 \cdot 1 = 0$	(P—2)
③	$1 \cdot 0 = 0$	(P—3)
④	$1 \cdot 1 = 1$	(P—4)
⑤	$0 + 0 = 0$	(P—5)
⑥	$0 + 1 = 1$	(P—6)
⑦	$1 + 0 = 1$	(P—7)
⑧	$1 + 1 = 1$	(P—8)
⑨	$0 = \bar{1}$	(P—9)
⑩	$1 = \bar{0}$	(P—10)

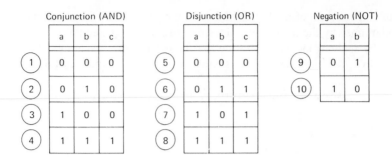

Figure 3–2   Basic logical operations

From these postulates, we have developed thirteen basic theorems of Boolean algebra.

Because we defined a variable as a quantity that can assume the value of any constant, the variables in our theorems are either 0 or 1 (they are the *only* constants available in the binary system).

THEOREM 1: $\boxed{A \cdot 0 = 0}$ (T–1)

Proof: If A is equal to 0, then Postulate 1 is proof.

$$A = 0, \quad \therefore \; 0 \cdot 0 = 0$$

If A is equal to 1, then Postulate 3 is proof.

$$A = 1, \quad \therefore \; 1 \cdot 0 = 0$$

THEOREM 2: $\boxed{0 \cdot A = 0}$ (T–2)

Proof: If A is equal to 0, then Postulate 1 is proof.

$$A = 0, \quad \therefore \; 0 \cdot 0 = 0$$

If A is equal to 1, then Postulate 2 is proof.

$$A = 1, \quad \therefore \; 0 \cdot 1 = 0$$

THEOREM 3: $\boxed{A \cdot 1 = A}$ (T–3)

Proof: If A is equal to 0, then Postulate 2 is proof.

$$A = 0, \quad \therefore \; 0 \cdot 1 = 0$$

If A is equal to 1, then Postulate 4 is proof.

$$A = 1, \quad \therefore \; 1 \cdot 1 = 1$$

THEOREM 4: $\boxed{1 \cdot A = A}$ (T–4)

Proof: If A is equal to 0, then Postulate 3 is proof.

$$A = 0, \quad \therefore \; 1 \cdot 0 = 0$$

If A is equal to 1, then Postulate 4 is proof.

$$A = 1, \quad \therefore \; 1 \cdot 1 = 1$$

THEOREM 5: $\boxed{A \cdot A = A}$ (T–5)

Proof: If A is equal to 0, then Postulate 1 is proof.

$$A = 0, \quad \therefore \; 0 \cdot 0 = 0$$

If A is equal to 1, then Postulate 4 is proof.

$$A = 1, \quad \therefore \; 1 \cdot 1 = 1$$

THEOREM 6: $\boxed{A \cdot \overline{A} = 0}$ (T–6)

Proof: If A is equal to 0, then $\overline{A}$ is equal to 1 by Postulate 9.

$$\left. \begin{array}{l} A = 0 \\ \overline{A} = 1 \end{array} \right\} \quad \therefore \; 0 \cdot 1 = 0$$

If A is equal to 1, then $\overline{A}$ is equal to 0 by Postulate 10.

$$\left. \begin{array}{l} A = 1 \\ \overline{A} = 0 \end{array} \right\} \quad \therefore \; 1 \cdot 0 = 0$$

THEOREM 7: $\boxed{A + 0 = A}$ (T–7)

Proof   If A is equal to 0, then Postulate 5 is proof.

$$A = 0, \quad \therefore \; 0 + 0 = 0$$

If A is equal to 1, then Postulate 7 is proof.

$$A = 1, \quad \therefore \; 1 + 0 = 1$$

THEOREM 8: $\boxed{0 + A = A}$ (T–8)

Proof: If A is equal to 0, then Postulate 5 is proof.

$$A = 0, \quad \therefore \; 0 + 0 = 0$$

If A is equal to 1, then Postulate 6 is proof.

$$A = 1, \quad \therefore \; 0 + 1 = 1$$

THEOREM 9: $\boxed{A + 1 = 1}$ (T–9)

Proof: If A is equal to 0, then Postulate 6 is proof.

$$A = 0, \quad \therefore \quad 0 + 1 = 1$$

If A is equal to 1, then Postulate 8 is proof.

$$A = 1, \quad \therefore \quad 1 + 1 = 1$$

THEOREM 10: $\boxed{1 + A = 1}$ (T–10)

Proof: If A is equal to 0, then Postulate 7 is proof.

$$A = 0, \quad \therefore \quad 1 + 0 = 1$$

If A is equal to 1, then Postulate 8 is proof.

$$A = 1, \quad \therefore \quad 1 + 1 = 1$$

THEOREM 11: $\boxed{A + A = A}$ (T–11)

Proof: If A is equal to 0, then Postulate 5 is proof.

$$A = 0, \quad \therefore \quad 0 + 0 = 0$$

If A is equal to 1, then Postulate 8 is proof.

$$A = 1, \quad \therefore \quad 1 + 1 = 1$$

THEOREM 12: $\boxed{A + \overline{A} = 1}$ (T–12)

Proof: If A is equal to 0, then $\overline{A}$ is equal to 1 by Postulate 9.

$$\begin{aligned} A &= 0 \\ \overline{A} &= 1 \end{aligned} \quad \therefore \quad 0 + 1 = 1$$

If A is equal to 1, then $\overline{A}$ is equal to 0 by Postulate 10.

$$\left. \begin{aligned} A &= 1 \\ \overline{A} &= 0 \end{aligned} \right\} \quad \therefore \quad 1 + 0 = 1$$

THEOREM 13: $\boxed{\overline{\overline{A}} = A}$ (T–13)

Proof: If A is equal to 0, then $\overline{A}$ is equal to 1 by Postulate 9. If $\overline{A}$ is equal to 1, then another negation operation $(\overline{\overline{A}})$ makes $\overline{\overline{A}}$ equal to 0 by Postulate 10.

$$\left.\begin{array}{l} A = 0 \\ \overline{\overline{A}} = 0 \end{array}\right\} \quad \therefore \ 0 = 0, \quad \text{proving} \ \ \overline{\overline{A}} = A.$$

Similarly, if A is equal to 1, then $\overline{A}$ is equal to 0 by Postulate 10. If $\overline{A}$ is equal to 0, then another negation operation $(\overline{\overline{A}})$ makes $\overline{\overline{A}}$ equal to 1 by Postulate 9.

$$\left.\begin{array}{l} A = 1 \\ \overline{\overline{A}} = 1 \end{array}\right\} \quad \therefore \ 1 = 1, \quad \text{proving} \ \ \overline{\overline{A}} = A.$$

Note that in each of these proofs, all possible values were examined for each variable. The fact that only *two* values exist greatly simplifies the task of proving theorems in Boolean algebra. Since most readers have labored wearily in deriving proofs for conventional and geometric algebraic theorems, where an infinite number of values must be assumed, the digital system offers an obvious mathematical advantage.

Symbology

Earlier, we devised a mathematical shorthand as a tool for digital-equipment analysis and troubleshooting. Now we will develop a *symbolic shorthand.* To begin, we must define the term *Boolean algebra expression.* Mathematically, an *expression* is a "string" of constants and/or variables connected by one or more arithmetic/algebraic operators. By revising this definition slightly, a Boolean algebra expression becomes *a string of constants and/or variables connected by one or more logical operators.*

Boolean algebra expressions can be represented graphically by symbols.* Such symbols can be directly related to the electrical/electronic gate and inverter circuits used in digital equipment. With such close correlation to the actual hardware, troubleshooting and analysis become much easier.

Figure 3–3 shows the MIL-STD 806B symbols that define the AND, OR, and NOT operations. In the AND gate, signal flow is always from the flat to the round side — the *round* side is the *output.* Signal flow in the OR gate is always from the concave to the convex side — the *convex* side is the *output.* The symbol defining NEGATION (the NOT operation) is unique because *it never appears by itself.* It *must* be connected to some other symbol (e.g., an AND

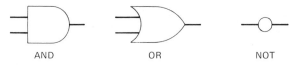

AND   OR   NOT

Figure 3–3   Combinational logic symbols

---

* Although numerous groups of symbols exist, MIL-STD-806B tends to be the standard governing a majority of the digital-equipment diagrams used by military and commercial organizations.

gate or an OR gate). Signal flow is conventionally from *left* to *right*. In various combinations, these three (AND, OR, NOT) symbols describe most of the combinational logic expressions commonly encountered.

Combinational logic elements are commonly called *gates*. In linear (analog) circuits, if a signal passes through or is blocked, depending on the characteristics of the controlling signal, one speaks of "linear gates," where output would be an approximate replica of at least one of the inputs. But for digital purposes, a *gate* is defined as a *circuit having two or more input terminals and one output terminal, where an output is present when and only when prescribed inputs are present.* The output of the digital (logic) gate may bear no resemblance to any of the inputs, except the fact that the output occurs during an interval selected by the control voltage(s).

The term *output* must be redefined for digital equipment. The output of an analog device might vary from minus infinity through zero to plus infinity. But, remember, digital equipment using the binary system can have only one of two outputs at any given time, and for all practical purposes, either of these two output values could be useful. Within this context, therefore, *a combinational logic device* (i.e., a *gate*), *produces an output whenever the gate performs the function denoted by the symbol.*

3–2 THE OR FUNCTION IN DIGITAL LOGIC

One aid to understanding a new concept (such as digital logic) is to construct a physical analogy that would be encountered in an everyday situation. Thus it is convenient — even necessary — for us to initially consider each logical operation as a physical entity. By using this approach, the reader should better comprehend the close correlation between the logic operations and the hardware that performs these operations.

A Physical Analogy

The conventional water faucet (Fig. 3–4) found in many homes provides an excellent analogy for a logic operation. Almost everyone has observed that if *both* the hot and cold water controls are turned OFF, no water comes from

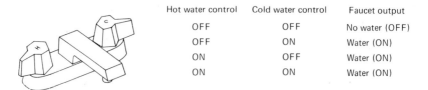

Hot water control	Cold water control	Faucet output
OFF	OFF	No water (OFF)
OFF	ON	Water (ON)
ON	OFF	Water (ON)
ON	ON	Water (ON)

Figure 3–4   A physical analogy of an OR gate

the faucet. If, however, *either* the hot or the cold water control (or both) are turned ON, a flow of water results. A table of all possible control combinations and the resultant flow of water or lack thereof is also shown in Fig. 3–4. Comparison of this table with the table for the logic operation DISJUNCTION (OR)* shows a remarkable resemblance. Apparently, the faucet performs the logic operation of DISJUNCTION.

Terms Associated with the OR Operation

Since this is a non-electronic text and since so many possible methods of implementing logic functions exist, actual values of voltage and/or current are not used descriptively when outputs and inputs are discussed. Instead, we use the HIGH (H) and LOW (L) notation. The H and L approach not only greatly simplifies logic circuit analysis but also makes it unnecessary to use actual numerical values for inputs and outputs.

The Boolean algebra expression for OR and the associated truth table have already been described. In addition, waveforms are commonly employed in the analysis of OR operations.†

Applications of the OR Operation

The device that performs the OR operation is called an OR *gate*. Physically, it may assume many forms; it may be constructed of discrete components or it may be in integrated circuit (IC) form. The number of inputs is not restricted to two, as implied in our earlier discussion. Input limitations are a matter of electronic design, and OR gates are commonly packaged with two, three, four, or eight inputs (an attempt to standardize production and logic design). Some manufacturers provide special configurations of OR gates with greater input capability. Physical packaging is similar to the examples shown in Fig. 2–10.

Many early applications of digital logic were in the area of *switching circuits*, although relays were used in place of actual switches. A *relay* is an electrically operated device that mechanically switches electrical circuits. Its symbol appears in Fig. 3–5a, which is a schematic diagram showing a relay-implemented OR circuit. The truth table in Fig. 3–5b defines the four possible combinations of inputs that can exist with two variables. Note that this table is identical to the truth table for the OR operation.

Relay logic is still used in some industrial and communication control applications. But such circuitry is rapidly being replaced by solid-state, digital-logic modules, since modern technology requires extreme reliability and small physical size in its complex system functions.

---

* The symbol for a device that performs the OR operation is shown in Fig. 3–3.

† To summarize the OR operation, the English-language statement, the Boolean algebra expression, the truth table, and typical waveforms are reproduced at the end of this chapter.

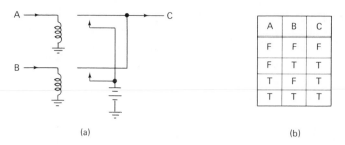

(a)   (b)

Figure 3—5   An OR circuit using relays

Theorems Using the OR Operation

Our Boolean algebra theorems using the OR operation can now be applied to the actual hardware. Again, a "true" (1) is considered to be HIGH and a "false" (0) is LOW. Theorem 7, $A + 0 = A$, can be viewed as two inputs to an OR gate. By definition, one input is a constant binary 0 (LOW). The other input, A, is a variable and can take on either binary 0 or binary 1 as a value. If A is a binary 0, it is at a LOW level; the OR truth table shows that when both inputs are LOW, the output is LOW. However if A is binary 1, then it is at a HIGH level; the OR truth table shows that these input conditions result in a HIGH output.

Actual demonstration of this theorem requires a *logic-signal source*, which has the capability of generating on command either a HIGH or LOW output. Such a device is relatively easy to implement, but for now, we will represent it by a simple rectangle. Control is provided for each signal source so that the output of the device will be either HIGH or LOW, as required. A logic diagram of the complete test set-up required to demonstrate Theorem 7 is shown in Fig. 3—6a with its associated waveforms and truth table.

One of the most valuable applications of the truth table is proving Boolean *identities*. If we can show that one column in a truth table (representing a variable) has a one-to-one correspondence with another column in the truth table (representing another variable or the results of a logical operation), then an "identity" exists between the variables or the results represented by each column. *One-to-one-correspondence* means that for each entry in a column in the truth table, an identical entry exists in the same row in another column. Thus in the short truth table used to define $A + 0 = A$, each entry under the A column has a corresponding entry in the OUT column, and identity is thereby established. That is, regardless of the logical value of A, when it is ORed with binary 0, the result will be the logic value of A.

Theorem 8, $0 + A = A$, may be easily demonstrated by interchanging the two inputs to the OR gate used in Theorem 7. Demonstration of Theorem 9, $A + 1 = 1$, is shown in Fig. 3—6b.

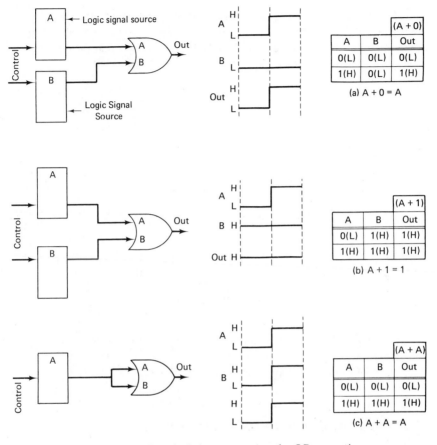

Figure 3—6   Proof of theorems using the OR operation

Theorem 10, $1 + A = 1$, may also be demonstrated by interchanging the two inputs to the OR gate used in Theorem 9. Theorem 11 is demonstrated in Fig. 3—6c.

3—3 THE AND FUNCTION IN DIGITAL LOGIC

A Physical Analogy

Another logic function, the AND operation, is seen in the form of a physical analogy in practically all homes. To illuminate a light both the associated switch *and* the circuit breaker must be on. If either the switch or the circuit breaker is off, the light will not illuminate. A simple schematic diagram showing this *switch implementation* of the AND operation is found (with its associated table of combinations) in Fig. 3–7. Comparison of this table with Table 3–1 (for the logic operation CONJUNCTION) shows that the circuit breaker/wall switch combination performs the AND* operation.

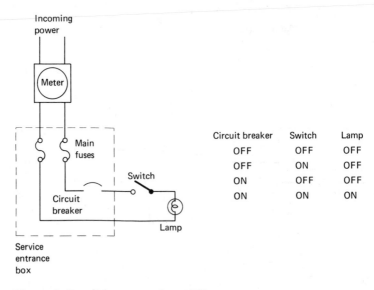

Circuit breaker	Switch	Lamp
OFF	OFF	OFF
OFF	ON	OFF
ON	OFF	OFF
ON	ON	ON

Figure 3–7   An analogy of an AND gate

Terms Associated with the AND Operation

The Boolean algebra expression and the related truth table for the AND operation have already been discussed. The third method of logic analysis, the waveform, is also adaptable to this operation. For easy reference, a synopsis of the important points relating to the AND operation is given in Sec. 3–8.

Applications of the AND Operation

The device that performs the AND operation is called an AND *gate*. Physically, it may assume may forms. Externally, the discrete component AND gate and the IC AND gate resemble the "packages" shown in Fig. 2–10. Like the OR gate, the number of inputs is restricted only by design requirements.

Switch implementation of the AND operation was shown in Fig. 3–7. Relay logic can also be employed in this operation. A typical relay operated AND circuit with its attendant truth table is shown in Fig. 3–8.

---

* The symbol for the hardware that performs the AND operation is shown in Fig. 3–3.

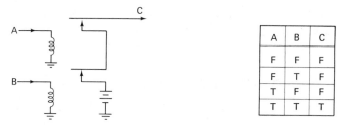

A	B	C
F	F	F
F	T	F
T	F	F
T	T	T

Figure 3–8   Relay AND circuit

Theorems Using the AND Operation

Basic Boolean algebra theorems also provide a means of demonstrating the AND gate in operation. Theorem 1, $A \cdot 0 = 0$, is implemented with hardware in Fig. 3–9a. Theorem 2 may be demonstrated by merely interchanging the two inputs to the AND gate.

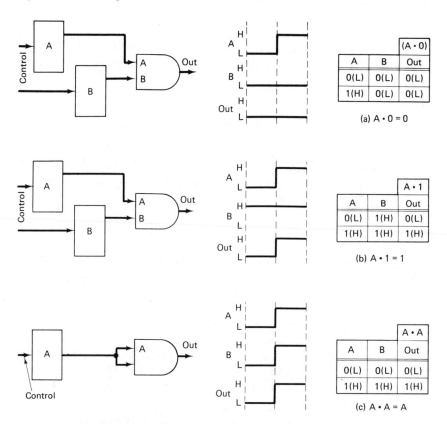

		(A · 0)
A	B	Out
0(L)	0(L)	0(L)
1(H)	0(L)	0(L)

(a) A · 0 = 0

		A · 1
A	B	Out
0(L)	1(H)	0(L)
1(H)	1(H)	1(H)

(b) A · 1 = 1

		A · A
A	B	Out
0(L)	0(L)	0(L)
1(H)	1(H)	1(H)

(c) A · A = A

Figure 3–9   Proof of theorems using the AND operation

Theorem 3, $A \cdot 1 = A$, is shown in Fig. 3–9b. Theorem 4, $1 \cdot A = A$, is represented simply by interchanging the inputs in Theorem 3.

Theorem 5, $A \cdot A = A$, is shown in the logic diagram in Fig. 3–9c.

3–4 THE NOT FUNCTION IN DIGITAL LOGIC

A Physical Analogy

The last of our fundamental logic operations, NEGATION (the NOT function), is not as easily visualized with a physical analogy as were the OR and the AND functions. Perhaps the engine oil pressure warning indicator system on a modern automobile comes as close as possible to a physical analogy. If oil pressure is within limits, the warning light is off. If oil pressure falls below required values, the warning light is on. All possible combinations of this system are shown here.

Oil Pressure	Warning Light
NORMAL	OFF
LOW	ON

Comparison of this list with Table 3–3 (for the logic operation NEGATION) indicates that the oil-pressure warning system performs in a manner similar to the NOT* function.

Terms Associated with the NOT Operation

The Boolean Algebra expression and the related truth table for the NOT operation have been discussed previously. Waveform analysis is also adaptable to this operation. A summary of the important points relating to the NOT operation is given in Sec. 3–8.

The Inverter

Before continuing the discussion of the NOT operation, we should introduce another logic symbol. Within digital logic systems, the need often arises to increase the current or power capabilities of a logic signal. A device called an *amplifier* does this (its symbol is shown in Fig. 3–10a). The amplifier's output signal may have greater power capabilities, but no *logic operation is performed* on the signal applied to the amplifier.

Depending on design, an amplifier† may or may not "invert" the incoming signal. The NOT symbol is commonly attached to the amplifier symbol (Fig. 3–10b, c) to show the inversion process; when this combination of symbols

---

* The symbol for the NOT operation is shown in Fig. 3–3.

† Means are provided within MIL-STD-806B to denote the inverting characteristics of amplifiers used in logic systems.

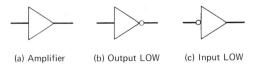

(a) Amplifier   (b) Output LOW   (c) Input LOW

Figure 3–10   The NOT symbol redefined

is used, it represents a device that performs the NOT operation. Such a device is commonly called an *inverter*. Physically, the inverter resembles other logic devices, so identification of the inverter is usually made by part number from the manufacturers' specification sheets.

Since the NOT symbol *cannot* appear alone, it *must* be physically connected to the amplifier symbol — *where* it connects is extremely important. The NOT symbol is placed adjacent to other logic symbols to indicate that a LOW signal either activates the logic operation on input or is a desired result of the logic operation on output. If the NOT symbol is connected at the output side of the amplifier symbol (Fig. 3–10b), it means that when the logic operation (NEGATION) is performed, the *output* will be LOW. LOW has tentatively been correlated with "false" (0), so the input must be HIGH ("true" or 1). If the NOT symbol is attached to the input side of the amplifier (Fig. 3–10c), then the *input* has to be LOW to provide activation of the NEGATION function. The NOT symbol,* therefore, is a *clue*, indicating what can be expected of the symbol to which it is attached.

Applications of the NOT Operation

A schematic diagram of the engine oil-pressure warning system (Fig. 3–11a) that we discussed earlier is one application of the NOT operation for those readers with an electrical/electronic background. Relay-logic implementation of the NOT function is described in Fig. 3–11b.

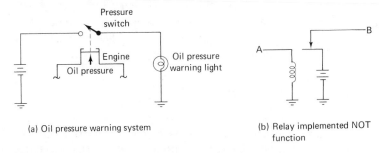

(a) Oil pressure warning system

(b) Relay implemented NOT function

Figure 3–11   Implementation of the NOT function

---

* Placement of the NOT symbol on the amplifier symbol has little meaning in the construction of hardware, but as we have just demonstrated, it does mean a great deal in terms of logic.

Theorems Using the NOT Operation

By the digital-logic definition of the NOT function, the output of the inverter is the inverse or, in Boolean algebra, the "other" value, when compared with the input. For example (see Fig. 3–12a), if the input to an inverter is the variable A, the output is $\overline{A}$. Extension of this application results in the proof for Theorem 13, $\overline{\overline{A}} = A$ (Fig. 3–12b).

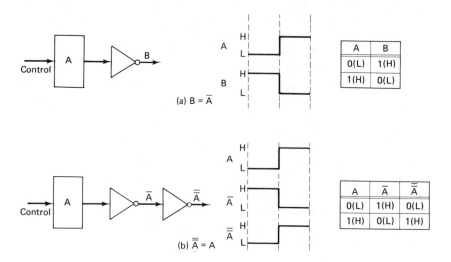

Figure 3–12   Proof of theorems using the NOT operation

3–5 SIMPLE, COMBINED FUNCTIONS

Logic operations are not limited to a *single* function (e.g., OR by itself). Nor does a limit exist on the number of simple logic operations that can be combined to form complex logic functions. So we shall examine various combinations of logic operations and their functions in this section.

Multiple OR Operations

OR operations may be combined on more than one level. For example, consider the Boolean expression $A + (B + C)$. The use of parentheses implies that A is to be ORed with the result of ORing B with C. Thus one OR gate is needed to disjunctively combine (OR) B with C, and another OR gate is needed to disjunctively combine (OR) A with the results. The logic diagram describing these operations is shown in Fig. 3–13a.

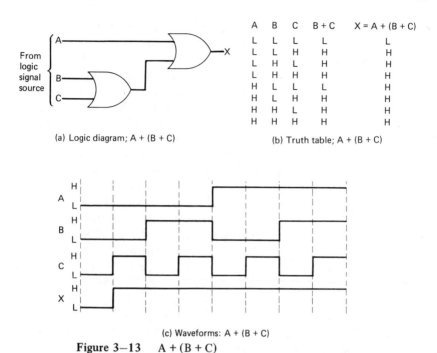

A	B	C	B + C	X = A + (B + C)
L	L	L	L	L
L	L	H	H	H
L	H	L	H	H
L	H	H	H	H
H	L	L	L	H
H	L	H	H	H
H	H	L	H	H
H	H	H	H	H

(a) Logic diagram; A + (B + C)   (b) Truth table; A + (B + C)

(c) Waveforms: A + (B + C)

Figure 3–13   A + (B + C)

If we want to find the logic value for this expression, the equation
$X = A + (B + C)$ can be written. A truth table can then be developed, and the
results of the total logic operation depicted in the X column. Note that in the
truth table (Fig. 3–13b), eight combinations of the three variables exist. When
evaluating logic functions, the total number of combinations of variables is
determined by taking the base of the binary number system (2) and raising it to
the power of the total number of variables being considered. Thus $2^3 = 8$ – the
total number of combinations to be examined. This concept may be generalized
by the following equation:

$$N = 2^M \qquad\qquad (3-1)$$

where N = the total number of combinations to the examined and M = the total
number of variables being considered.

Waveforms (Fig. 3–13c) showing all possible combinations of the three
variables may also be drawn. Like the truth table, all eight possible combinations
must be examined.

Multiple AND Operations

AND operations may also be combined on more than one level. For
example, consider the Boolean expression $A \cdot (B \cdot C)$. The use of parentheses
implies that A is to be ANDed with the result of ANDing B with C. Thus an

AND gate is needed to conjunctively combine (AND) B with C, and another AND gate is needed to AND A with the results. The logic diagram describing these operations is shown in Fig. 3–14a.

If we want to find the logic value for this expression, the equation $X = A \cdot (B \cdot C)$ can be written; a truth table can be developed; and the results of the total logical operation can be shown in the X column. Note that in the truth table (Fig. 3–14b), eight combinations of the three variables exist. Applying Eq. (3–1), $2^3 = 8$, which is the total number of combinations to be examined.

Waveforms showing all possible combinations of the three variables may also be drawn. Like the truth tables, all eight possible combinations must be examined (see Fig. 3–14c).

Until now, we have considered waveforms to be present for a given period of time. However one of the most common applications of multilevel gate logic is the *selection of given time periods* in which a relatively short "pulse" of information is allowed through. A good example would be a digital system that requires a "clock" which provides timing signals of a specific nature (usually short pulses at regular intervals). Some operations may not require pulses at the basic clock interval but only at specific times, depending on logic conditions elsewhere in the system. By ANDing these logic conditions with clock pulses, specific pulses can be selected and routed to perform their functions. The logic diagram and representative waveforms shown in Fig. 3–15 represent such a circuit. Here, clock pulses are ANDed with $A, B,$ and C to select only every eighth pulse.

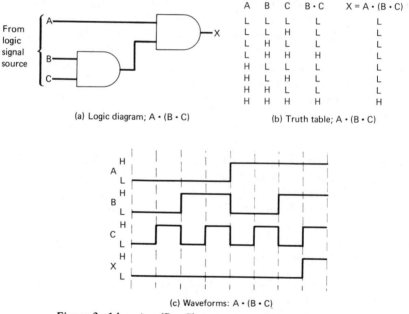

A	B	C	B · C	X = A · (B · C)
L	L	L	L	L
L	L	H	L	L
L	H	L	L	L
L	H	H	H	L
H	L	L	L	L
H	L	H	L	L
H	H	L	L	L
H	H	H	H	H

(a) Logic diagram; A · (B · C)   (b) Truth table; A · (B · C)

(c) Waveforms: A · (B · C)

Figure 3–14   A · (B · C)

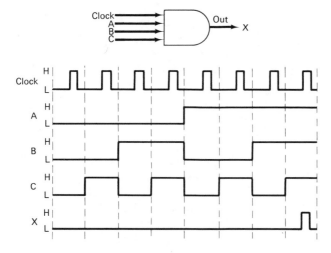

Figure 3—15   A "clocked" AND gate

AND/OR Operations

Both AND and OR operations may be combined. The Boolean expression
$A \cdot (B + C)$ is an example; it implies that A is to be ANDed with the results of
ORing B with C. Thus an OR gate is needed to OR B with C, and an AND gate
is needed to AND the results with the A. The logic diagram and truth table
resulting from changing the expression to an equation $[X = A \cdot (B + C)]$ is shown
in Fig. 3—16. Typical waveform analysis is also provided. Inputs are from a
"logic-signal" source.

OR/AND Operations

Another simple example may be illustrated by the Boolean equation
$X = A + (B \cdot C)$. A is to be ORed with the results of ANDing B with C. An AND
gate is needed to AND B with C, and an OR gate to OR A with the results. The
logic diagram, truth table, and waveforms are provided in Fig. 3—17. Again,
inputs are from a "logic-signal" source.

Order of Operations and Grouping Signs

The *order* in which Boolean algebra operations are *performed* is also very
important. It affects the *logic value of an expression* — just as the order of
arithmetic operations in conventional arithmetic affects the arithmetic value of
an expression. If the total value of the expression 100 x 25 + 2000 is to be
determined, the procedure involves first the multiplication of 100 x 25 and then

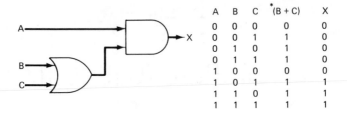

A	B	C	*(B + C)	X
0	0	0	0	0
0	0	1	1	0
0	1	0	1	0
0	1	1	1	0
1	0	0	0	0
1	0	1	1	1
1	1	0	1	1
1	1	1	1	1

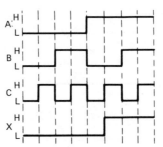

*Note the use of O as LOW and 1 as HIGH

Figure 3–16 $X = A \cdot (B + C)$

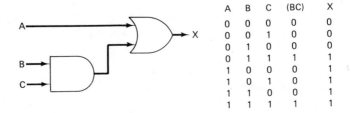

A	B	C	(BC)	X
0	0	0	0	0
0	0	1	0	0
0	1	0	0	0
0	1	1	1	1
1	0	0	0	1
1	0	1	0	1
1	1	0	0	1
1	1	1	1	1

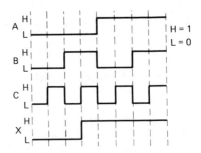

Figure 3–17 $X = A + (B \cdot C)$

the addition of 2000. The result is the answer 4500. A review of arithmetic rules discloses that, unless otherwise indicated, *multiplication and division of numbers within terms is performed before addition and subtraction.* However if the above expression is written as 100 × (25 + 2000), then 202, 500 is the correct answer. Arithmetic rules state that *operations within parentheses are performed before operations outside of parentheses.*

Similar discrepancies arise in Boolean algebra. Consider the three variables A, B, and C. Combining them in the expression $A + B \cdot C$ can result in different logic values, depending on the order in which the operations are performed. Comparison of the effect of performing the OR operation prior to the AND operation $[(A + B) \cdot C]$ with performing the AND operation prior to the OR operation $[A + (B \cdot C)]$ is shown in Table 3–6.

The order of priority in Boolean algebra is NOT (*inversion*) first, AND second, and OR last, *unless otherwise indicated by grouping signs*, such as parentheses, brackets, braces, or the vinculum. Following these rules, the previous expression $A + B \cdot C$ can be evaluated *only* as $A + (B \cdot C)$ with the existence of the parentheses implicitly understood. In fact, knowing this, the expression can be further simplified to $A + BC$. (As a matter of convention, the AND operator is usually omitted in Boolean algebra.) BC is also considered to be grouped naturally because the letters are written together and are separated from A by the OR operator. Both of the conventions ($B \cdot C$ and BC) are used, but BC is preferred (unless clarity demands showing the logic operator).

When working with logic expressions, misuse of grouping signs is a common occurrence. Too many grouping signs do not usually generate problems, as long as they occur in pairs. But this is valid *only* when the grouping signs have been placed properly. Inserting a pair of parentheses to enclose $A + B$ in the previous expression obviously changes the complete meaning of the original expression. However enclosing $B \cdot C$ in parentheses has no effect and is, in fact, an unnecessary step — grouping signs are *not* necessary when ANDed variables or constants are combined with other variables or constants by the OR operator. An insufficient number of parentheses usually results in the wrong interpreta-

Table 3–6 $(A + B) \cdot C \neq A + (B \cdot C)$

A	B	C	$A + B$	$(A + B) \cdot C$	$B \cdot C$	$A + (B \cdot C)$
L	L	L	L	L	L	L
L	L	H	L	L	L	L
L	H	L	H	L	L	L
L	H	H	H	H	H	H
H	L	L	H	L	L	H
H	L	H	H	H	L	H
H	H	L	H	L	L	H
H	H	H	H	H	H	H

└──────Not equal──────┘

tion of logic expressions. Thus if we wish to AND $B + C$ with A, we must enclose $B + C$ in parentheses. Otherwise, the expression would be evaluated as $A \cdot B + C$, and the rules for order of priority of operations would AND B with A — prior to ORing with C. Table 3–7 clearly shows that $A \cdot B + C$ is not the same as $A(B + C)$.

If an expression already contains *parentheses* ( ) and additional signs of grouping are required, first *brackets* [ ] and then *braces* { } are used. The *vinculum* ($^-$) is also a grouping sign — it groups the portion(s) of an expression to be inverted. Remember, letters and expressions are also considered to be *grouped* when they are written without a connecting operator. Here is a complex expression containing various examples of common grouping practices.

$$Y + W \cdot [W(X\overline{Z} + YZ)] + [WX\overline{Y}Z + \overline{X + Y + WZ}]$$

Table 3–7 $(A \cdot B) + C \neq A \cdot (B + C)$

A	B	C	$A \cdot B$	$(A \cdot B) + C$	$B + C$	$A \cdot (B + C)$
L	L	L	L	L	L	L
L	L	H	L	H	H	L
L	H	L	L	L	H	L
L	H	H	L	H	H	L
H	L	L	L	L	L	L
H	L	H	L	H	H	H
H	H	L	H	H	H	H
H	H	H	H	H	H	H

└────Not equal────┘

AND/OR/NOT Operations

The NOT operation may be combined with the AND operation, the OR operation, and the combined AND/OR operations. Logic functions using combinations of OR, AND, and NOT range from the very simple to the very complex. Certain Boolean algebra theorems are well suited to a demonstration of such logic functions. Theorem 12, $A + \overline{A} = 1$, for example, requires the use of an OR gate and the NOT operation. When A is applied to the input of an inverter (NOT), the output is $\overline{A}$. The same variable, A, may also be applied as an input to an OR gate. Thus both required inputs, A and $\overline{A}$, are available. The logic diagram in Fig. 3–18a illustrates one method of proving that $A + \overline{A} = 1$; the attendant truth table also provides proof of the theorem.

Theorem 6, $A \cdot \overline{A} = 0$, combines the NOT operation with the AND operation. Both A and $\overline{A}$ can be made available in the same manner as they were in Theorem 12. The logic diagram and accompanying truth table appear in Fig. 3–18b.

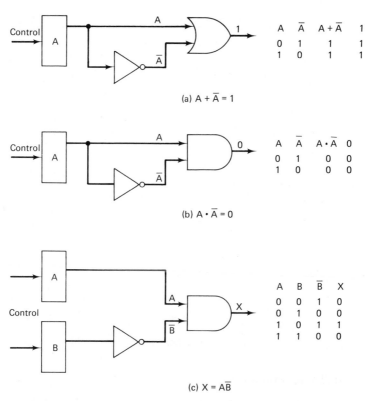

Figure 3—18 Simple AND/OR/NOT operations

Figures 3—18a & b are typical examples of single-level, relatively simple logic circuits using AND/OR/NOT operations. Extending the inputs to two variables, *A* and *B*, does not especially complicate matters, as Fig. 3—18 c shows.

With two variables, only four possible combinations exist, so no real problem should be encountered in analyzing the operation. Also, all four locations of input inverters should be examined, and the effect of each configuration on the truth table should be carefully noted by comparing the information in Fig. 3—19.

3—6 DERIVED FUNCTIONS

In our preceding discussion, AND/OR operations were combined with the NOT operation. In each case, the inverter, which implemented the NOT operation, extended the capabilities of either the AND or the OR gate. Two special cases exist, however, that we will now introduce to illustrate the versatility of simple digital logic.

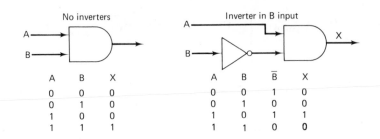

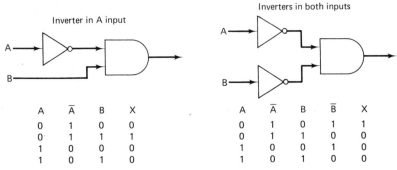

Figure 3—19   Variations of A and B combined conjunctively

The NAND Function

The first case is one in which both inputs to an OR gate are fed through inverters. Under these circumstances, the OR gate functions in an *opposite* (complementary) manner. With inverted inputs, the OR operation is present when one or the other (or both) inputs are *not* present. In other words, *the negated input* OR *gate is activated when one or both inputs are not present.* Compare the negated input OR operation shown in Fig. 3—20a with the normal OR operation in Fig. 3—2 and determine the differences.

Next we will examine an AND operation with negated output. The inverter appears at the output rather than at the input of the AND gate, as in the OR gate example. Thus the AND operation is performed *before* negation. The logic diagram and resulting truth table are shown in Fig. 3—20b.

At this point you will notice a rather startling resemblance between the truth tables resulting from the negated input OR operation and the negated output AND operation. In fact, they are *identical*; that is, *when two truth tables with identical inputs have the same outputs, the logic operations described by the truth tables are equal (identical).* This is one of the statements of *DeMorgan's Laws* and is also one of the most important statements of equality, $\overline{A} + \overline{B} = \overline{AB}$, in Boolean algebra. The conventional method (i.e., truth tables) of indicating equality of Boolean algebra functions is shown in Table 3—8.

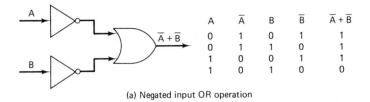

A	$\bar{A}$	B	$\bar{B}$	$\bar{A} + \bar{B}$
0	1	0	1	1
0	1	1	0	1
1	0	0	1	1
1	0	1	0	0

(a) Negated input OR operation

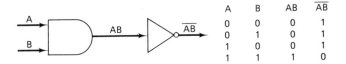

A	B	AB	$\overline{AB}$
0	0	0	1
0	1	0	1
1	0	0	1
1	1	1	0

(b) Negated output AND operation

Figure 3—20   The NAND function

Table 3—8 $\bar{A} + \bar{B} = \overline{AB}$

A	$\bar{A}$	B	$\bar{B}$	$\bar{A} + \bar{B}$	AB	$\overline{AB}$
L	H	L	H	H	L	H
L	H	H	L	H	L	H
H	L	L	H	H	L	H
H	L	H	L	L	H	L

—Equality—

The NOR Function

The second special case is one in which both inputs to an AND gate are fed through inverters. This configuration allows the AND gate to operate in the *complementary* (opposite) manner, just as the OR gate did in the first case. With inverted inputs, the AND operation is present only when both inputs are not present. In other words, *the negated input* AND *gate is activated only when both inputs are not present.* Compare the negated input AND operation shown in Fig. 3—21a with the normal AND operation in Fig. 3—2 to determine the differences.

Now we will examine an OR operation with negated output. The inverter appears at the output rather than at the input of the OR gate, as in the AND gate example. Thus the OR operation is performed *before* negation. The logic diagram and resulting truth table are shown in Fig. 3—21b.

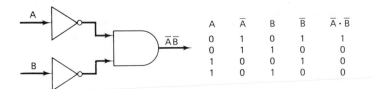

A	$\bar{A}$	B	$\bar{B}$	$\bar{A}\cdot\bar{B}$
0	1	0	1	1
0	1	1	0	0
1	0	0	1	0
1	0	1	0	0

(a) Negated input AND operation

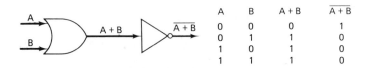

A	B	A + B	$\overline{A+B}$
0	0	0	1
0	1	1	0
1	0	1	0
1	1	1	0

(b) Negated output OR operation

Figure 3—21   The NOR function

Once again, the resemblance between the negated-input AND operation truth table and the negated-output OR operation truth table is obvious. Since the truth tables are identical, the logic functions are identical. This is the second of DeMorgan's Laws, which in Boolean algebra becomes $\bar{A}\cdot\bar{B} = \overline{A+B}$. Comparative truth tables are shown in Table 3—9.

While we do not plan a detailed discussion of derived logic operations in this chapter, we would like to point out that the logic operations resulting from DeMorgan's Laws are commonly implemented in *hardware form*. A device called the NAND (NOT AND) gate (Fig. 3—22a) implements the first law, and the NOR (NOT OR) gate (Fig. 3—22b) implements the second law. Since both AND and OR operations may be derived from NAND and NOR gates, many manufacturers furnish only the latter types of logic elements. Economically, it may be more feasible for a logic designer to use a minimum number of different logic elements (e.g., all NAND and all NOR) rather than mixed AND, OR, NAND, NOR, and inverter elements. In some cases, such designs may actually require more of one type of logic assembly, but the saving in quantity purchases and reduced inventory will reduce overall system costs.

Table 3—9 $\bar{A}\cdot\bar{B} = \overline{A+B}$

A	$\bar{A}$	B	$\bar{B}$	$\bar{A}\cdot\bar{B}$	$A+B$	$\overline{A+B}$
L	H	L	H	H	L	H
L	H	H	L	L	H	L
H	L	L	H	L	H	L
H	L	H	L	L	H	L

└───Equality───┘

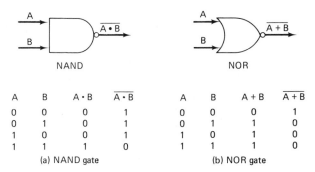

A	B	A · B	$\overline{A \cdot B}$		A	B	A + B	$\overline{A + B}$
0	0	0	1		0	0	0	1
0	1	0	1		0	1	1	0
1	0	0	1		1	0	1	0
1	1	1	0		1	1	1	0
		(a) NAND gate					(b) NOR gate	

Figure 3—22   NAND and NOR gates

The symbology and truth tables for both the NAND and NOR operations*
are reproduced in Fig. 3—22.

Using the Logic Signal Source To Prove Identities

Our introduction of DeMorgan's Laws and an increasing use of truth-table
identity proofs leads us to a more detailed discussion of the *logic signal source.*
Since the identity of two truth tables is considered proof of logic identity, a
device that would speed up such proofs by mechanizing the truth tables is
highly desirable. Such a device, the logic signal source, has already been used
to examine some of the Boolean algebra theorems. In Sec. 2—6, the FF was
discussed in conjunction with counting. If we examine the counting table
developed for the four-stage binary counter, we see that the entries match
directly with the truth table developed for the logic signal source. Essentially,
therefore the logic signal source is *a simple binary counter with outputs arranged
and named so that they can be used to mechanize the standard truth table.*
The two tables (3—10) are identical, thus establishing the identity of the counter
and the logic signal source.

Table 3—10   Binary counter versus logic-signal source

Counter table				*Logic-signal source*		
A	*B*	*C*	*Decimal Value*	*A*	*B*	*C*
0	0	0	0	0	0	0
0	0	1	1	0	0	1
0	1	0	2	0	1	0
0	1	1	3	0	1	1
1	0	0	4	1	0	0
1	0	1	5	1	0	1
1	1	0	6	1	1	0
1	1	1	7	1	1	1

Note: 0 = LOW and 1 = HIGH

* Derivation of the AND and OR operations from NAND and NOR elements is
explained in Chapter 4.

One additional fact: as previously stated, the total number of combinations to be evaluated when examining the operations of combinational logic circuits is 2^n, where n is the number of variables. Thus with two variables, 2^2 or four combinations must be evaluated. You might randomly determine the four possible combinations, but it is more efficient to organize the combinations into a usable list. By arranging the truth table entries in ascending order (starting with 0 and ending with $2^n - 1$), all possible combinations will occur in an easily remembered and verifiable order. In a two-variable truth table (e.g., Table 3—11),

Table 3—11   2-variable truth table

Decimal value	A	B
0	0	0
1	0	1
2	1	0
3	1	1

Note: 0 = LOW and 1 = HIGH

2^2 entries (starting with 0 and ending with $2^2 - 1$ or decimal 3) are made. If three variables are to be examined, then 2^3 entries (starting with 0 and ending with $2^3 - 1$ or decimal 7) must be made. (See Tables 3—10 and 3—11.)

The technique, mentioned in Chapter 2, that advocated visual representation of logic states is also extremely useful with mechanized truth tables. If the *light-on* is "true" (1 or HIGH) and the *light-off* is "false" (0 or LOW), then either a waveform display or any electrical indications are unnecessary in determining logic states. In actual practice, a special type of amplifier called a *lamp driver* (Fig. 3—23) provides sufficient power to operate the indicator lamp.

Figure 3—23   Lamp Driver symbol

Each stage of the logic-signal source has an indicator, which will determine its state. In addition, indicators are attached to the output of each logic circuit being tested. The logic-signal source is then started through its sequence, observing the states of each stage and the outputs of the logic circuits. If the indicators attached to the outputs of the logic circuits being tested are in the same configuration for each input step, then those outputs are identical. Thus if the output indicators stay the same (i.e., changing together throughout all possible input conditions), then *identity* is established. Logic diagrams of typical Boolean algebra identity proofs, using both of DeMorgan's Laws, are shown in Fig. 3—24. Lamp drivers are connected so as to display the logic state of each circuit.

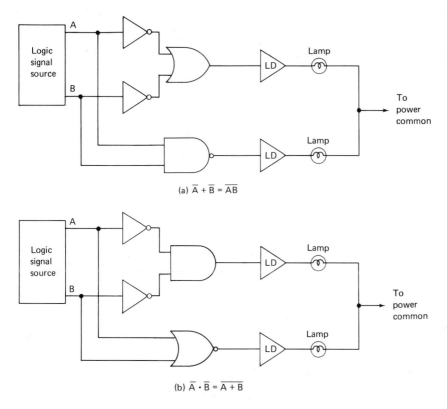

Figure 3—24   Use of Lamp Drivers in logic circuits to indicate proof of DeMorgan's Laws

3—7   A SIMPLE APPLICATION OF BASIC COMBINATIONAL LOGIC FUNCTIONS

The Logic Circuit

A practical application of combinational logic is the *count-decoding* method. Many logic applications require the counting of objects, events, timing pulses, and so forth. For example, consider the logic circuit in Fig. 3—25. Although the binary counter shown is sequential in nature (a concept that has not yet been discussed), the reader should be able to understand its operation from both the counting-situation and the logic-signal source discussions. Specifically, the number of input pulses counted may be determined by examining the

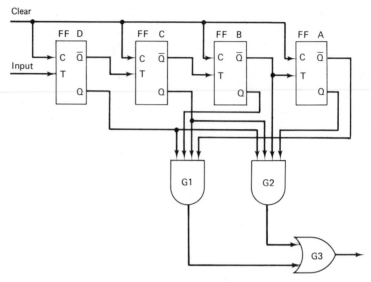

Figure 3—25   A "Count Decoding" logic circuit

states of the outputs of each flip-flop (FF). If we can assume that the FF interconnections provide the capability to count, then the connection of the FF
outputs to G_1 and G_2 inputs *could* be used to decode specific counter configurations. G_3 should provide an output when either G_1, G_2, or both are
activated. Now comes the problem: when will an active output appear at G_3,
and what form will it take?

Waveform Analysis

One approach uses waveform analysis. The waveforms for a 4-FF counter,
developed in Ch. 2, are reproduced in Fig. 3—26 along with the CLEAR and
INPUT waveforms.

To activate the AND gate G_1 (i.e., cause its output to go HIGH), all inputs
to G_1 must be HIGH. The inputs to G_1 are furnished by FFs D, C, B, and A, so
each of these inputs must be HIGH to activate the gate. Thus the Q output of
D, C, and B and the $\overline{Q}$ output of A must all be HIGH at the same time. The rules
of operation for FFs state that the $\overline{Q}$ output is HIGH when the FF is in the
CLEAR (RESET) state. G_1 will therefore be activated when D, C, and B are
SET and A is RESET.

Waveform analysis* of this type of logic circuit usually begins at the most-
significant-digit (MSD) FF and works backward. Investigation of the $\overline{Q}$ output
of FF A (one of the inputs to G_1) will show that it is HIGH from the time of the

---

* If any doubt exists as to how the waveforms are obtained, review the FF-counter
section of Ch. 2.

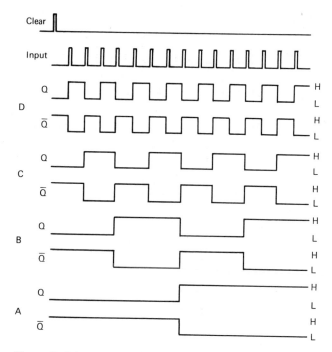

Figure 3—26 Counter waveforms

CLEAR pulse application through (and including) count 7. Thus G_1 *must* be decoding a count between 0 and 7. Observing the Q output of FF B (another input to G_1) indicates that it is HIGH only from count 4 through (and including) count 7. The count is now narrowed to the range between 4 and 7. The Q output of FF C is HIGH during counts 2 and 3, and counts 6 and 7. Since the Q output of B is LOW during counts 2 and 3, G_1 cannot be activated at that time. However the B FF Q output does provide an enabling input during counts 6 and 7. G_1 activation is now narrowed to either count 6 or 7. The D FF Q output is HIGH during count 7, so G_1 becomes enabled approximately from the time input pulse 7 arrives until the arrival of input pulse 8. In other words, G_1 is *decoding* count 7. (G_1 output is shown in the waveform diagram.) Of course, since any HIGH input to an OR gate results in a HIGH output, gate G_3 also provides an indication at count 7. Another output from G_3 occurs when G_2 is enabled. The same analytical approach may be applied to G_2 to show that it is decoding count 11. G_3 output then occurs at either count 7 *or* count 11. Figure 3—27 shows all the important waveforms.

Since waveform analysis provides a visual representation of logic-circuit functions, some readers may prefer to use this method. However as logic circuits become more complex, waveform drawings become more cumbersome, so Boolean algebra notations are preferred.

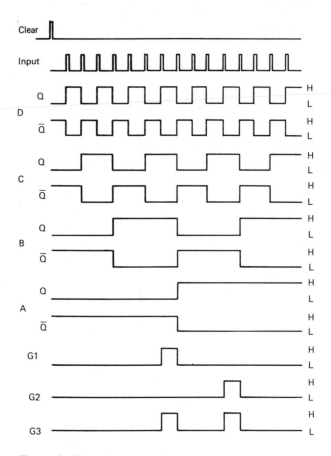

Figure 3—27   Counter and decoder waveforms

A few words must be said at this point to clarify the use of Boolean notation when discussing FFs. The standard FF has two outputs, commonly called
the Q and the $\overline{Q}$ outputs. When the FF is in the SET state, the Q output is
HIGH and the $\overline{Q}$ output is LOW. When FFs are named, (e.g., D, C, B, A), the
SET state is considered to be the *true* state, and the CLEAR (RESET) state
becomes the *false* state. For example, if FF D is in the SET state, it is commonly denoted as D — in the RESET state as $\overline{D}$.

Evaluation of the decode 7 gate (G_1) shows that $D, C,$ and B must be in
the SET state and A must be in the RESET state (because the $\overline{Q}$ output rather
than the Q output is used as an input to G_1). The Boolean expression for
enabling G_1 then becomes $D \cdot C \cdot B \cdot \overline{A}$. G_2 is enabled by FFs $D, C,$ and A in the
SET state and B in the RESET state. The resulting expression is $D \cdot C \cdot \overline{B} \cdot A$.
Remember, when the Boolean expression is written for an AND gate (under
MIL-STD-806B rules), the variables shown must *all* be HIGH to activate the gate.

Writing $D \cdot C \cdot B \cdot \overline{A}$ means that all four variables must be HIGH. Although $\overline{A}$ appears, this merely means that whatever the $\overline{A}$ input comes from must provide a HIGH output when A is in the *false* condition. This is accomplished in a FF by using the $\overline{Q}$ output.

Until now, we have written expressions describing gate outputs in the order of FF positions as they appear on the logic diagram. Actually, it makes no difference in what *order* the variables are ANDed. Thus it is more convenient to list the variables in alphabetical order (i.e., $A \cdot B \cdot C \cdot D$). Unfortunately, logic diagrams showing FFs are conventionally arranged with the LSD on the left, since information flows from LSD to MSD in such a circuit. The truth table, however, is arranged with the MSD on the left, which is convenient if all combinations of variables are placed in ascending decimal order (see Fig. 3–28).

Rather than break completely with these conventions, the reader should merely adapt a bit. Since much of our work is done by referring to logic diagrams, it is advisable to retain the diagram convention of data flow (from left to right) and to adapt the truth table to fit the situation. The table showing the sequential states of the three FFs is actually a "mirror image" of the truth table. Keeping this in mind, the advantages of both conventions can be retained.

When the outputs of G_1 and G_2 are combined in the OR gate (G_3), the resulting expression $\overline{A} \cdot B \cdot C \cdot D + A \cdot \overline{B} \cdot C \cdot D$ becomes the output of the multiple-count decoding circuit. Assignment of numerical values to each variable, based on the position of the FFs in the counter circuit, produces the actual decoded counts. The expression $\overline{A} \cdot B \cdot C \cdot D$ is equivalent to $\overline{8}$ (which is 0) plus 4 plus 2 plus 1 equals 7. $A \cdot \overline{B} \cdot C \cdot D$ means 8 plus $\overline{4}$ (0) plus 2 plus 1 equals 11. Study the Boolean expressions and the waveforms thoroughly to establish the correct correlations and relationships.

Since we had no intention of explaining all the complexities of binary counters — they were used here merely to develop input signals for the gates — please refer to Chapter 11 for further details about the counter's operation.

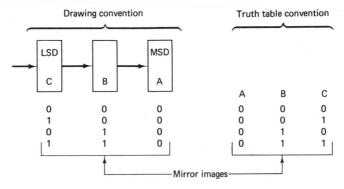

Figure 3–28   Drawing convention compared with truth table convention

Truth-Table Analysis

Once the inputs to the gates have been defined, truth-table techniques can be used to analyze combinational logic circuits. In this particular case, four input variables are used. Applying Eq. (3–1), we see that a total of 16 possible combinations of the four variables must be examined. Arranging the 16 combinations in ascending decimal order results in Table 3–12. (Since negated values are used in both expressions, they are included in the table.) Now we merely examine each of the combinations for the *binary value* (truth value) of each expression. $\overline{A} \cdot B \cdot C \cdot D$ is true only at count 7, and $A \cdot \overline{B} \cdot C \cdot D$ is true only at count 11. The truth table thus proves the original analysis.

Table 3–12   4-variable truth table used to analyze count decoder

A	$\overline{A}$	B	$\overline{B}$	C	D	$\overline{A}BCD$	$A\overline{B}CD$	$\overline{A}BCD + A\overline{B}CD$
0	1	0	1	0	0	0	0	0
0	1	0	1	0	1	0	0	0
0	1	0	1	1	0	0	0	0
0	1	0	1	1	1	0	0	0
0	1	1	0	0	0	0	0	0
0	1	1	0	0	1	0	0	0
0	1	1	0	1	0	0	0	0
0	1	1	0	1	1	1	0	1
1	0	0	1	0	0	0	0	0
1	0	0	1	0	1	0	0	0
1	0	0	1	1	0	0	0	0
1	0	0	1	1	1	0	1	1
1	0	1	0	0	0	0	0	0
1	0	1	0	0	1	0	0	0
1	0	1	0	1	0	0	0	0
1	0	1	0	1	1	0	0	0

3–8 SUMMARY

The OR Operation

$$A + B$$

"If A is true, or B is true, or A and B are true together, then A OR B is true." From a more practical standpoint: "If A is HIGH, or B is HIGH, or A and B are HIGH together, then A OR B is HIGH."
(See Figure at top left of page 83.)

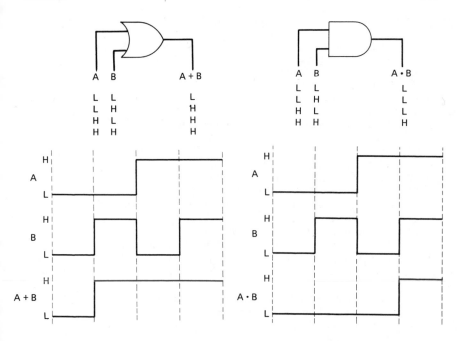

The AND Operation

$$A \cdot B$$

"If A and B are both true at the same time, then $A \cdot B$ is true." Stated in more common language: "If A and B are both HIGH at the same time, then $A \cdot B$ is HIGH." (See Figure above right.)

The NOT Operation

$$A = \bar{B}, \quad \bar{A} = B$$

"A is true if B is false; A is false if B is true." Stated differently: "A is HIGH when B is LOW; A is LOW when B is HIGH." (See Figure at right.)

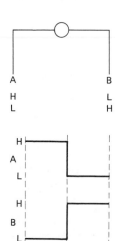

PROBLEM APPLICATIONS

3—1.   What are the two general types of logic circuits? Define them.
3—2.   What is a logic operation?
3—3.   Define CONJUNCTION.
3—4.   Define DISJUNCTION.
3—5.   Define NEGATION.
3—6.   List all of the Boolean constants.
3—7.   What is the maximum number of constant values a Boolean variable can assume?
3—8.   Define truth table.
3—9.   What is the purpose of a truth table?
3—10. Why is Boolean algebra less complex than conventional algebra?
3—11. Draw the MIL-STD-806B symbols for devices that will perform the CONJUNCTION operation, the DISJUNCTION operation, the NEGATION operation.
3—12. In combinational logic terms, what is a gate?
3—13. Show the proper symbol, construct a truth table, and draw the waveforms for the Boolean expression $A + B + C$.
3—14. Explain at least three ways in which the OR operation may be put to practical use.
3—15. Show the proper symbol, construct a truth table, and draw the waveforms for the Boolean's expression ABC.
3—16. Explain at least three ways in which the AND operation may be put to practical use.
3—17. What are the restrictions on the use of the NOT symbol?
3—18. What is the most convenient method for examining multiple-combinational logic operations?
3—19. In the Boolean expression $A + B + C + D$, how many combinations of the four variables exist?
3—20. What is the order of performing operations in a complex Boolean expression?
3—21. What is a NAND gate?
3—22. What is a NOR gate?

4

Advanced Combinational Logic

Our investigation of the postulates and theorems of Boolean algebra in Ch. 3 has prepared us for new territory — *the laws and identities of Boolean algebra*. In developing techniques for logic-circuit analysis, we must continually broaden our knowledge and understanding of the mathematics of logic as well as such other analytical "tools" as truth tables and waveform analysis.

Each logic diagram we encounter presents a new challenge. Some are relatively simple, so the use of logic symbols and the application of "logical-thought processes" will provide an adequate understanding of the diagram. Initially, due to inexperience, the reader will find few diagrams that he can submit to this method of analysis. Instead, as increasingly complex logic circuits are encountered, he will be forced to compile truth tables to decipher the intricacies of the logic functions. A reader with an extensive background in electronic troubleshooting, on the other hand, may find waveform analysis a more useful tool. For the mathematically oriented reader, the use of Boolean algebra is probably most appropriate. Undoubtedly, however, any combinational logic circuit will require the application of more than one (and perhaps even all) of these tools.

In this chapter, we will learn how to apply basic laws and identities of Boolean algebra to these various methods of combinational logic diagram analysis.

4–1 BASIC LAWS OF BOOLEAN ALGEBRA

In conventional mathematics, it is often difficult to provide sufficient non-mathematical proofs to help simplify the cornerstones — *the laws* — of the system. In Boolean algebra, however, truth tables and waveforms supply visual clues and explanations that fix the basic principles in the readers' minds. In addition, whenever it is appropriate, the actual mathematical manipulations are shown. Thus the reader can progress from simple to complex laws with a minimum of effort.

The Law of Identity

The *Law of Identity* states that *any letter, number, or expression is equal to itself.* Thus in Boolean algebra,

$$A = A \quad \text{and} \quad \overline{A} = \overline{A} \tag{4-1}$$

Some additional examples of this law are

$$B = B, \quad 0 = 0, \quad AB = AB, \quad \overline{X} = \overline{X}, \quad \text{and so on.}$$

It is unnecessary to prove the Laws of Identity by such methods as truth tables — the proof is implicit in the statement of the law.

The Commutative Law

The first part of the *Commutative Law* states that *when inputs to a logic symbol are ANDed, the order in which they are written does not affect the binary value of the output.* Stated algebraically,

$$AB = BA \tag{4-2}$$

The proof of this statement is easily established by truth tables. Furthermore, mechanization of the truth tables and the use of indicator lamps (discussed in Sec. 3–6) are employed in most of the proofs presented here. These logic circuits may be constructed by using any of the digital-logic training aids available. And we strongly recommend that each of the logic circuits be investigated by using actual hardware.

When the output level is HIGH (1), the lamp is ON. If the lamp is not ON, the output level is LOW (0). As each input combination is established (each line on the truth table), note the condition of the indicator lamps. If both lamps are in the same condition (*either* OFF or ON), both circuits are reacting in the same manner to the specific input combination. The logic-signal source can then be set to the *next* input combination (the next line on the truth table) and the process repeated. If both lamps are in the same condition (*either* OFF or ON) for *all* possible input combinations, then the two circuits are identical in logic functions. Of course, the condition of the lamps determines whether the output is 1 or 0, and the truth table may be constructed directly by recording the lamp conditions for each input condition. Figure 4–1a shows the logic diagram and truth table proofs for the identity $AB = BA$.

In logic diagrams, the power connections are assumed to be made and are therefore *not* shown. In the case of the lamp drivers, when the input is LOW, the output is LOW (as defined by MIL-STD-806B), so the lamps are OFF. Con-

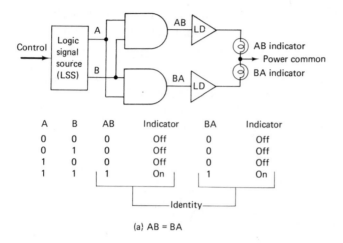

A	B	AB	Indicator	BA	Indicator
0	0	0	Off	0	Off
0	1	0	Off	0	Off
1	0	0	Off	0	Off
1	1	1	On	1	On

Identity

(a) AB = BA

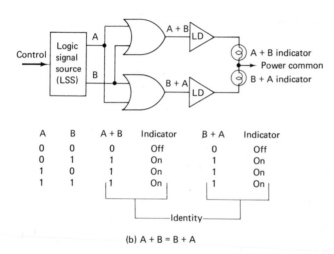

A	B	A + B	Indicator	B + A	Indicator
0	0	0	Off	0	Off
0	1	1	On	1	On
1	0	1	On	1	On
1	1	1	On	1	On

Identity

(b) A + B = B + A

Figure 4−1 Logic diagram and truth table proofs for the Commutative Laws

versely, HIGH input causes a HIGH output, so the lamps are ON. The lamp drivers act as *switches*, connecting power when the inputs are HIGH and removing power when the inputs are LOW.

The second part of the *Commutative Law* states that *when inputs to a logic symbol are ORed, the order in which they are written does not affect the binary value of the output.* In terms of Boolean algebra, we would write

$$A + B = B + A \qquad (4-3)$$

A proof (Fig. 4−1b) similar to that used for the first part of the Commutative Law will suffice.

The Commutative Law is *not* limited to two variables. With three inputs (A, C, and D) provided to an AND gate, the output could be

$$A \cdot C \cdot D, A \cdot D \cdot C, C \cdot D \cdot A, C \cdot A \cdot D, D \cdot C \cdot A \quad \text{or} \quad D \cdot A \cdot C$$

Furthermore, the expression $R \cdot (S + T)$ may also be written as $(T + S) \cdot R$ — or in numerous other ways. $W + X + YZ$ and $ZY + X + W$ are the same, although the order of *both* the ANDed and ORed variables have been changed. The Commutative Law still applies, as proven in Fig. 4–2.

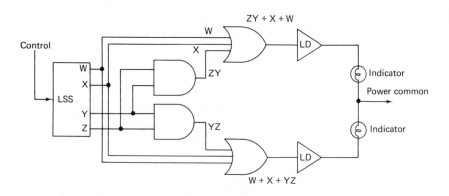

W	X	Y	Z	YZ	ZY	W + X + YZ	Indicator	ZY + X + W	Indicator
0	0	0	0	0	0	0	Off	0	Off
0	0	0	1	0	0	0	Off	0	Off
0	0	1	0	0	0	0	Off	0	Off
0	0	1	1	1	1	1	On	1	On
0	1	0	0	0	0	1	On	1	On
0	1	0	1	0	0	1	On	1	On
0	1	1	0	0	0	1	On	1	On
0	1	1	1	1	1	1	On	1	On
1	0	0	0	0	0	1	On	1	On
1	0	0	1	0	0	1	On	1	On
1	0	1	0	0	0	1	On	1	On
1	0	1	1	1	1	1	On	1	On
1	1	0	0	0	0	1	On	1	On
1	1	0	1	0	0	1	On	1	On
1	1	1	0	0	0	1	On	1	On
1	1	1	1	1	1	1	On	1	On

Identity

Figure 4–2   Logic diagram and truth table proofs W + X + YZ = ZY + X + W

Note that the 16 possible combinations of four variables are arranged in the truth table in ascending decimal order.

The procedure for developing a logic diagram from simple Boolean expressions consists of four steps.

1. Identify the general type of circuit.
2. Begin with the output expression, drawing the logic symbol that represents the last (final) operation.
3. Work toward the input, drawing logic symbols for each operation until all inputs are accounted for.
4. Label inputs and outputs of all required gates.

Naturally, as expressions become more complex, we will be forced to expand this basic procedure.

The identity just investigated may also be demonstrated by using waveforms. Figure 4–3, for example, is a waveform representation of the logic diagram and truth table of Fig. 4–2. Waveform diagrams show quite adequately the logic conditions defined by Boolean algebra, logic diagrams, and truth tables. HIGH and LOW conventions, discussed in previous chapters, are used in this chapter in all waveform diagrams. Many of the identities that will follow are explained by using waveform diagrams, in addition to the truth tables and algebraic proofs of Boolean algebra laws.

$(J + L)K$ is *not* the same as $(J + K)L$. The Commutative Law applies *only* to inputs to the *same* logic symbol. Note that the logic values of the outputs do

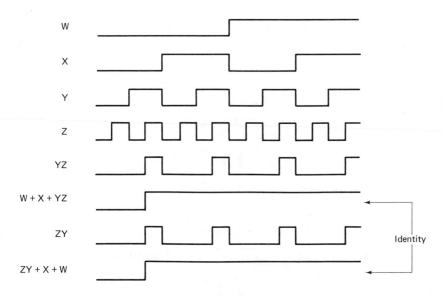

Figure 4–3   Waveform drawing for the identity W + X + YZ = ZY + X + W

not match, thus showing that the two expressions are *not* identical. A logic diagram, truth table, and waveform proofs for $(J + L)K \neq (J + K)L$ are shown in Fig. 4—4.

A composite restatement of the Commutative Law might therefore say: "In a simple AND or OR relationship, the positions of the input signals may be interchanged with no change in the output signal."

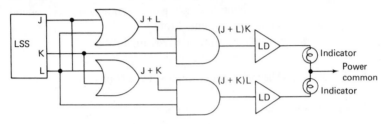

J	K	L	J + L	(J + L) K	Indicator	J + K	(J + K)L	Indicator
0	0	0	0	0	Off	0	0	Off
0	0	1	1	0	Off	0	0	Off
0	1	0	0	0	Off	1	0	Off
0	1	1	1	1	On	1	1	On
1	0	0	1	0	Off	1	0	Off
1	0	1	1	0	Off	1	1	On
1	1	0	1	1	On	1	0	Off
1	1	1	1	1	On	1	1	On

Not identical

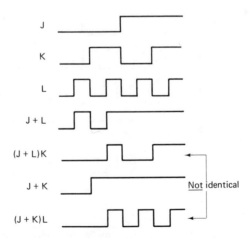

Figure 4—4 $(J + L)K \neq (J + K)L$ – truth table, logic diagram, and waveforms

The Associative Law

The *Associative Law* simply states that *it makes no difference in what order parts of an* AND *or an* OR *expression are combined.* Thus

$$A(BC) = ABC \qquad (4\text{-}4)$$

(See Fig. 4–5.)

and $\qquad A + (B + C) = A + B + C \qquad (4\text{-}5)$

(See Fig. 4–6.)

A	B	C	(BC)	A(BC)	Indicator	ABC	Indicator
0	0	0	0	0	Off	0	Off
0	0	1	0	0	Off	0	Off
0	1	0	0	0	Off	0	Off
0	1	1	1	0	Off	0	Off
1	0	0	0	0	Off	0	Off
1	0	1	0	0	Off	0	Off
1	1	0	0	0	Off	0	Off
1	1	1	1	1	On	1	On

Identity

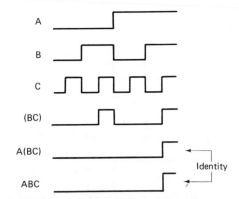

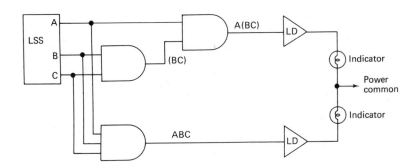

Figure 4–5 Truth table, logic diagram, and waveform proofs for A(BC) = ABC (Associative Law)

The reader may wonder why something as simple as the Associative Law gets so much attention. If so, consider the hardware implementation of $A + (B + C)$. Note that one 2-input OR gate is required to combine B with C, and another 2-input OR gate is needed to combine A with the results of ORing B with C. Implementing $A + B + C$ requires only one 3-input OR gate. Thus we have saved one gate! Even in the simpler digital devices, hundreds of such expressions may be required, so any simplification operations produce a great saving.

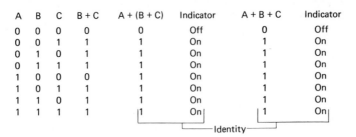

A	B	C	B + C	A + (B + C)	Indicator	A + B + C	Indicator
0	0	0	0	0	Off	0	Off
0	0	1	1	1	On	1	On
0	1	0	1	1	On	1	On
0	1	1	1	1	On	1	On
1	0	0	0	1	On	1	On
1	0	1	1	1	On	1	On
1	1	0	1	1	On	1	On
1	1	1	1	1	On	1	On

Identity

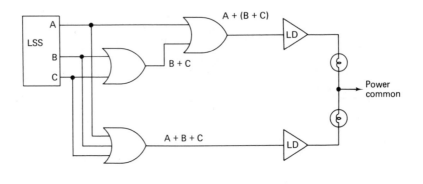

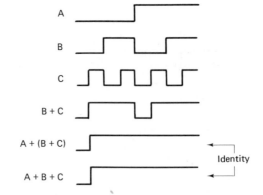

Figure 4–6   Truth table, logic diagram, and waveform proofs for A + (B + C) = A + B + C (Associative Law)

The Idempotent Laws

The *Idempotent Law* states that *if a binary variable is* ANDed *with itself or* ORed *with itself, the binary value of the output will be that of the input variable.*

$$AA = A \quad \text{and} \quad A + A = A \tag{4-6}$$

These expressions are merely restatements of Theorems 5 and 11, respectively, whose proofs were shown in Chapter 3.

The Distributive Law

The first identity of the Distributive Law is

$$A(B + C) = AB + AC \tag{4-7}$$

Replace our conventional methods of proof (i.e., logic diagram and the truth table) and try to "think this one out." Thus for the binary value of $A(B + C)$ to be 1, A must be 1 *and* either B or C must be 1. If B were 1, the AB combination would yield a 1. Likewise, if C were 1, the AC combination would also yield a 1. This type of approach — the *logical thought process* — is sometimes an effective method of simplification for Boolean expressions.

The truth table proof of $A(B + C) = AB + AC$ and the associated logic diagram and waveforms are shown in Fig. 4–7.

This first identity of the Distributive Law may be easily proven by applying the familiar rules of *logical multiplication* (ANDing). Recall that when variables inside parentheses are logically multiplied by a variable outside the parentheses, the result is the logical product of the "outside" variable and each of the "inside" variables, joined by the operator connecting the "inside" variables. Thus by logical multiplication, $A(B + C) = AB + AC$.

The second identity of the Distributive Law is

$$A + BC = (A + B)(A + C) \tag{4-8}$$

This identity is valid in Boolean algebra but *not* in ordinary algebra.

The logic diagram, truth table, and waveform diagrams proving this identity are shown in Fig. 4–8. A proof of this identity using the postulates and theorems of Boolean algebra is also given.

Steps	Justification
$(A + B)(A + C) = AA + AC + AB + BC$	Logical Multiplication
$= A + AC + AB + BC$	$AA = A$
$= [A(1 + C)] + AB + BC$	Factoring
$= A + AB + BC$	$1 + A = 1$
$= [A(1 + B)] + BC$	Factoring
$= A + BC$	$1 + A = 1$
$\therefore \ (A + B)(A + C) = A + BC$	

A	B	C	B + C	A(B + C)	Indicator	AB	AC	AB + AC	Indicator
0	0	0	0	0	Off	0	0	0	Off
0	0	1	1	0	Off	0	0	0	Off
0	1	0	1	0	Off	0	0	0	Off
0	1	1	1	0	Off	0	0	0	Off
1	0	0	0	0	Off	0	0	0	Off
1	0	1	1	1	On	0	1	1	On
1	1	0	1	1	On	1	0	1	On
1	1	1	1	1	On	1	1	1	On

————Identity————

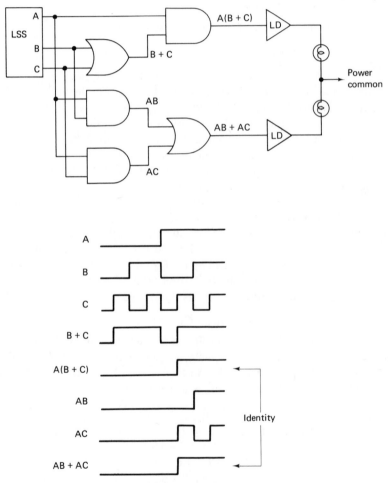

Figure 4—7   Truth table, logic diagram, and waveform proofs for A(B + C) = AB + AC (Distributive Law)

A	B	C	BC	A + BC	Indicator	A + B	A + C	(A + B)(A + C)	Indicator
0	0	0	0	0	Off	0	0	0	Off
0	0	1	0	0	Off	0	1	0	Off
0	1	0	0	0	Off	1	0	0	Off
0	1	1	1	1	On	1	1	1	On
1	0	0	0	1	On	1	1	1	On
1	0	1	0	1	On	1	1	1	On
1	1	0	0	1	On	1	1	1	On
1	1	1	1	1	On	1	1	1	On

———————————— Identity ————————————

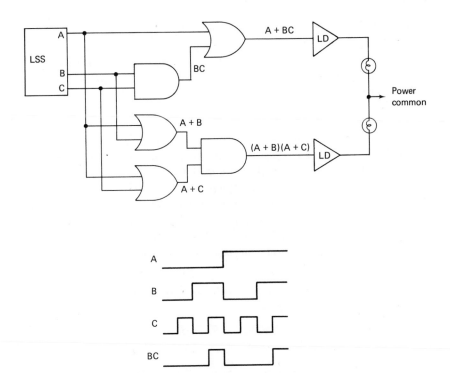

Figure 4–8   Truth table, logic diagram, and waveform proofs
for A + BC = (A + B) (A + C) (Distributive Law)

This is only one form that a Boolean algebra proof can assume. Many other approaches (some with less steps) yield the same results. *But we have selected proofs for this and all the laws and identities that will demonstrate the most important uses of Boolean algebra.* The reader is encouraged to investigate other forms of algebraic proof.

Application of the Distributive Law* to simplification operations significantly reduces hardware requirements and increases analytic capabilities.

DeMorgan's Law

DeMorgan's Law expresses a relationship between the OR and the AND operations in Boolean algebra. Stated in simple terms: *DeMorgan's Law says that inversion or negation of a Boolean expression may be accomplished by negating each variable/constant and by changing each AND to OR and OR to AND.* Equation (4–9) generalizes DeMorgan's Law.

$$\overline{A \cdot B} = \overline{A} + \overline{B} \quad \text{and} \quad \overline{A + B} = \overline{A} \cdot \overline{B} \tag{4–9}$$

Be sure to note the *difference* between the expression $\overline{A + B}$ and the expression $\overline{A} + \overline{B}$. In the former, A and B are combined in an OR relationship; then the output is inverted. In the latter, both A and B are first inverted; then the inverted variables are combined in the OR relationship. Hence *the two expressions are not equal*, which is also evident from the truth tables and the waveforms in Fig. 4–9.

All these laws are powerful tools that can be used to simplify Boolean expressions and to minimize the number of logic devices necessary to perform a particular function, as the following examples illustrate.

Example 4–1. Simplify the expression $(\overline{A} + \overline{B} + \overline{C})D$. Then draw the logic diagrams for the original and the simplified expressions.

Solution. $(\overline{A} + \overline{B} + \overline{C})D = (\overline{ABC})D$ DeMorgan's Laws

See Fig. 4–10.

Example 4–2. Simplify the expression $R + S + (TV) + (\overline{RS})$. Then draw the logic diagrams for the original and the simplified expressions and show truth-table proof of the equivalency.

Solution. $R + S + (TV) + (\overline{RS}) = R + S + (TV) + (\overline{R} + \overline{S})$ DeMorgan's Law
$\phantom{Solution. R + S + (TV) + (\overline{RS})} = R + S + (\overline{R} + \overline{S}) + TV$ Commutative Law
$\phantom{Solution. R + S + (TV) + (\overline{RS})} = R + S + \overline{R} + \overline{S} + TV$ Associative Law
$\phantom{Solution. R + S + (TV) + (\overline{RS})} = R + \overline{R} + S + \overline{S} + TV$ Commutative Law
$\phantom{Solution. R + S + (TV) + (\overline{RS})} = 1 + 1 + TV$ Theorem 12
$\phantom{Solution. R + S + (TV) + (\overline{RS})} = 1$ Theorem 10,
$\phantom{Solution. R + S + (TV) + (\overline{RS}) = 1}$ Postulate 8.

---

* The Distributive Law will be examined in greater detail in subsequent chapters.

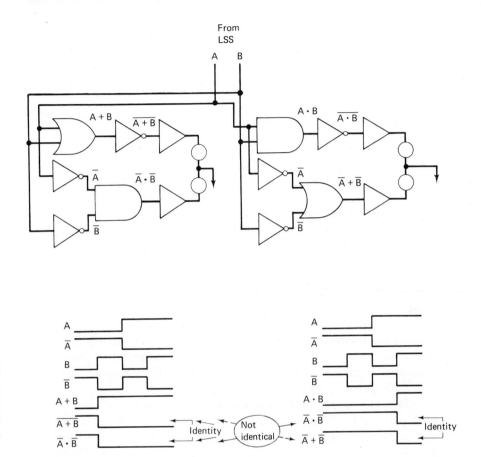

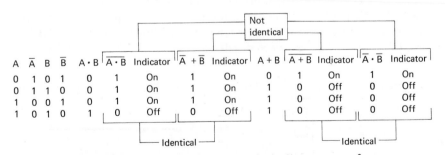

A	$\overline{A}$	B	$\overline{B}$	$A \cdot B$	$\overline{A \cdot B}$	Indicator	$\overline{A} + \overline{B}$	Indicator	$A + B$	$\overline{A + B}$	Indicator	$\overline{A} \cdot \overline{B}$	Indicator
						Not identical							
0	1	0	1	0	1	On	1	On	0	1	On	1	On
0	1	1	0	0	1	On	1	On	1	0	Off	0	Off
1	0	0	1	0	1	On	1	On	1	0	Off	0	Off
1	0	1	0	1	0	Off	0	Off	1	0	Off	0	Off
					Identical					Identical			

Figure 4—9   DeMorgan's Laws — logic diagram, waveforms, and truth tables

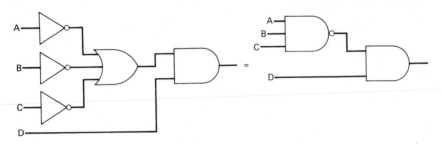

Figure 4—10   Logic diagram, Example 4—1

See Fig. 4—11 for the logic diagrams and Table 4—1 for the truth-table proof.

DeMorgan's Law also states mathematically the *equivalence of certain gate forms.* Since $\overline{A \cdot B} = \overline{A} + \overline{B}$, the hardware implementation of each side of the equation will provide identical results (see Fig. 4—9). Thus the negated input OR gate is functionally equivalent to the negated output AND (NAND) gate. Similarly, since $\overline{A} \cdot \overline{B} = \overline{A + B}$, the negated input AND gate is equivalent to negated output OR (NOR) gate.

As we mentioned in Ch. 3, many hardware manufacturers supply only certain types of gates — in the interests of cost. By applying DeMorgan's Law and the Distributive Law, we can show how *all* combinational functions implemented by the AND, OR, and NOT method may *also* be implemented by the NOR method *alone* or by the NAND method *alone.*

Section 3—6 defines *a NOR gate as an OR gate followed by an inverter,* which performs the NOT function. The truth table for the OR gate (see Fig. 3—2) reveals that when any one or more of the inputs to an OR gate is HIGH, the output is HIGH and that when all inputs are LOW, the output is LOW. Terminating all but one of the gate inputs — as recommended by the manufacturer's specification sheet — results in a 1-input OR gate. The 1-input OR gate still behaves according to Fig. 3—2; i.e., a HIGH input results in a HIGH output, a LOW input results in a LOW output. By using the 1-input OR gate followed by an inverter (a 1-input NOR gate), the "straight-through" operation of the 1-input OR gate is inverted, resulting in the logic operation of NOT. Thus it *is* possible to obtain the NOT operation by using only NOR gates.

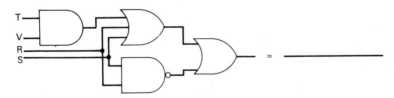

Figure 4—11   Logic diagram, Example 4—2

Table 4–1   Truth table $R + S + (TV) + (\overline{RS})$

R	$\overline{R}$	S	$\overline{S}$	T	$\overline{T}$	V	$\overline{V}$	$T \cdot V$	$R \cdot S$	$\overline{R \cdot S}$	$\overline{R}+\overline{S}$	$R+S+(TV)+(\overline{RS})$	$R+\overline{R}+S+\overline{S}+(TV)$
0	1	0	1	0	1	0	1	0	0	1	1	1	1
0	1	0	1	0	1	1	0	0	0	1	1	1	1
0	1	0	1	1	0	0	1	0	0	1	1	1	1
0	1	0	1	1	0	1	0	1	0	1	1	1	1
0	1	1	0	0	1	0	1	0	0	1	1	1	1
0	1	1	0	0	1	1	0	0	0	1	1	1	1
0	1	1	0	1	0	0	1	0	0	1	1	1	1
0	1	1	0	1	0	1	0	1	0	1	1	1	1
1	0	0	1	0	1	0	1	0	0	1	1	1	1
1	0	0	1	0	1	1	0	0	0	1	1	1	1
1	0	0	1	1	0	0	1	0	0	1	1	1	1
1	0	0	1	1	0	1	0	1	0	1	1	1	1
1	0	1	0	0	1	0	1	0	1	0	0	1	1
1	0	1	0	0	1	1	0	0	1	0	0	1	1
1	0	1	0	1	0	0	1	0	1	0	0	1	1
1	0	1	0	1	0	1	0	1	1	0	0	1	1

Figure 4–12a shows how NOR gates can be combined to perform the AND operation. The truth table accompanying the diagram verifies both the equality and the NOR gate implementation of the 2-input AND gate.

The OR operation may also be implemented using only NOR gates. The logic diagram and substantiating truth table appear in Fig. 4–12b.

At this point, we should mention some alternate types of logic diagrams. In some cases a *block diagram* form is used, since it more closely resembles the physical configuration of the devices performing the logic functions. A typical illustration is the 914 IC (integrated circuit) form, where two NOR gates are packaged in a TO-5 case that is approximately the same size as a single transistor. Its diagram may assume the form of either Fig. 4–13a or 4–13b and is equivalent to the logic diagram in Fig. 4–12.

NAND gates may also be used to perform the AND, OR, and NOT operations. *A* NAND *gate has been previously defined as an* AND *gate followed by an inverter* (NOT). The NAND gate's output is LOW when all inputs are HIGH. If any one input is LOW, its output* is HIGH. A 1-input AND may be derived in a

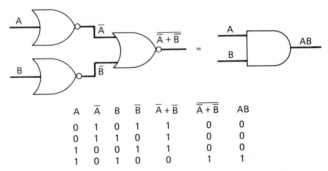

A	$\overline{A}$	B	$\overline{B}$	$\overline{A}+\overline{B}$	$\overline{\overline{A}+\overline{B}}$	AB
0	1	0	1	1	0	0
0	1	1	0	1	0	0
1	0	0	1	1	0	0
1	0	1	0	0	1	1

(a) AND operation using NOR gates—Logic diagram and truth tables

A	B	A + B	$\overline{A+B}$	$\overline{\overline{A+B}}$	A + B
0	0	0	1	0	0
0	1	1	0	1	1
1	0	1	0	1	1
1	1	1	0	1	1

(b) OR operation using NOR gates—Logic diagram and truth table

Figure 4–12   AND and OR operations using NOR gates – logic diagram and truth tables

_____
* An AND gate, in comparison, has a HIGH output when (and only when) all inputs are HIGH.

manner similar to the 1-input OR. Termination of unused inputs must be in accordance with the manufacturer's specifications. The 1-input AND performs the same function as the 1-input OR; that is, HIGH in –HIGH out, and LOW in–LOW out. Combining the 1-input AND with the inverter (NOT) results in a 1-input NAND that performs the NOT operation.

The AND operation may be obtained from a NAND gate with two inputs A and B, followed by a 1-input NAND gate. Justification and diagrams appear in Fig. 4–14a.

Another commonly encountered form of logic diagram that is device-oriented rather than logic-operation oriented is shown in Fig. 4–14b. This is known as the *dual in-line IC package*. Physically, it is approximately 0.28 inch wide, 0.785 inch long, and 0.18 inch thick. Common configurations include 14-pin (shown) and 16-pin versions.

As the complexity of logic functions within a single IC package increases, it becomes less and less practical to use the form of diagram in Fig. 4–14a. It is therefore quite common to provide the capability to perform complete arithmetic operations, such as addition, on a single IC "chip." Obviously the logic diagram of this type of operation would not fit inside the symbol provided and still be readable; hence, the form in Fig. 4–14b is used. Complete manufacturer's logic descriptions and diagrams must be available to perform logic analysis with this type of diagram. Figure 4–14b is logically equivalent to Fig. 4–14a.

If the output $A + B$ is desired, it can also be obtained from NAND gates. DeMorgan's Law states that $\overline{A + B}$ is equivalent to $\overline{A} \cdot \overline{B}$, which is the output of a NAND gate with inputs $\overline{A}$ and $\overline{B}$. These inputs may be easily obtained by two 1-input NANDs being fed by A and B, respectively. Logic diagrams and the truth table are provided in Fig. 4–14c.

Thus any of the three basic operations of Boolean Algebra – *conjunction* (AND), *disjunction* (OR), and *negation* (NOT) – may be obtained either from NOR or NAND gates. As an aid to using and remembering the three basic logic

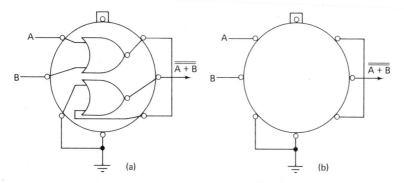

Figure 4–13   Physically orientated logic diagram (TO-5) Case

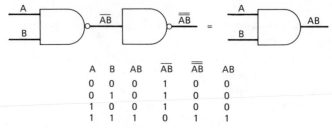

A	B	AB	$\overline{AB}$	$\overline{\overline{AB}}$	AB
0	0	0	1	0	0
0	1	0	1	0	0
1	0	0	1	0	0
1	1	1	0	1	1

(a) AND operation using NAND gates—Logic diagram and truth table

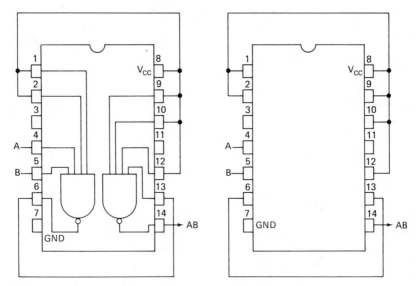

(b) Physically oriented logic diagram (DIP case)

A	$\overline{A}$	B	$\overline{B}$	$\overline{A} \cdot \overline{B}$	$\overline{\overline{A} \cdot \overline{B}}$	A + B
0	1	0	1	1	0	0
0	1	1	0	0	1	1
1	0	0	1	0	1	1
1	0	1	0	0	1	1

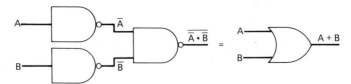

(c) OR operation using NAND gates—Logic diagram and truth table

Figure 4–14 AND and OR operations using NAND gates — logic diagrams and truth tables

operations and the two derived logic operations, Table 4–2 provides the logic characteristics of each in an easily comparable format. Innumerable combinations of AND, OR, NOT, NAND, and NOR operations are possible. Extensive coverage of all possible configurations is not intended. The basic fundamentals have been provided and will be expanded as examinations of more complex logic circuits are undertaken.

Table 4–2   AND-OR/NAND-NOR equivalences

A	$\overline{A}$	B	$\overline{B}$	$A \cdot B$	$A + B$	$\overline{A \cdot B}$	$\overline{A} + \overline{B}$	$\overline{A + B}$	$\overline{A} \cdot \overline{B}$
0	1	0	1	0	0	1	1	1	1
0	1	1	0	0	1	1	1	0	0
1	0	0	1	0	1	1	1	0	0
1	0	1	0	1	1	0	0	0	0

The Law of Absorption

The first identity of the *Law of Absorption* is

$$A + AB = A \qquad\qquad (4-10)$$

Actually, this identity is nothing more than common sense. Whenever A is 1, the value of $A + AB$ is 1. It is really unnecessary to consider what happens to the value of the expression when both A and B are 1. Regardless of the value of A and B together, the value of the expression is determined by the value of A alone. The truth table and diagrams in Fig. 4–15a verify this statement.

The second identity of the Law of Absorption

$$A(A + B) = A \qquad\qquad (4-11)$$

is demonstrated in Fig. 4–15b.

The Law of Expansion

The first identity of the *Law of Expansion*

$$AB + A\overline{B} = A \qquad\qquad (4-12)$$

can be proven by any of the methods employed so far. For example, the following is an algebraic proof.

$$
\begin{aligned}
AB + A\overline{B} &= A(B + \overline{B}) &&\text{Factoring} \\
&= A &&B + \overline{B} = 1 \\
\therefore\ AB + A\overline{B} &= A
\end{aligned}
$$

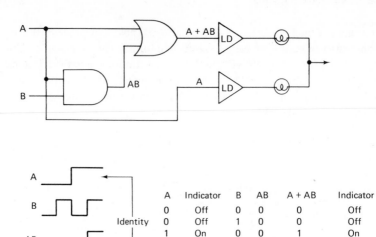

A	Indicator	B	AB	A + AB	Indicator
0	Off	0	0	0	Off
0	Off	1	0	0	Off
1	On	0	0	1	On
1	On	1	1	1	On

Identity

(a) A + AB = A

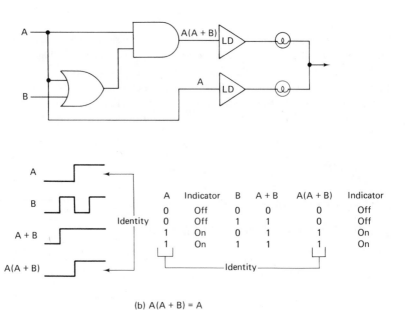

A	Indicator	B	A + B	A(A + B)	Indicator
0	Off	0	0	0	Off
0	Off	1	1	0	Off
1	On	0	1	1	On
1	On	1	1	1	On

Identity

(b) A(A + B) = A

Figure 4—15   The Laws of Absorption

The logic diagram, waveforms, and accompanying truth table in Fig. 4–16a further demonstrate the Law of Expansion.

Truth-table proof of the second identity of the Law of Expansion

$$(A + B)(A + \overline{B}) = A \tag{4-13}$$

appears in Fig. 4–16b, along with the logic diagram, waveforms, and algebraic proofs.

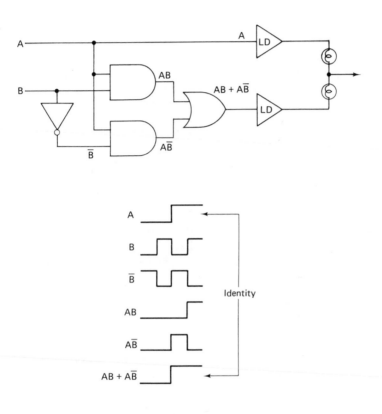

A	B	$\overline{B}$	AB	$A\overline{B}$	$AB + A\overline{B}$	Indicator	A	Indicator
0	0	1	0	0	0	Off	0	Off
0	1	0	0	0	0	Off	0	Off
1	0	1	0	1	1	On	1	On
1	1	0	1	0	1	On	1	On

Identity

Figure 4–16 Truth table, logic diagram, and waveform proofs – Law of Expansion: (a) $AB + A\overline{B} = A$;

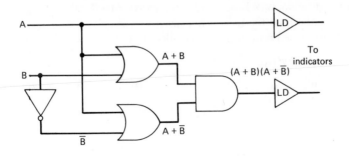

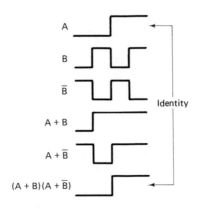

A	B	$\overline{B}$	A + B	A + $\overline{B}$	(A + B)(A + $\overline{B}$)	Indicator	A	Indicator
0	0	1	0	1	0	Off	0	Off
0	1	0	1	0	0	Off	0	Off
1	0	1	1	1	1	On	1	On
1	1	0	1	1	1	On	1	On

Identity

Figure 4–16 (cont'd)   (b) (A + B) (A + $\overline{B}$) = A

Duality

Our investigation of the postulates, theorems, and laws of Boolean algebra points out an interesting and useful relationship. The Commutative Law, $AB = BA$ and $A + B = B + A$, may be used as an example. These identities as well as the other laws, theorems, and postulates, appear in *pairs*. One pair is usually in a *conjunctive form* (the ANDing of variables/constants), while the other is in a *disjunctive form* (the ORing of variables/constants). Both the conjunctive and the disjunctive versions of the Commutative Law have already been proven separately. In fact, perusal of the other laws, theorems, and postulates reveals that if the conjunctive version has been proven, the disjunctive version is also valid and vice versa. It can therefore be stated that *the conjunctive version of a postulate, theorem, or law is the dual of the disjunctive version* and that *the disjunctive version is the dual of the conjunctive version.*

The *symmetry* (duality) of Boolean algebra can be more formally stated.

> Given any postulate, theorem, or law in Boolean algebra,
> if each 0 is replaced by 1, each 1 by 0, each + by ·, and
> each · by +, the result is the *dual* of the selected postu-
> late, theorem, or law.

This effectively reduces by half the task of investigation of the theorems and laws of Boolean algebra. The mathematical derivation of the duality relationship is itself another subject and is not investigated in this text.

4–2 COMMON IDENTITIES OF BOOLEAN ALGEBRA

Considerable justification for the investigation of the identities of Boolean algebra exists from an economic viewpoint. In conventional mathematics, identities are used to facilitate the *mathematical simplification* of equations and expressions; in Boolean algebra such simplifications usually result not only in simplified forms of equations and expressions but also in *reduced requirements for hardware* to implement them. Only the most common identities encountered in everyday work are discussed here. Each of the identities is proven by at least two methods previously employed. The remaining proofs are left to the reader as exercises at the end of the chapter.

The Identity

$$A(\overline{A} + B) = AB \tag{4–14}$$

This identity is easily analyzed by logical thinking. For $A(\overline{A} + B) = AB$ to be true (logical 1), both A and B must be 1. Confirmation is made in the following manner: If A is 0, it doesn't matter what value B possesses, since 0 ANDed with *anything* is 0. If A is 1 and B is 0, then the value of $A(\overline{A} + B)$ is 0, that is $1(0 + 0) = 0$. However, with $B = 1$ and $A = 1$, the value of $A(\overline{A} + B)$ becomes 1, since $1(0 + 1) = 1$. Note that the *logical thought* method of analysis closely resembles the truth-table method. If this "English-language" method is more easily understood by some readers, it should be applied wherever possible. Usually, however, the truth-table method provides the same information in a more organized manner. Figure 4–17 shows the proofs of this identity.

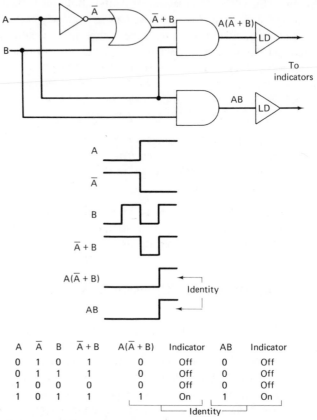

Figure 4—17   Logic diagram, truth table, and waveform proofs of the identity $A(\overline{A} + B) = AB$

The Identity

$$A + \overline{A}B = A + B \tag{4-15}$$

Logically thinking, if $A = 1$, it doesn't matter whether $B = 0$ or $B = 1$, since 1 ORed with *anything* is 1. If $A = 0$, then $\overline{A} = 1$. B must equal 1 for $\overline{A}B = 1$. $\overline{A}B = 1$ so that $A + \overline{A}B = 1$, that is, $0 + (1 \cdot 1) = 1$. The logic diagram and algebraic proofs appear in Fig. 4—18.

Another Identity

$$AB(A + B) = AB \tag{4-16}$$

The identity in Eq. (4—16) is often encountered in digital logic circuits. Recognition of this identity allows the same operation to be performed with one AND gate instead of one OR gate *and* two AND gates. Proofs of the identity are shown in Fig. 4—19.

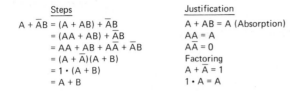

Steps	Justification
$A + \overline{A}B = (A + AB) + \overline{A}B$	$A + AB = A$ (Absorption)
$= (AA + AB) + \overline{A}B$	$AA = A$
$= AA + AB + A\overline{A} + \overline{A}B$	$A\overline{A} = 0$
$= (A + \overline{A})(A + B)$	Factoring
$= 1 \cdot (A + B)$	$A + \overline{A} = 1$
$= A + B$	$1 \cdot A = A$

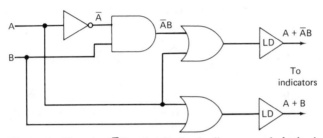

Figure 4–18 $A + \overline{A}B = A + B$: Logic diagram and algebraic proofs

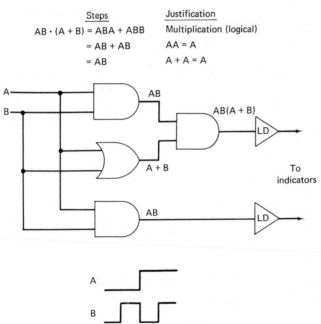

Steps	Justification
$AB \cdot (A + B) = ABA + ABB$	Multiplication (logical)
$= AB + AB$	$AA = A$
$= AB$	$A + A = A$

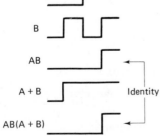

Figure 4–19 $AB(A + B) = AB$

The EXCLUSIVE—OR Identity

$$\overline{AB}(A + B) = \overline{A}B + A\overline{B} \qquad (4\text{--}17)$$

Investigation of the right-hand side of the equation shows that it can have a value of 1 only when A is 1 and B is 0 or when A is 0 and B is 1. This specific relationship is a special case of disjunction, called the EXCLUSIVE—OR *operation*.

It has a special symbol and a special operating sign $\oplus$.

This operation is commonly encountered in digital logic and can appear in many different forms. One of these forms is shown on the left-hand side of the equation.

As expressions and equations, such as the EXCLUSIVE—OR, become more complex, it becomes more difficult to use the logical-thought method of analysis. However increased experience with Boolean forms results in greater insight and capability. Although the approach may not appear immediately obvious, it will be shown here as an example of a more advanced form of logical-thought.

For the value of the left-hand side of the equation to be 1, $\overline{AB}$ must be 1 and $(A + B)$ must be 1. For $\overline{AB}$ to be 1, AB must be 0. For $\overline{AB}$ to be 0, A must be 0, B must be 0, or both must be 0. For $(A + B)$ to be 1, A must be 1, B must be 1, or both must be 1. Note the discrepancy in the requirements for $\overline{AB}$ and $(A + B)$ to be 1.

If A and B are both 0, making $\overline{AB} = 1$, then $(A + B)$ is 0. $\overline{AB}(A + B)$ is then equal to 0, since $1 \cdot 0 = 0$. When A and B are both 1, $\overline{AB} = 0$, while $(A + B) = 1$. The combination is again equal to 0, since $0 \cdot 1 = 0$. The only condition under which the combination is equal to 1 is either when $A = 1$ and $B = 0$ or when $A = 0$ and $B = 1$. Once again, the use of *grouping marks* (vinculum and parentheses) aids in developing the logic diagram. The parentheses imply that $A + B$ is to be implemented by itself, and the vinculum implies that A is to be ANDed with B and negated as a single term. When these operations are completed, $\overline{AB}$ is to be ANDed with $(A + B)$. The truth table and logic diagram may be seen in Fig. 4–20.

The EXCLUSIVE—NOR Identity

$$\overline{A\overline{B} + \overline{A}B} = A \cdot B + \overline{A} \cdot \overline{B} \qquad (4\text{--}18)$$

This identity shows that the inverted form of the EXCLUSIVE—OR operation, at left side of the equal sign, is the same as the disjunction of the

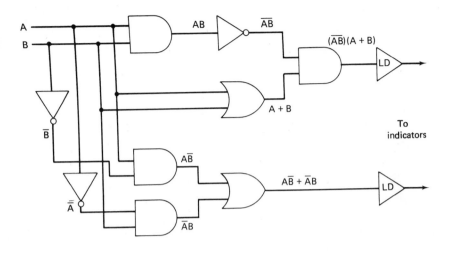

A	$\overline{A}$	B	$\overline{B}$	AB	$\overline{AB}$	A + B	$(\overline{AB})(A + B)$	$A\overline{B}$	$\overline{A}B$	$A\overline{B} + \overline{A}B$
0	1	0	1	0	1	0	0	0	0	0
0	1	1	0	0	1	1	1	0	1	1
1	0	0	1	0	1	1	1	1	0	1
1	0	1	0	1	0	1	0	0	0	0

Identity

Figure 4–20 $(\overline{AB})(A + B) = A\overline{B} + \overline{A}B$

conjunction of the true form *and* the false form of the variables. In other words, by ORing A AND B with $\overline{A}$ AND $\overline{B}$, the inverted form of the EXCLUSIVE–OR operation is obtained. Although this identity may be examined using the logical-thought method, this approach becomes unwieldy when either many variables or numerous negations are encountered. It is recommended that truth-table or other forms of analysis be used in this and remaining identities. (See Fig. 4–21.)

A Three-variable Identity

$$(A + B)(B + C)(A + C) = AB + BC + AC \quad (4-19)$$

The commonly encountered identities are not limited to two variables. *Three variables*, arranged in the manner shown in Eq. (4–19), are often seen in digital logic circuits. Figure 4–22 shows the logic diagram and waveform proofs for this identity.

A	$\bar{A}$	B	$\bar{B}$	$A\bar{B}$	$\bar{A}B$	$A\bar{B}+\bar{A}B$	$\overline{A\bar{B}+\bar{A}B}$	AB	$\bar{A}\bar{B}$	$AB+\bar{A}\bar{B}$
0	1	0	1	0	0	0	1	0	1	1
0	1	1	0	0	1	1	0	0	0	0
1	0	0	1	1	0	1	0	0	0	0
1	0	1	0	0	0	0	1	1	0	1

Identity

$$\overline{A\bar{B}+\bar{A}B} = \overline{(A\bar{B})}\,\overline{(\bar{A}B)}$$
$$= (\bar{A}+B)(A+\bar{B})$$
$$= A\bar{A} + \bar{A}\bar{B} + AB + B\bar{B}$$
$$= O + \bar{A}\bar{B} + AB + O$$
$$= \bar{A}\bar{B} + AB$$
$$= AB + \bar{A}\bar{B}$$

DeMorgan's Law
DeMorgan's Law
Multiplication
$A\bar{A} = O$
$A + O = A$
Commutation

Figure 4–21 $\overline{A\bar{B}+\bar{A}B} = AB + \bar{A}\bar{B}$

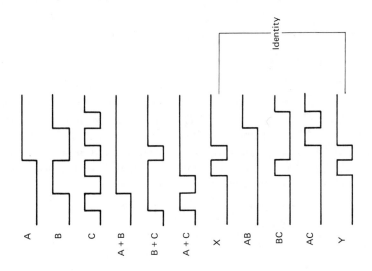

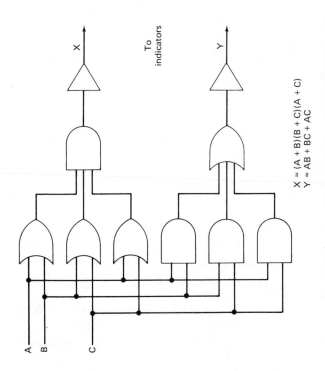

$X = (A + B)(B + C)(A + C)$
$Y = AB + BC + AC$

To indicators

Figure 4—22 $(A + B)(B + C)(A + C) = AB + BC + AC$

A Complex Three-variable Identity

$$(A + B)(\overline{A} + C) = AC + \overline{A}B \tag{4-20}$$

For this identity to be true, at least one variable inside each set of parentheses on the left-hand side of the equal sign must be equivalent to 1. If $A = 1$, then $\overline{A} = 0$ and C must be equal to 1. The value of B is not important, since if $A = 1$, then $1 + B = 1$. Therefore AC is one of the combinations required. If $A = 0$, then $\overline{A} = 1$ and B must be equal to 1. The value of C is not important, since if $\overline{A} = 1$, then $1 + C = 1$. Therefore $\overline{A}B$ is the other combination required. The following is the algebraic proof.

$$
\begin{aligned}
(A + B)(\overline{A} + C) &= A\overline{A} + AC + \overline{A}B + BC \\
&= 0 + AC + \overline{A}B + BC \\
&= AC + \overline{A}B + BC \\
&= \overline{A}B + AC + (A + \overline{A})(BC) \\
&= \overline{A}B + AC + ABC + \overline{A}BC \\
&= \overline{A}B + \overline{A}BC + AC + ABC \\
&= \overline{A}B(1 + C) + AC(1 + B) \\
&= (\overline{A}B \cdot 1) + (AC \cdot 1) \\
&= \overline{A}B + AC \\
&= AC + \overline{A}B \\
\therefore \quad (A + B)(\overline{A} + C) &= AC + \overline{A}B
\end{aligned}
$$

Justification for each step in this proof should be determined by the reader. The truth table for the identity $(A + B)(\overline{A} + C) = AC + \overline{A}B$ may be seen in Table 4–3.

Table 4–3   Truth Table $(A + B)(\overline{A} + C) = AC + A\overline{B}$

A	$\overline{A}$	B	$\overline{B}$	C	$\overline{C}$	$A+B$	$\overline{A}+C$	$(A+B)(\overline{A}+C)$	AC	$\overline{A}B$	$AC+\overline{A}B$
0	1	0	1	0	1	0	1	0	0	0	0
0	1	0	1	1	0	0	1	0	0	0	0
0	1	1	0	0	1	1	1	1	0	1	1
0	1	1	0	1	0	1	1	1	0	1	1
1	0	0	1	0	1	1	0	0	0	0	0
1	0	0	1	1	0	1	1	1	1	0	1
1	0	1	0	0	1	1	0	0	0	0	0
1	0	1	0	1	0	1	1	1	1	0	1

————Identity————

AN OR–AND Identity

$$AC + AB + B\overline{C} = AC + B\overline{C} \tag{4-21}$$

Logic diagrams and waveforms (Fig. 4–23) are used to prove this identity. Equation (4–21) may be considered to be an example of the Law of Absorption in more complex form. The term AB is effectively "absorbed" by the first term AC, which contains the variable A, and the third term $B\overline{C}$, which contains the variable B. Development of the algebraic proof (left to the reader as an exercise at the end of the chapter) makes this viewpoint more apparent.

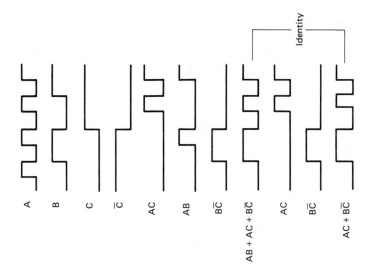

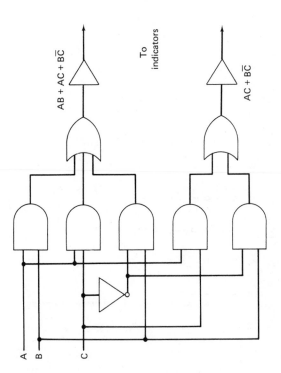

Figure 4–23 $AC + AB + B\overline{C} = AC + B\overline{C}$

Table 4–4 Truth table $(A + B)(B + C)(\bar{A} + C) = (A + B)(\bar{A} + C)$

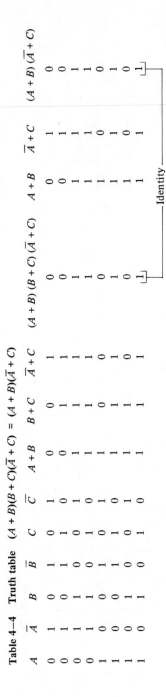

A	$\bar{A}$	B	$\bar{B}$	C	$\bar{C}$	$A+B$	$B+C$	$\bar{A}+C$	$(A+B)(B+C)(\bar{A}+C)$	$A+B$	$\bar{A}+C$	$(A+B)(\bar{A}+C)$
0	1	0	1	0	1	0	0	1	0	0	1	0
0	1	0	1	1	0	0	1	1	0	0	1	0
0	1	1	0	0	1	1	1	1	1	1	1	1
0	1	1	0	1	0	1	1	1	1	1	1	1
1	0	0	1	0	1	1	0	0	0	1	0	0
1	0	0	1	1	0	1	1	1	1	1	1	1
1	0	1	0	0	1	1	1	0	0	1	0	0
1	0	1	0	1	0	1	1	1	1	1	1	1

Identity

AN AND–OR Identity

$$(A + B)(B + C)(\overline{A} + C) = (A + B)(\overline{A} + C) \tag{4–22}$$

Truth-table proof of this identity is shown in Table 4–4, while the associated algebraic proof follows.

$$
\begin{aligned}
(A + B)(B + C)(\overline{A} + C) &= [(A + B)(B + C)]\,[\overline{A} + C] \\
&= [AB + AC + BB + BC]\,[\overline{A} + C] \\
&= \overline{A}AB + \overline{A}AC + \overline{A}BB + \overline{A}BC + ABC + ACC + BBC + BCC \\
&= 0 + 0 + \overline{A}B + \overline{A}BC + ABC + AC + BC + BC \\
&= \overline{A}B + \overline{A}BC + ABC + AC + BC \\
&= \overline{A}B + BC(\overline{A} + A) + AC + BC \\
&= \overline{A}B + BC + AC
\end{aligned}
$$

$$\therefore \quad (A + B)(B + C)(\overline{A} + C) = \overline{A}B + BC + AC = (A + B)(\overline{A} + C)$$

Once again the reader is encouraged to determine the justification for each step in this proof.

4–3 SUMMARY OF ALGEBRAIC TO LOGIC CONVERSION

Although they have not been specifically pointed out, the major techniques for converting algebraic expressions to equivalent logic diagrams have been discussed in this chapter. Collecting the various techniques and listing them in sequence of execution provides the reader with a valuable aid in the conversion of algebraic expressions to logic diagrams. A list of suggested procedures is provided in this section, along with explanatory notes.

1. *Identify the general type of circuit.* Except in very complex cases, a logic circuit may be catagorized as either an AND–OR or an OR–AND type. The AND–OR type consists of two or more ANDed operations combined together in an OR operation. For example, $AB + CD$ is an AND–OR type of logic expression; it would be implemented with an AND gate to combine A and B, an AND gate to combine C and D, and an OR gate to combine the results of the two ANDed operations.

   The OR–AND logic circuit consists of two or more ORed operations combined together in an AND operation. For example $(A + B)(C + D)$ is a typical OR–AND logic expression; it would be implemented with an OR gate to combine A and B, an OR gate to combine C and D, and an AND gate to combine the results of the two ORing operations.

2. *Starting with the overall expression, draw the logic symbol that represents the last logic operation.* The final logic operation can be identified after deciding what the general type of circuit does. If it is an AND–OR circuit, the last operation is an OR operation; if it is an OR–AND circuit, the last operation is an AND operation. As logic circuits become more complex, it may not be easy to determine the final operation. Experience with logic-circuit analysis provides the reader with the necessary skills to make this determination.

3. *Working toward the input, draw the logic symbol for each operation until all inputs are accounted for.* In the simple AND–OR example $(AB + CD)$, it was shown that the last logic operation is an OR operation. Each of the inputs to the OR gate will have to come from AND gates, since the two terms being ORed are ANDed terms. Thus one of the inputs would come from an AND gate with A and B inputs; the other OR input would come from another AND gate with C and D inputs. The logic diagram is now complete, since all inputs have been accounted for.

Quite often, *apparent* simplifications to the logic diagram being constructed may be noted. Until added experience is gained, however, such random simplifications* are discouraged.

4–4   A COMBINATIONAL LOGIC CIRCUIT (BINARY HALF-ADDER)

Combinational logic problems appear in many different forms – most of which have been presented in this and preceding chapters. Now a partial combinational logic circuit, borrowed from a computer application, will be described using "English-language" statements. From this description, various other methods of presenting the same information will be shown.

Binary Addition

Digital computers use logic circuits called *adders* to obtain the *sum* of binary numbers. One of the added numbers is called the *augend*, which is the number to which the other number is to be added. The other number is called the *addend.* Conventionally, the augend is written first (on the top line), and

---

* Chapter 5 deals with algebraic methods of logic-circuit simplification where an *organized* method for reducing logic circuit requirements is discussed. Conversion techniques are also expanded in Ch. 5.

the addend is written next (below the augend). The rules for arithmetic addition of positive binary integers are as follows.

0 *plus* 0 = 0,   0 *plus* 1 = 1,   1 *plus* 0 = 1,   1 *plus* 1 = 0 *carry* 1

A logic circuit that performs these operations is called a *half-adder.* Although the derivation of the term "half-adder" is somewhat obscure, investigation shows that such a circuit must be capable of adding two single-digit binary numbers and providing the sum and carry outputs. An example of binary addition might prove helpful in clarifying the meaning of the term. When the binary addition to be performed concerns only two single-digit binary numbers, the half-adder quite adequately performs the function. Thus

	0		0		1		1
plus	0	*plus*	1	*plus*	0	*plus*	1
sum	0		1		1		0
carry	0		0		0		1

It seems reasonable, therefore, that if the preceding addition is performed on the least significant digits of multidigit numbers, it is necessary to provide only one other half-adder for each order of the numbers. To make it simple: consider the addition of the 2-digit binary numbers 11 and 11. The carry generated in the least-significant order must be added to the next-most-significant order. But this results in a three input requirement, and the half-adder has only two inputs. Hence the name half-adder, since it can only do *half* the job. A *full-adder* is required in all but the least-significant positions to accommodate the carry inputs from previous orders. Full-adders are discussed in subsequent chapters.

Analysis of the Half-Adder

The operation of the half-adder logic circuit is described by the following definition.

> If the augend is zero and the addend is one or if the
> augend is one and the addend is zero, the sum is one.
> If the augend is one and the addend is one, the carry
> output is one.

This turns out to be a relatively complex statement, subject to misinterpretation and difficult to visualize. Perhaps substituting letter names for each of the variable names in the "English-language" statement would clarify the meaning. Thus the augend can be named A, the addend can be B, the sum can be S, and the carry can be C. Now the statement reads: "If A is zero and B is one or if A is one and B is zero, S is one. If A is one and B is one, C is one." By arbitrarily assigning the logical value 0 to the numerical value zero, and the logical value 1 to the numerical value one, another approach can be initiated.

Consider logical 0 as a LOW and logical 1 as a HIGH. This statement is shown in truth-table form in Table 4—5a. Note that two variables are used in this table. Only three of the four possible combinations of the two variables are shown. But we desire to expand the truth table to include all four combinations so that the table will more closely resemble standard truth tables (see Table 4—5b).

Remember that the designation of a variable as LOW (0) is usually considered to be the false (negated) form, and HIGH (1) is the true form. The two terms that result in a HIGH S output and the one term that results in a HIGH C output can be written in Boolean algebra equations. Thus

$$S = \bar{A}B + A\bar{B}$$
$$C = AB \tag{4-23}$$

Applying what was discussed earlier in this chapter allows this equation to be displayed in symbolic diagram form (Fig. 4—24a) and, finally, as a waveform (Fig. 4—24b).

Table 4—5   Half-adder truth tables

A	B	S	C
0	1	1	0
1	0	1	0
1	1	0	1

(a) Direct conversion of half-adder description to tabular form

A	B	S	C
0	0	0	0
0	1	1	0
1	0	1	0
1	1	0	1

(b) Conventional half-adder truth table

Summary of Analysis Methods

The same combinational logic circuit has been described by five different methods. Descriptive methods to be used depend not only on the ultimate user of the material but also on the complexity of the description. The "English-language" description is commonly encountered in material written for non-technical people, since the combinational logic circuit can be presented in a relatively nontechnical manner.

Algebraic and truth-table methods provide an excellent means of presenting circuit-function information to engineering groups, who may use these mathematical forms to develop actual logic circuits. These methods are also useful to technicians in analyzing and troubleshooting logic circuits. Symbolic diagrams are used by both engineers and technicians as a visual aid to understanding circuit operation. Waveform drawings and photographs assist trouble-

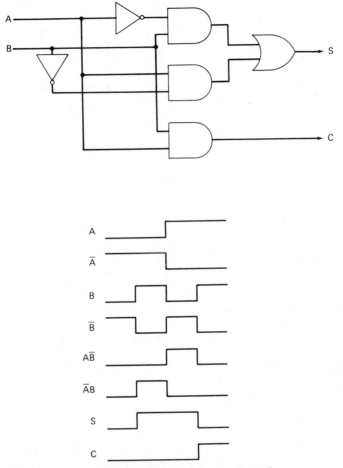

Figure 4—24   Half-adder logic diagram and waveforms

shooting during the detection and repair of malfunctions in digital equipment.
Of course, the use of any of these descriptive methods is not limited to a specific
group. Comments made herein merely provide a general picture for those who
want to find which method would be most useful.

PROBLEM APPLICATIONS

4—1.   List four combinational logic circuit analysis methods.
4—2.   Explain a common application of the Commutative Law in combinational
logic.
4—3.   Can the Associative Law be used to save hardware when implementing
logic equations? Explain.

4-4.   Why is the Distributive Law sometimes considered one of the most powerful of the laws of Boolean algebra?

4-5.   What is a NOR gate? Show how it can be developed from AND, OR, and NOT hardware elements.

4-6.   What is a NAND gate? Show how it can be developed from NOR hardware elements.

4-7.   Explain the second identity of the Law of Absorption, $A(A + B) = A$, using the "logical thought" method.

4-8.   What is duality? Why is it a valuable attribute of Boolean algebra?

4-9.   Prove the identity $A(\overline{A} + B) = AB$ using the postulates, theorems, and laws of Boolean algebra.

4-10.   Develop the truth table and waveform diagrams to prove the identity $A + \overline{A}B = A + B$.

4-11.   Prove the identity $AB(A + B) = AB$ using the "logical-thought" method.

4-12.   Using the postulates, theorems, and laws of Boolean algebra, prove the identity $\overline{AB}(A + B) = A\overline{B} + \overline{A}B$.

4-13.   Show the waveform diagram that represents the identity in Problem 12.

4-14.   Draw a logic diagram and show the waveforms that represent the identity $A\overline{B} + \overline{A}B = A \cdot B + \overline{A} \cdot \overline{B}$.

4-15.   Develop the truth table and show the Boolean algebra proof for the identity $(A + B)(B + C)(A + C) = AB + BC + AC$.

4-16.   Show the waveforms and logic diagram that represent the identity $(A + B)(\overline{A} + C) = AC + \overline{A}B$.

4-17.   Develop the Boolean algebra proof for the identity $AC + AB + B\overline{C} = AC + B\overline{C}$.

4-18.   Draw the waveforms that represent the identity $(A + B)(B + C)(\overline{A} + C) = (A + B)(\overline{A} + C)$.

Given the following logic circuit description, develop the

4-19.   Boolean Algebra equation,

4-20.   truth table,

4-21.   logic diagram, and

4-22.   waveform diagram.

> The light will illuminate if the switch at the top of the stairs is on and the switch at the bottom of the stairs is off, or if the switch at the top of the stairs is off and the switch at the bottom of the stairs is on.

Given the truth table on the right, develop the

4-23.   Boolean algebra equation,

4-24.   logic diagram, and

4-25.   waveform diagram.

A	B	C	X
0	0	0	0
0	0	1	0
0	1	0	1
0	1	1	1
1	0	0	1
1	0	1	1
1	1	0	0
1	1	1	0

5

Algebraic Simplification Methods

5-1 WHAT IS LOGIC SIMPLIFICATION?

Logic simplification definitions are multitudinous — there are almost as many definitions as there are people to make them! In general, however, the word *simplification* implies *a reduction in complexity*. Since there are so many different ways to describe a logic circuit in terms of ANDs, ORs, NOTs, NANDs, and NORs, some arbitrary definition of which circuit is the simplest must be determined.

In some cases (e.g., many engineering situations), the basis for choosing a particular circuit is cost. But with minimum cost as a criterion, other parameters come into play. Can a relationship between the cost of simple diode gates and the number of inputs per gate be discovered? How much do NAND or NOR gates* cost in quantity compared to AND—OR—NOT circuits? Is it advantageous to purchase a "prepackaged" logic function in integrated circuit (IC) form? These are but a few of the many questions that must be considered when logic simplification is to result in minimum cost requirements. The increasing use of digital computers has allowed designers to consider these variables in the design phase of logic systems; hence cost-effective designs are resulting in a rapid expansion of digital applications to society's everyday needs.

But cost-effective designs† are basically a problem for the design engineer. Since this text is using logic simplification merely as a way to develop skill in logic analysis and troubleshooting, a more appropriate definition must be developed. Sometimes logic expressions are simplified so that *a minimum*

---

* Economic reasons frequently limit the type(s) of gates available to implement the logic. Use of NAND gates or NOR gates may simplify the manufacturing process due to fewer *types* of modules in the system.

† Of course, cost cannot be disregarded completely. Implementing any logic function without regard to cost would be financial suicide.

number of logic elements (gates, inverters, etc.) are required for implementation. Often this is equivalent to finding the "simplest" form for the logic function. Other simplification techniques result in Boolean expressions containing a minimum number of terms with a minimum number of variables in each term. As we mentioned earlier, it all depends on the definition of the "simplest form."

Two such definitions are used in this text. First, after applying the procedures for simplification that are detailed in this chapter, *the algebraic form that results in the smallest number of logic elements is considered the simplest form for the Boolean expression.* It may be necessary to examine several different alternatives to determine the final form.

Second, the Boolean expression should also be stated in the AND–OR form in such a way that *no variables are enclosed in parentheses and no vincula extend over more than one variable.* This second form may not necessarily be the form with the smallest number of logic elements, but we should be familiar with it in preparation for the chapter on graphic simplification.

The Study of Logic Simplification

Although some people find the study of a subject fascinating just for the "sheer enjoyment of learning something new," this may not be enough motivation for everyone. *From a strictly economic viewpoint, logic simplification is a useful undertaking.* We have already mentioned that considerable savings are realized if the capability to simplify logic is achieved. If the task of the reader is to design logic circuits, then there is no doubt that such a capability is valuable.

However many of our readers are involved in analysis and troubleshooting, both of which require *the recognition of many different logic operations and functions.* Besides performing the analysis and troubleshooting on actual equipment, perhaps one of the most satisfactory ways to recognize these logic circuits is to work with them on paper. And perhaps the best way to work with them on paper is to study the various methods of logic simplification.

As numerous logic circuits performing the same function are observed, *the many different ways of performing the same operations become obvious.* As a desirable by-product, the reader experiences actual logic simplification work and should therefore develop *a certain amount of capability in the area of simplification techniques.* Simplification is one of the steps toward a *design capability.* Such a capability, with the actual hardware experience that many technical personnel already possess, is and always will be in demand in the labor force.

Methods of Logic Simplification

Two general methods of logic simplification are in common use today. One method (the subject of this chapter) uses the postulates, theorems, laws, and identities of Boolean algebra. The other method uses graphic techniques and is detailed in Ch. 6. Each has its advantages and disadvantages — its easy roads and pitfalls. Certain similarities become obvious as the simplification methods are developed, and mastery of both methods is a big step toward a complete understanding of *combinational logic circuits.*

5–2 ALGEBRAIC SIMPLIFICATION PROCEDURES

A General Procedure

Every combinational logic circuit has its own peculiarities and requires its own analysis procedure. However enough similarities exist to develop a general procedure for all combinational logic circuits and then modify the procedure slightly for each new circuit encountered. Such a procedure is listed here.

1. Obtain the Boolean expression for the original circuit:
   (a) from the English language description (Sec. 3–1);
   (b) from a truth table (Sec. 4–1);
   (c) from a logic diagram (Sec. 4–3);
   (d) from the algebraic description (already in proper form).
2. Simplify algebraically:
   (a) apply postulates, theorems, laws, and identities as required (Ch. 3);
   (b) factor, substitute, expand, modify, and convert as required (Ch. 5);
   (c) try all over again to make sure the simplest form has been obtained.
3. Compare the new expression with the original expression by using truth tables (Sec. 4–1);
4. Draw the simplified logic diagram (Sec. 4–3);

Example 5–1. Algebraically simplify $A + ABC + AB + \overline{A}BC$.
Solution.

$$
\begin{aligned}
A + ABC + AB + \overline{A}BC &= A + AB + ABC + \overline{A}BC && \text{Commutative Law}\\
&= (A + AB) + (\overline{A}BC + ABC) && \text{Associative Law}\\
&= (A \cdot 1 + AB) + (ABC + \overline{A}BC) && \text{Theorem 3}\\
&= A(1 + B) + BC(A + \overline{A}) && \text{Factoring}\\
&= A \cdot 1 + BC(A + \overline{A}) && \text{Theorem 9}\\
&= A + BC(A + \overline{A}) && \text{Theorem 3}\\
&= A + BC \cdot 1 && \text{Theorem 12}\\
\therefore \quad A + ABC + AB + \overline{A}BC &= A + BC && \text{Theorem 3}
\end{aligned}
$$

As experience is gained, the number of steps required to reach a simplified version of an original expression diminishes. Once the factoring is accomplished (as in Ex. 5—1), the experienced analyst probably recognizes that both $(1 + B)$ and $(A + \bar{A})$ are equivalent to 1 and that any variable ANDed with 1 is equal to that variable. In other words, the step following the factoring step *could* have been $A + BC$ directly. But don't jump steps until it is obvious that such jumps can be taken. Many traps exist between steps, and a jump in the wrong direction may end up in a "deep hole!"

Some New Terminology

Another word of caution is justified at this time. The use of the multiplication dot (·) and the plus sign (+) in Boolean algebra is a carryover from conventional arithmetic and algebra. Since the plus sign is used to designate disjunction in Boolean algebra, it is only reasonable that disjunction be called *logical addition*. Similarly, the dot's designation as a conjunctive operator results in conjunction being called *logical multiplication*. Thus sums and products exist in Boolean algebra. Of course, these are "logical sums" and "logical products," but the name "logical" is dropped by many users. It is not at all uncommon to see the result of ANDing called a "product" and the result of ORing called a "sum." But please remember that these terms are *logical operations* – *not* conventional arithmetic operations. When such terms are used in this text, they will be specifically expressed as *logical sum* and *logical product*.

In Boolean algebra, it is generally recognized that *any logic expression* can be stated as either *a conjunction of disjunctive terms* or *a disjunction of conjunctive terms*. In other words, any expression can be stated either as OR operations connected by AND operations or AND operations connected by OR operations. The expression $(A + B)(C + D)$ is an example of the first case and is said to be in the OR—AND form or the "logical-product-of-logical-sums" form. This same expression may be converted to the AND—OR form by "multiplying" the terms out to form $AC + AD + BC + BD$. Now the expression is said to be in the "logical-sum-of-logical-products" form. Be careful, though, because it is not uncommon to see these expressions called product of sums and sum of products. If logical operations are being performed at the time, be sure to recognize that these are logical products of logical sums and logical sums of logical products.

Procedural Summary

Once a simplified form is reached, there is no guarantee that it is the simplest form available. This is one of the main disadvantages of the algebraic-simplification method. Until considerable experience has been gained, much time may be wasted in trying different forms to determine which is the simplest. And there is still no guarantee that all of the forms have been tried!

Upon completion of the simplification steps, the new expression should be tested by using a truth table. Comparison of this new truth table with the truth table for the original expression assures us that both expressions are equivalent. It is not impossible to make a mistake while taking a short cut or to misinterpret a portion of the expression and end up with nonequivalence. Once equivalence is determined, the final step is to draw the simplified logic diagram and bask in the glory of a successful reduction in hardware requirements!

5–3 1-LEVEL LOGIC SIMPLIFICATION

In a 1-*level logic circuit*, a given logic variable goes through only *one logic element* between source and output. Such circuits (discussed in earlier chapters) have only one logic operation, either AND or OR. However, so that a starting point can be established in this chapter, some of the basic ideas of logic symbology and Boolean algebra notation are reviewed.

The common AND gate is an example of a 1-level logic circuit. Consider such a simple circuit with four inputs A, B, C, and D and one output X. The definition of the AND operation states that if and only if inputs A, B, C, and D are all HIGH *simultaneously*, the output X will be HIGH. This statement is written in Boolean notation as $A \cdot B \cdot C \cdot D = X$ or, more simply, as $ABCD = X$. The AND operator ($\cdot$) is conventionally omitted, unless it is needed to clarify the expression.

Now consider a logic circuit that is defined by the equation $ABAB = X$. The logic diagram is drawn in Fig. 5–1 along with the equivalent truth table. Inspection of the truth table (Fig. 5–1b) should be sufficient to determine that the 4-input AND gate (Fig. 5–1a) required by the equation can be replaced by a 2-input AND gate with A and B each appearing once as inputs. However the steps in the algebraic simplification are shown to start developing the techniques to be used in more complex simplifications.

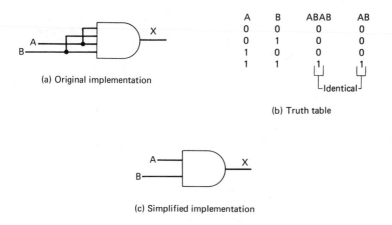

(a) Original implementation

A	B	ABAB	AB
0	0	0	0
0	1	0	0
1	0	0	0
1	1	1	1

└─Identical─┘

(b) Truth table

(c) Simplified implementation

Figure 5–1   Logic diagram and truth table, ABAB = AB

$$ABAB = AABB \qquad \text{Commutative Law}$$
$$= (AA)(BB) \qquad \text{Associative Law}$$
$$= A(BB) \qquad \text{Idempotent Law}$$
$$= AB \qquad \text{Idempotent Law}$$
$$\therefore \ ABAB = AB$$

The truth table and simplified logic diagrams are drawn in Figs. 5–1b and and c. Truth table comparison will reveal that $ABAB = AB$ is indeed valid and that the 4-input AND gate is replaceable by a 2-input AND gate. If the number of inputs to an AND gate is a direct cost factor, then the circuit has been simplified, and the cost cut in half.

Another example appears in Fig. 5–2. The logic diagram (Fig. 5–2a) shows that the variables A, B, $\overline{A}$, and $\overline{B}$ are being ORed together. This may be expressed as $A + B + \overline{A} + \overline{B} = X$. By developing the truth table (Fig. 5–2b), it is apparent that $X = 1$ – no matter what value any of the variables take on. Therefore the 4-input OR gate can be replaced by a direct connection (Fig. 5–2c), eliminating a logic element – that is a *real* simplification! The algebraic simplification steps to justify this replacement are shown here.

$$A + B + \overline{A} + \overline{B} = A + \overline{A} + B + \overline{B} \qquad \text{Commutative Law}$$
$$= (A + \overline{A}) + (\overline{B} + B) \qquad \text{Associative Law}$$
$$= 1 + (\overline{B} + B) \qquad \text{Theorem 12}$$
$$= 1 + 1 \qquad \text{Theorem 12}$$
$$= 1 \qquad \text{Postulate 8}$$
$$\therefore \ A + B + \overline{A} + \overline{B} = 1$$

So although it might appear that very little simplification could take place in a 1-level logic circuit, we have shown that the recognition of certain relationships of variables has considerably reduced the requirements for hardware.

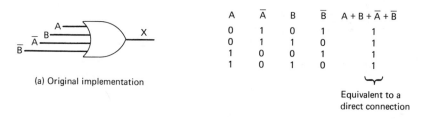

A	$\overline{A}$	B	$\overline{B}$	$A + B + \overline{A} + \overline{B}$
0	1	0	1	1
0	1	1	0	1
1	0	0	1	1
1	0	1	0	1

(a) Original implementation

Equivalent to a
direct connection

(b) Truth table

(c) Simplified implementation
(a direct connection)

Figure 5–2   Logic diagram and truth table, $A + B + \overline{A} + \overline{B} = 1$

5—4 2-LEVEL LOGIC SIMPLIFICATION

In *2-level logic circuits*, a variable passes through *two cascaded logic elements* between input and output. It is assumed that both the true and the complemented values of variables are available as inputs to the circuits being simplified. However, when inverters are required to negate variables within the logic circuit, these inverters contribute to the overall level of the logic circuit. The most common configuration of 2-level logic circuits is AND circuits feeding OR circuits or vice versa.

A Simplification Starting with an English-Language Description

Our first example of a 2-level logic circuit is presented in terms of an English-language statement, and the procedural steps necessary to prepare the problem for simplification are demonstrated. This example is typical of industrial-control applications of logic. It is only a partial statement of an overall requirement, but it is sufficiently complete in itself to demonstrate how simplification is performed.

The operation under consideration deals with a conveyor belt in an industrial plant. The belt carries empty boxes that are to be filled with breakfast cereal. A portion of the operational requirement is to have an alarm sound if a dangerous situation or a situation requiring immediate attention should occur. The systems designer has provided the following information: An alarm should sound if a box is available and the cereal hopper is empty, or if no box is available and electric power is lost, or if the cereal hopper is empty and the electric power is lost. This description of circuit operation should be investigated to determine if any simplification can be made prior to system construction. Reference to the general procedure for logic simplification shows that the first thing to do is to develop the Boolean expression. Thus we assign variable names to each of the conditions mentioned in the English-language description so that a truth table, a logic diagram, and the Boolean expression can be developed. The following assignments are made.

B = Box available
$\overline{B}$ = No box available
C = Cereal hopper empty
$\overline{C}$ = Cereal hopper not empty
P = Electric power available
$\overline{P}$ = Electric power lost
X = Alarm sounds

Substituting, the Boolean expression becomes

$$X = B \cdot C + \overline{B} \cdot \overline{P} + C \cdot \overline{P}$$

Note that the optional method of showing conjunctive terms is used in this equation to reduce the possibility of misinterpreting the term $\overline{B} \cdot \overline{P}$.

The truth table for $X = B \cdot C + \bar{B} \cdot \bar{P} + C \cdot \bar{P}$ appears in Fig. 5–3a and the equivalent logic diagram in Fig. 5–3b. Algebraic proof is shown in Example 5–2.

Example 5–2. Simplify the expression $B \cdot C + \bar{B} \cdot \bar{P} + C \cdot \bar{P}$.
Solution.

$$
\begin{aligned}
B \cdot C + \bar{B} \cdot \bar{P} + C \cdot \bar{P} &= B \cdot C + \bar{B} \cdot \bar{P} + C \cdot \bar{P}(B + \bar{B}) && \text{Theorem 3} \\
&= B \cdot C + \bar{B} \cdot \bar{P} + B \cdot C \cdot \bar{P} + \bar{B} \cdot C \cdot \bar{P} && \text{Distributive Law} \\
&= B \cdot C + B \cdot C \cdot \bar{P} + \bar{B} \cdot \bar{P} + \bar{B} \cdot C \cdot \bar{P} && \text{Commutative Law} \\
&= (B \cdot C + B \cdot C \cdot \bar{P}) + \bar{B} \cdot \bar{P} + \bar{B} \cdot C \cdot \bar{P} && \text{Associative Law} \\
&= BC(1 + \bar{P}) + \bar{B} \cdot \bar{P} + \bar{B} \cdot C \cdot \bar{P} && \text{Factoring} \\
&= BC + \bar{B} \cdot \bar{P} + \bar{B} \cdot C \cdot \bar{P} && \text{Theorems 10 \& 3} \\
&= BC + (\bar{B} \cdot \bar{P} + \bar{B} \cdot C \cdot \bar{P}) && \text{Associative Law} \\
&= BC + \bar{B} \cdot \bar{P}(1 + C) && \text{Factoring} \\
\therefore \quad B \cdot C + \bar{B} \cdot \bar{P} + C \cdot \bar{P} &= BC + \bar{B} \cdot \bar{P} && \text{Theorems 10 \& 3}
\end{aligned}
$$

Note that the third term of the original expression has been eliminated. Apparently it was not necessary to completely specify the expression. This may be verified by constructing the truth table for the expression (Fig. 5–3c) and comparing it with the truth table for the original expression (Fig. 5–3a). The truth tables are identical, thus verifying the equality of the original and the new expression. The logic diagram for the new expression (Fig. 5–3d) uses one less

B	B̄	C	C̄	P	P̄	BC	B̄P̄	CP̄	BC + B̄P̄ + CP̄	BC	B̄P̄	BC + B̄P̄
0	1	0	1	0	1	0	1	0	1	0	1	1
0	1	0	1	1	0	0	0	0	0	0	0	0
0	1	1	0	0	1	0	1	1	1	0	1	1
0	1	1	0	1	0	0	0	0	0	0	0	0
1	0	0	1	0	1	0	0	0	0	0	0	0
1	0	0	1	1	0	0	0	0	0	0	0	0
1	0	1	0	0	1	1	0	1	1	1	0	1
1	0	1	0	1	0	1	0	0	1	1	0	1

Identical

(a) Original truth table   (b) Simplified truth table

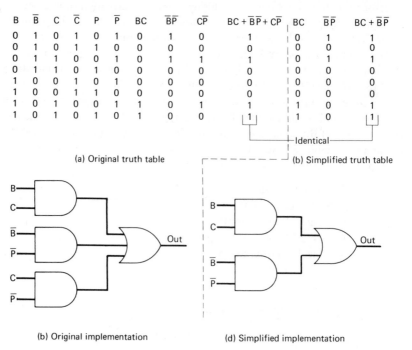

(b) Original implementation   (d) Simplified implementation

Figure 5–3   Logic diagram and truth tables $BC + \bar{B}\bar{P} + C\bar{P} = BC + \bar{B}\bar{P}$

AND gate than the original logic diagram (Fig. 5—3b), so a reduction in number of gates has been accomplished. Additional attempts at simplification may be tried, but the form $B \cdot C + \bar{B} \cdot \bar{P}$ is probably the simplest. It meets the definition of "simplest" provided earlier: the smallest number of logic elements to perform the task apparently has resulted, and the expression is in the AND—OR form with no variables enclosed in parentheses and no vincula extending over more than one variable.

In the process of performing this simplification, English-language statements have been equated with variable names; truth tables and logic diagrams have been constructed; and theorems and laws of Boolean algebra have been applied. Although all simplification attempts may not require such a complete analysis, the steps performed in this example are typical, and the analyst should be prepared to apply any or all of these techniques.

A Simplification Starting with a Logic Diagram

Often it is necessary to convert a logic expression to a form that is readily implemented by the type of logic elements available. For example, a logic circuit may have been originally implemented strictly with NOR elements, due to a particular designer's choice or for economic reasons. As we mentioned before, when a logic system is designed exclusively with one type of logic element, cost reductions are realized in the design and fabrication processes. For the sake of discussion, however, assume that the logic diagram given in Fig. 5—4a must be implemented with AND—OR elements — not NOR elements as shown.

The first step is to determine the logic expression that this diagram represents. Derivation of the logic expression starts with inputs and moves toward the output — one gate or logic element at a time. The output at $\boxed{1}$ is $\overline{A + B}$. At $\boxed{2}$ the output is $\overline{C + D}$. Combining these two expressions in the NOR gate results in the expression $\overline{\overline{A + B} + \overline{C + D}}$ at $\boxed{3}$. Inversion of the expression at $\boxed{3}$ by the single input NOR gate provides the final output expression

$$\overline{\overline{\overline{A + B} + \overline{C + D}}}$$

at $\boxed{4}$. The double negation may be removed by applying Theorem 13, and the final expression is usually written as $\overline{A + B} + \overline{C + D}$. Truth-table representation of this expression is provided in Fig. 5—4b and algebraic proof in Example 5—3.

Example 5—3. Convert the NOR form expression $\overline{\overline{A + B} + \overline{C + D}}$ to the AND—OR form.

Solution. $\overline{\overline{A + B} + \overline{C + D}} = \overline{\overline{A + B}} \cdot \overline{\overline{C + D}}$ DeMorgan's Law

$= AB + \overline{C + D}$ Theorem 13

$= AB + \overline{C} \cdot \overline{D}$ DeMorgan's Law

$= AB + \overline{C}D$ Theorem 13

$\therefore \overline{\overline{A + B} + \overline{C + D}} = AB + \overline{C}D$

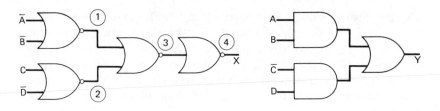

(a) Original implementation   (c) Simplified implementation

A	$\bar{A}$	B	$\bar{B}$	C	$\bar{C}$	D	$\bar{D}$	$\bar{A}+\bar{B}$	$\overline{\bar{A}+\bar{B}}$	$C+\bar{D}$	$\overline{C+\bar{D}}$	X	AB	$\bar{C}D$	Y
0	1	0	1	0	1	0	1	1	0	1	0	0	0	0	0
0	1	0	1	0	1	1	0	1	0	0	1	1	0	1	1
0	1	0	1	1	0	0	1	1	0	1	0	0	0	0	0
0	1	0	1	1	0	1	0	1	0	1	0	0	0	0	0
0	1	1	0	0	1	0	1	1	0	1	0	0	0	0	0
0	1	1	0	0	1	1	0	1	0	0	1	1	0	1	1
0	1	1	0	1	0	0	1	1	0	1	0	0	0	0	0
0	1	1	0	1	0	1	0	1	0	1	0	0	0	0	0
1	0	0	1	0	1	0	1	1	0	1	0	0	0	0	0
1	0	0	1	0	1	1	0	1	0	0	1	1	0	1	1
1	0	0	1	1	0	0	1	1	0	1	0	0	0	0	0
1	0	0	1	1	0	1	0	1	0	1	0	0	0	0	0
1	0	1	0	0	1	0	1	0	1	1	0	1	1	0	1
1	0	1	0	0	1	1	0	0	1	0	1	1	1	1	1
1	0	1	0	1	0	0	1	0	1	1	0	1	1	0	1
1	0	1	0	1	0	1	0	0	1	1	0	1	1	0	1

—Identical—

(b) and (d) Original and simplified truth tables

$$X = \overline{\bar{A}+\bar{B}} + \overline{C+\bar{D}}$$
$$Y = AB + \bar{C}D$$

Figure 5—4   Logic diagrams and truth tables $\overline{\bar{A}+\bar{B}} + \overline{C+\bar{D}} = AB + \bar{C}D$

Comparison of the truth table for the simplified expression (Fig. 5—4d) demonstrates the equivalence of the NOR form and the AND—OR form in this example. Conversion from NOR form to AND—OR form is a commonly encountered situation in digital logic. The logic diagram of the AND—OR form in Fig. 5—4c shows a reduction in gate count from four NOR gates to two AND and one OR gate. More importantly, derivation of logic expressions from logic diagrams and application of DeMorgan's Law have been demonstrated.

A Simplification Starting with a Boolean Expression

Example 5–4. Simplify $(A + \overline{C})(A + D)(B + \overline{C})(B + D)$.
Solution.

$(A + \overline{C})(A + D)(B + \overline{C})(B + D) = (\overline{C} + A)(D + A)(\overline{C} + B)(D + B)$
Commutative Law

$= [(\overline{C} + A)(\overline{C} + B)] \, [(D + A)(D + B)]$
Associative Law

$= (\overline{C} \cdot \overline{C} + \overline{C}B + A\overline{C} + AB)(DD + DB + AD + AB)$
Distributive Law

$= [\overline{C}(1 + B + A) + AB] \, [D(1 + B + A) + AB]$
Factoring

$= (\overline{C}1 + AB)(D1 + AB)$ Theorem 9

$= (\overline{C} + AB)(D + AB)$ Theorem 3

$= \overline{C}D + \overline{C}AB + ABD + ABAB$
Distributive Law

$= \overline{C}D + AB(\overline{C} + D + 1)$ Factoring

$= \overline{C}D + AB(1)$ Theorem 9

$= \overline{C}D + AB$ Theorem 3

$= AB + \overline{C}D$ Commutative Law

$\therefore \; (A + \overline{C})(A + D)(B + \overline{C})(B + D) = AB + \overline{C}D$

Once again, no major steps are omitted. By now, it should be apparent that some steps may be combined, and as experience is gained, the reader may find it advantageous to do so. Comparative logic diagrams are shown in Figs. 5–5a and b, and a truth table showing both expressions is provided in Fig. 5–5c. This example demonstrates conversion from OR–AND to AND–OR form.

5–5 3-LEVEL LOGIC SIMPLIFICATION

A Simplification Starting with a Boolean Expression

When *any* variable in a Boolean expression must pass through *three cascaded logic elements* between input and output, the circuit is considered to be a *3-level logic circuit.* The Boolean expression $A(A + B) + B(\overline{A} + B)$ is a relatively simple example in which at least one of the variables in the expression passes through three logic elements.

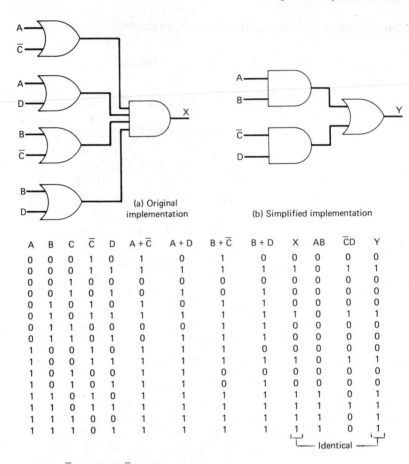

A	B	C	C̄	D	A + C̄	A + D	B + C̄	B + D	X	AB	C̄D	Y
0	0	0	1	0	1	0	1	0	0	0	0	0
0	0	0	1	1	1	1	1	1	1	0	1	1
0	0	1	0	0	0	0	0	0	0	0	0	0
0	0	1	0	1	0	1	0	1	0	0	0	0
0	1	0	1	0	1	0	1	1	0	0	0	0
0	1	0	1	1	1	1	1	1	1	0	1	1
0	1	1	0	0	0	0	1	1	0	0	0	0
0	1	1	0	1	0	1	1	1	0	0	0	0
1	0	0	1	0	1	1	1	0	0	0	0	0
1	0	0	1	1	1	1	1	1	1	0	1	1
1	0	1	0	0	1	1	0	0	0	0	0	0
1	0	1	0	1	1	1	0	1	0	0	0	0
1	1	0	1	0	1	1	1	1	1	1	0	1
1	1	0	1	1	1	1	1	1	1	1	1	1
1	1	1	0	0	1	1	1	1	1	1	0	1
1	1	1	0	1	1	1	1	1	1	1	0	1

Identical

$X = (A + \bar{C})(A + D)(B + \bar{C})(B + D)$
$Y = AB + \bar{C}D$

(a) Original implementation   (b) Simplified implementation   (c) Truth table

Figure 5–5 $(A + \bar{C})(A + D)(B + \bar{C})(B + D) = AB + \bar{C}D$

Example 5–5. Simplify $A(A + B) + B(\bar{A} + B)$.
Solution.

$$
\begin{aligned}
A(A + B) + B(\bar{A} + B) &= AA + AB + \bar{A}B + BB & \text{Distributive Law} \\
&= A + AB + \bar{A}B + BB & \text{Theorem 5} \\
&= A + AB + \bar{A}B + B & \text{Theorem 5} \\
&= A(1 + B) + \bar{A}B + B & \text{Factoring} \\
&= A(1) + \bar{A}B + B & \text{Theorem 10} \\
&= A + \bar{A}B + B & \text{Theorem 3} \\
&= A + B(1 + \bar{A}) & \text{Factor} \\
&= A + B(1) & \text{Theorem 10} \\
&= A + B & \text{Theorem 3}
\end{aligned}
$$

$\therefore\ A(A + B) + B(\bar{A} + B) = A + B$

The logic diagrams for both the original expression and the simplified expression appear in Fig. 5–6a, while the truth tables that are used to demonstrate equivalence are shown in Fig. 5–6b.

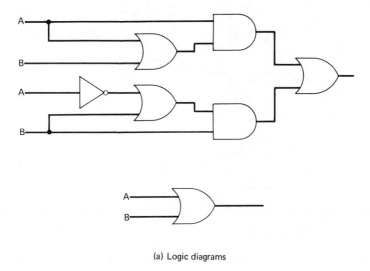

(a) Logic diagrams

A	$\overline{A}$	B	A + B	A(A + B)	$\overline{A}$ + B	B($\overline{A}$ + B)	A(A + B) + B($\overline{A}$ + B)
0	1	0	0	0	1	0	0
0	1	1	1	0	1	1	1
1	0	0	1	1	0	0	1
1	0	1	1	1	1	1	1

└──────────────── Identical ────────────────┘

(b) Truth tables

Figure 5–6   Logic diagrams and truth tables A(A + B) + B($\overline{A}$ + B) = A + B

A Simplification Starting with a Logic Diagram

Another 3-level logic circuit is shown in Fig. 5–7a. The straightforward derivation of the Boolean expression is shown on the logic diagram. In the algebraic simplification that follows, note that as the *number of variables* in an expression increases, more and more of the theorems, laws, and identities come into play. Numerous exercises are provided at the end of this chapter to supply practice in the use of logic simplification techniques.

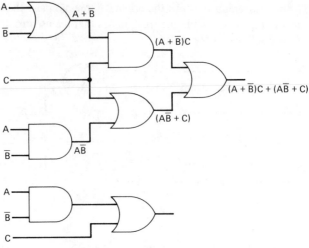

(a) and (b) Logic diagrams

A	B	$\bar{B}$	C	$A+\bar{B}$	$(A+\bar{B})C$	$A\bar{B}$	$(A\bar{B}+C)$	$(A+\bar{B})C+(A\bar{B}+C)$	$A\bar{B}+C$	
0	0	1	0	1	0	0	0	0	0	
0	0	1	1	1	1	0	1	1	1	1
0	1	0	0	0	0	0	0	0	0	
0	1	0	1	0	0	0	1	1	1	
1	0	1	0	1	0	1	1	1	1	
1	0	1	1	1	1	1	1	1	1	
1	1	0	0	1	0	0	0	0	0	
1	1	0	1	1	1	0	1	1	1	

(c) Truth tables
Identical

Figure 5−7   Logic diagrams and truth tables, $(A+\bar{B})C+$
$(A\bar{B}+C)=A\bar{B}+C$

Example 5−6.   Simplify $(A+\bar{B})C+(A\bar{B}+C)$.
Solution.

$$
\begin{aligned}
(A+\bar{B})C+(A\bar{B}+C) &= (AC+\bar{B}C)+(A\bar{B}+C) && \text{Distributive Law}\\
&= AC+\bar{B}C+A\bar{B}+C && \text{Associative Law}\\
&= AC+C+\bar{B}C+A\bar{B} && \text{Commutative Law}\\
&= C(A+1)+\bar{B}C+A\bar{B} && \text{Factoring}\\
&= C(1)+\bar{B}C+A\bar{B} && \text{Theorem 9}\\
&= C+\bar{B}C+A\bar{B} && \text{Theorem 3}\\
&= C(1+\bar{B})+A\bar{B} && \text{Factoring}\\
&= C(1)+A\bar{B} && \text{Theorem 10}\\
&= C+A\bar{B} && \text{Theorem 3}\\
&= A\bar{B}+C && \text{Commutative Law}
\end{aligned}
$$

$\therefore (A+\bar{B})C+(A\bar{B}+C)=A\bar{B}+C$

The simplified logic diagram is in Fig. 5−7b. Comparison of the original
expression and the simplified expression is shown in the truth tables of Fig. 5−7c.

Simplification of a Complex 3-Level Expression

Complexity does *not* have to increase as the level of the circuit increases, but in many cases it does. Consider the 3-level expression $(D + E)(F + E) + (\overline{F + E})(\overline{D + E})$. In the previous example, the tendency was to simplify the expression by converting the entire expression to another form. As we shall see in the following algebraic simplification, such an approach becomes quite cumbersome. Since this problem is very comprehensive, each of the steps is discussed separately so that the philosophy of each operation is more apparent.

Example 5–7. Simplify $(D + E)(F + E) + (\overline{F + E})(\overline{D + E})$.
Solution.

$$\underbrace{(D + E)}_{X}\underbrace{(F + E)}_{Y} + \underbrace{(\overline{F + E})}_{\overline{Y}}\underbrace{(\overline{D + E})}_{\overline{X}}$$

Step 1. The technique of *substituting single variables for groups of variables* is demonstrated in this step. As we shall shortly see, it is easier to visualize the Boolean algebra form of an expression when a lesser number of variables are concerned. For a few steps, then, $(D + E)$ will be replaced by X; $(F + E)$ will be replaced by Y; $(\overline{F + E})$ will be replaced by $\overline{Y}$; and $(\overline{D + E})$ will be replaced by $\overline{X}$. The expression to be simplified is now $X \cdot Y + \overline{Y} \cdot \overline{X}$. Later in the procedure the original terms will be resubstituted.

$$\underbrace{XY}_{A} + \overline{X} \cdot \overline{Y}$$
$$\quad\; A\; +\underset{\downarrow}{B}\;\; \underset{\downarrow}{C}$$

Step 2. The Commutative Law is applied to the second term in the new expression $(\overline{Y} \cdot \overline{X})$, making it $(\overline{X} \cdot \overline{Y})$. This portion of Step 2 is merely for "housekeeping" purposes. Variables are conventionally arranged in alphabetical order. But another action takes place in this step. Sometimes it is helpful to *regroup variables* so that the algebraic form of the expression is more readily recognized. Regrouping is allowable *only* if it does not change the logical meaning and value of the expression. If any doubt exists as to the validity of the regrouping, both expressions should be compared with truth tables. In this case, XY are considered one variable, but $\overline{X}$ and $\overline{Y}$ are considered two variables. The expression then takes on the form $A + BC$, i.e., $(XY) + \overline{X} \cdot \overline{Y}$, which is one of the common forms seen in the Distributive Law.

Step 3. A basic identity of the Distributive Law states that $A + BC = AB + AC$. Application of the Distributive Law to the expression in Step 2 then yields the expression

$$(XY + \overline{X})(XY + \overline{Y})$$

Step 4. Once again, the Commutative Law is applied. This time another convention is demonstrated. Single variable terms in an expression are usually written *before* two variable terms, two variable terms before three variable terms, and so forth.

$$(\overline{X} + XY)(\overline{Y} + XY)$$

Step 5. If the terms $(\overline{X} + XY)$ and $(\overline{Y} + XY)$ are to be of the form $A + BC$, then the Distributive Law may be used to expand both terms. The expression becomes $(\overline{X} + X)(X + Y)(\overline{Y} + X)(\overline{Y} + Y)$. Note, however, that the original expression, which started out in the logical sum of logical products form, is approaching the logical product of logical sums form. Such a simplification approach is very valuable, particularly since some expressions that may not be adaptable to simplification in one form may be easily adaptable to simplification in the other form.

$$(\overline{X} + X)(\overline{X} + Y)(\overline{Y} + X)(\overline{Y} + Y)$$

Step 6. Theorem 12 is used twice in this step. Both $(\overline{X} + X)$ and $(\overline{Y} + Y)$ are equated to 1. Now the expression appears in the following form:

$$(1)(\overline{X} + Y)(\overline{Y} + X)(1)$$

Step 7. To obtain $(\overline{X} + Y)(\overline{Y} + X)$ from the expression in Step 6, a number of substeps must be considered. First, the (1) at the right-hand side of the expression is moved to the left-hand side of the expression, thus giving $(1)(1)(\overline{X} + Y)(\overline{Y} + X)$. The Commutative Law allows this operation. Separating the expression into two parts by applying the Associative Law, i.e., $[(1)(1)]$ $[(\overline{X} + Y)(\overline{Y} + X)]$, reveals that Postulate 4 allows replacement of the first bracketed term with (1). Again, using the Associative Law to separate terms will result in $[(1)(\overline{X} + Y)] [(\overline{Y} + X)]$. Theorem 4 replaces the first bracketed term with $(\overline{X} + Y)$. Removal of the brackets by the Associative Law brings about the final expression,

$$(\overline{X} + Y)(\overline{Y} + X)$$

Step 8. The original groups of variables are resubstituted for the single variables in this step.

$$[(\overline{F + E}) + (D + E)] [(\overline{D + E}) + (F + E)]$$

To show the advantage of substituting single variables for groups of variables, Steps 1–8 are shown in their original form.

1. $(D + E)(F + E) + (\overline{F + E})(\overline{D + E})$
2. $(D + E)(F + E) + (\overline{D + E})(\overline{F + E})$
3. $[(D + E)(F + E) + (\overline{D + E})] [(D + E)(F + E) + (\overline{F + E})]$
4. $[(\overline{D + E}) + (D + E)(F + E)] [(\overline{F + E}) + (D + E)(F + E)]$
5. $[(\overline{D + E}) + (D + E)] [(\overline{D + E}) + (F + E)] [(\overline{F + E}) + (D + E)] [(\overline{F + E}) + (F + E)]$
6. $[1] [(\overline{D + E}) + (F + E)] [(\overline{F + E}) + (D + E)] [1]$
7. $[(\overline{D + E}) + (F + E)] [(\overline{F + E}) + (D + E)]$
8. Same as Step 7.

Note that although one step is saved, the complexity of the other steps tends to make visualization of the algebraic form more difficult.

Step 9. Removal of brackets is accomplished by using the Associative Law.

$$(\overline{F + E} + D + E)(\overline{D + E} + F + E)$$

Step. 10. Since the variable E is contained in both terms of the expression, it may be factored out to obtain the expression

$$E + (\overline{F + E} + D)(\overline{D + E} + F)$$

Step 11. DeMorgan's Law is used to break the vincula in both terms, since one requirement for a simplified form is that *no vincula extend over more than one term*. In addition, the two terms in parentheses are regrouped to aid in recognition of the algebraic form used in the Distributive Law. (Brackets are added for clarity.)

$$E + [(\underbrace{\overline{F} \cdot \overline{E} + D})(\underbrace{\overline{D} \cdot \overline{E}} + F)] \\ H \phantom{\cdot \overline{E} + D)(}J \phantom{\cdot \overline{E})} K$$

Step 12. Expansion of the expression in Step 11 by application of the Distributive Law yields

$$E + [(\underbrace{\overline{F} \cdot \overline{E} + D})(\underbrace{\overline{D} \cdot \overline{E}}) + (\underbrace{\overline{F} \cdot \overline{E} + D})(\underbrace{F})] \\ H \phantom{\cdot \overline{E} + D)(}J H K$$

Step 13. Removal of the brackets added in Step 11 is accomplished by application of the Associative Law.

$$E + (\overline{F} \cdot \overline{E} + D)(\overline{D} \cdot \overline{E}) + (\overline{F} \cdot \overline{E} + D)(F)$$

Step 14. The last two terms in the expression of Step 13 are expanded by use of the Distributive Law.

$$E + \overline{D} \cdot \overline{E} \cdot \overline{F} \cdot \overline{E} + \overline{D} \cdot E \cdot D + F \cdot \overline{F} \cdot E + F \cdot D$$

Step 15. The second term of the expression in Step 14 is simplified by Theorem 5 and the third and fourth terms by Theorem 6, accompanied by liberal application of the Commutative Law.

$$E + \overline{D} \cdot \overline{E} \cdot \overline{F} + 0 + 0 + F \cdot D$$

Step 16. Postulate 1 and Theorems 7 or 8 are used to extract this next expression from Step 15. The Associative Law permits writing this expression with parentheses in the following manner:

$$(E + \overline{D} \cdot \overline{E} \cdot \overline{F}) + F \cdot D$$

Step 17. Expansion of the terms $(E + \overline{D} \cdot \overline{E} \cdot \overline{F})$ by using the Distributive Law helps to derive this expression.

$$(E + \overline{D})(E + \overline{E})(E + \overline{F}) + FD$$

Step 18. $(E + \overline{E})$ is removed by application of Theorem 12 and either Theorem 3 or 4. Brackets are added to aid in the recognition of the distributive form. So

$$[(E + \overline{D})(E + \overline{F})] + FD$$

Step 19. The Distributive Law is used on the bracketed terms to simplify the expression to $E + \overline{D} \cdot \overline{F} + FD$. The Commutative Law changes the last term from FD to DF, thus resulting in the final simplified expression

$$E + \overline{D} \cdot \overline{F} + D \cdot F$$

Examination of the simplification process just completed may show other methods by which the same or a similar final expression can be obtained. Our approach demonstrates many postulates, theorems, and laws, but the Distributive Law applied to a complete expression was the most common method used throughout the example. Furthermore, a number of steps are saved by applying the Distributive Law to *part* of an expression. Note the following reduction in number of steps that is brought about by applying this method to the same expression.

1. $(D + E)(F + E) + (\overline{F + E})(\overline{D + E})$
2. $(E + DF) + (\overline{F+E})(\overline{D+E})$ Distributive Law to $(D+E)(F+E)$
3. $(E + DF) + \overline{F} \cdot \overline{E} \cdot \overline{D} \cdot \overline{E}$ DeMorgan's and Associative Laws
4. $E + DF + \overline{F} \cdot \overline{E} \cdot \overline{D}$ Associative Law and Theorem 5
5. $DF + (E + \overline{F} \cdot \overline{E} \cdot \overline{D})$ Commutative and Associative Laws
6. $DF + (E + \overline{F})(E + \overline{E})(E + \overline{D})$ Distributive Law to $(E + \overline{F} \cdot \overline{E} \cdot \overline{D})$
7. $DF + (E + \overline{F})(1)(E + \overline{D})$ Theorem 12
8. $DF + [(E + \overline{F})(E + \overline{D})]$ Theorem 3 or 4 and Associative Law
9. $DF + E + \overline{D} \cdot \overline{F}$ Distributive Law to $(E + \overline{F})(E + \overline{D})$
10. $E + \overline{D} \cdot \overline{F} + DF$ Commutative Law

The Distributive Law may be applied to part of an expression, provided that parentheses or brackets can be placed around that part without changing the value of the expression. In the previous example, the Distributive Law was applied to this part of the expression:

$$\boxed{(D + E)(F + E)} + (\overline{F + E})(\overline{D + E})$$

The expression could *not* have been divided into

$$(D+E)\boxed{(F+E) + (\overline{F+E})}(\overline{D+E}) \quad \text{or} \quad \boxed{(D+E)(F+E) + (\overline{F+E})}(\overline{D+E})$$

Whether to apply the Distributive Law to the entire expression or to just part of it becomes a matter of experience. *Think ahead!* Try to locate variables that will become the complement of or the same as other variables so that simplifying theorems may be applied along with laws and identities.

Obviously, the diagram in Fig. 5–8b (the simplified form) uses fewer logic elements than the original form in Fig. 5–8a. A reduction in hardware has been accomplished, and the outputs of the two circuits are identical, as shown in the logic truth table in Table 5–1.

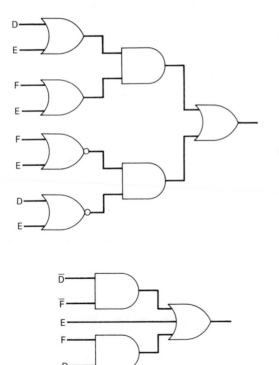

Figure 5–8 Logic diagram $(D + E)(F + E) + (\overline{F + E})(\overline{D + E}) = E + \overline{D}\overline{F} + DF$

Table 5–1 Truth table $(D+E)(F+E) + (\overline{F+E})(\overline{D+E}) = E + \overline{D}\,\overline{F} + DF$

D	$\overline{D}$	E	$\overline{E}$	F	$\overline{F}$	D+E	$\overline{D+E}$	F+E	$\overline{F+E}$	(D+E)(F+E)	$(\overline{F+E})(\overline{D+E})$	$[(D+E)(F+E)+(\overline{F+E})(\overline{D+E})]$	$\overline{D}\,\overline{F}$	FD	$E+\overline{D}\,\overline{F}+FD$
0	1	0	1	0	1	0	1	0	1	0	1	1	1	0	1
0	1	0	1	1	0	0	1	1	0	0	0	0	0	0	0
0	1	1	0	0	1	1	0	1	0	1	0	1	1	0	1
0	1	1	0	1	0	1	0	1	0	1	0	1	0	0	1
1	0	0	1	0	1	1	0	0	1	0	0	0	0	0	0
1	0	0	1	1	0	1	0	1	0	1	0	1	0	1	1
1	0	1	0	0	1	1	0	1	0	1	0	1	0	0	1
1	0	1	0	1	0	1	0	1	0	1	0	1	0	1	1

Identity

5–6 4-LEVEL LOGIC SIMPLIFICATION

Simplification Starting with a Logic Diagram

A path through *four cascaded logic elements* for any variable in a Boolean expression is the criterion for classifying a logic circuit as "4-level." Figure 5–9 is a logic diagram of a 4-level logic circuit. It represents a logic function that has been implemented strictly with NAND gates. But for the purpose of discussion, assume that it is necessary to implement this same function with AND–OR–INVERTER type logic elements. The first step is to determine the Boolean expression at each numbered output. Since two input variables A and B are present at the input to the first NAND gate, its output (1) is $\overline{AB}$. The single-input NAND gate serves as an inverter, resulting in $\overline{\overline{AB}}$ at (2). $\overline{\overline{AB}}$ is combined with $\overline{C}$ in the third NAND gate, giving $\overline{(\overline{\overline{AB}})\overline{C}}$ at (3). Finally, the last single-input NAND gate, acting as an inverter, provides $\overline{\overline{(\overline{AB})\overline{C}}}$ at (4), the circuit output. Simplification of the output expression of this circuit is fairly easy.

Example 5–8. Simplify $\overline{\overline{(\overline{AB})\overline{C}}}$.

Solution. $\overline{\overline{(\overline{AB})\overline{C}}} = (\overline{\overline{AB}})\overline{C}$ Theorem 13
$= (AB)\overline{\overline{C}}$ Theorem 13
$= AB\overline{\overline{C}}$

One simple 3-input AND gate performs the logic function previously requiring four NAND gates. From the viewpoint of the simplification criteria used in this text, a saving of three gates has been realized. Again, however, in extremely large and complex logic systems, a cost savings might still have been realized by implementing this logic function with NAND gates. It is not uncommon to purchase logic elements in quantities of 1000 or more units at half the single-unit price. This savings, coupled with simplifications in manufacturing processes when only one type of logic element is used, can quite often "swing the balance" toward single-element implementation of a logic system.

The reader should be prepared to encounter many NAND or NOR implemented systems. Recognition of the actual logic functions being performed may not be as easy as with AND–OR–INVERTER implemented systems. This situation calls for application of the techniques discussed in this chapter. Conversion of the logic expressions to straightforward AND–OR–INVERTER form can quite often make an apparently senseless logic circuit easily understandable.

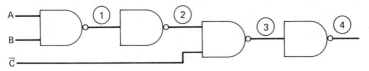

Figure 5–9   Logic diagrams (AB)C = ABC

Simplification Starting with a Boolean Expression

A Boolean expression that can be simplified by using one of the common identities of Boolean algebra is $[A\overline{B} + (\overline{A} + B)\overline{C}] + C$. Many of the theorems and laws of Boolean algebra are applied, as well as the technique of substituting variables for the purpose of clarification of an expression.

Example 5–9. Simplify $[A\overline{B} + (\overline{A} + B)\overline{C}] + C$.
Solution.

$[A\overline{B} + (\overline{A} + B)\overline{C}] + C$	Original problem
$[A\overline{B} + \overline{A} \cdot \overline{C} + B\overline{C}] + C$	Distributive Law
$A\overline{B} + (\overline{A} \cdot \overline{C} + B\overline{C}) + C$	Associative Law
$A\overline{B} + \overline{C}(\overline{A} + B) + C$	Factoring
$A\overline{B} + [C + \overline{C}(\overline{A} + B)]$	Commutative/Associative Laws
$A\overline{B} + [C + (\overline{A} + B)]$	Identity – Eq. (4–15)
$\overline{A} + A\overline{B} + B + C$	Associative/Commutative Laws
$\overline{A} + \overline{B} + B + C$	Identity – Eq. (4–15)
$\overline{A} + 1 + C$	Associative Law, Theorem 12
$1 + C$	Theorem 9
1	Theorem 10

Thus two AND and three OR gates are replaced by a *direct connection.* There should be no doubt that the simplified expression meets all criteria for the "least complex" implementation of the original expression: less logic elements are used; the cost of a direct connection is surely less than five logic gates; and no further algebraic simplification is possible. Comparative logic diagrams and truth tables appear in Fig. 5–10.

A	$\overline{A}$	B	$\overline{B}$	C	$\overline{C}$	$A\overline{B}$	$\overline{A} + B$	$\overline{C}(\overline{A} + B)$	$[A\overline{B} + (\overline{A} + B)\overline{C}]$	$[A\overline{B} + (\overline{A} + B)\overline{C}] + C$
0	1	0	1	0	1	0	1	1	1	1
0	1	0	1	1	0	0	1	0	0	1
0	1	1	0	0	1	0	1	1	1	1
0	1	1	0	1	0	0	1	0	0	1
1	0	0	1	0	1	1	0	0	1	1
1	0	0	1	1	0	1	0	0	1	1
1	0	1	0	0	1	0	1	1	1	1
1	0	1	0	1	0	0	1	0	0	1

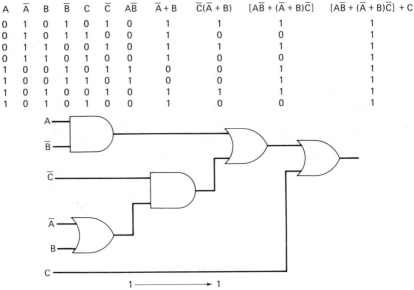

Figure 5–10 Logic diagram and truth table $[A\overline{B} + (\overline{A} + B)\overline{C}] + C = 1$

Recognition of Alternate Simplification Methods

The final example of 4-level logic circuit simplification demonstrates the use and application of many of the theorems and laws of Boolean algebra. In addition, the existence of alternate solution methods is mentioned, and the reader is provided with necessary clues to determine the simplification approach best suited to his own knowledge and experience.

Example 5–10. Simplify $[AB(C + D + E) + DBA][\overline{A} + \overline{B} + \overline{E}]$.
Solution.

$[AB(C + D + E) + DBA][\overline{A} + \overline{B} + \overline{E}]$	Original problem
$[(ABC + ABD + ABE) + DBA][\overline{A} + \overline{B} + \overline{E}]$	Distributive Law
$[ABC + ABD + ABE + DBA][\overline{A} + \overline{B} + \overline{E}]$	Associative Law
$(ABC + ABD + ABE)(\overline{A} + \overline{B} + \overline{E})$	Commutative Law, Theorem 11
$(ABC + ABD + ABE)(\overline{ABE})$	DeMorgan's Law
$\overline{ABE}ABC + \overline{ABE}ABD + \overline{ABE}ABE$	Distributive Law
$\overline{ABE}ABC + \overline{ABE}ABD + 0$	Theorem 6
$\overline{ABE}ABC + \overline{ABE}ABD$	Theorem 7
$\overline{ABE}AB(C + D)$	Factoring
$AB(\overline{ABE})(C + D)$	Commutative/Associative Laws
$AB(\overline{AB} + \overline{E})(C + D)$	DeMorgan's Law
$(AB\overline{AB} + AB\overline{E})(C + D)$	Distributive Law
$(0 + AB\overline{E})(C + D)$	Theorem 6
$(AB\overline{E})(C + D)$	Theorem 8
$AB\overline{E}C + AB\overline{E}D$	Distributive Law
$ABC\overline{E} + ABD\overline{E}$	Commutative Law

An alternate method of solution to this simplification may be observed at the fourth and fifth steps in Example 5–10. The method selected in Example 5–10 involved recognizing the equivalency of the terms $(\overline{A} + \overline{B} + \overline{E})$ and $\overline{ABE}$ and using the $\overline{ABE}$ term to cancel the ABE term. However, at the fourth step, the expression could be changed to the form $(\overline{A} + \overline{B} + \overline{E})(ABC + ABD + ABE)$. By assigning single-variable names to replace certain multiple-variable names and by applying the Distributive Law, a means of expansion results that removes the parentheses and allows further simplification. Let

$$W = (\overline{A} + \overline{B} + \overline{E}), \quad X = ABC, \quad Y = ABD, \quad \text{and} \quad Z = ABE.$$

Expansion is accomplished by using the equality

$$W(X + Y + Z) = WX + WY + WZ$$

This approach is left to the reader as an exercise in simplification.

Either method of simplification reduces the requirement for three AND and three OR gates to two AND and one OR gate. Comparative logic diagrams appear in Figs. 5–11a and b. Truth-table proof of the equivalency of the original and final expression is left to the reader.

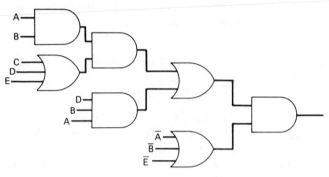

(a) Original implementation

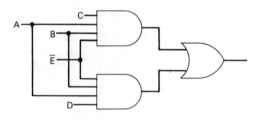

(b) Simplified implementation

Figure 5–11   Logic diagram $[AB(C + D + E) + DBA]$
$[\overline{A} + \overline{B} + \overline{E}] = ABC\overline{E} + ABD\overline{E}$

5–7 ALGEBRAICALLY SIMPLIFYING THE BINARY ADDER

When a digital computer must perform arithmetic addition, the half-adder (discussed in Ch. 4) is inadequate. A *full adder* is required in all but the least-significant positions to accommodate the carry input from previous orders. The full adder has two outputs, *SUM* and *CARRY*, just like the half-adder. However, it must handle not only the augend and the addend (as in the half-adder), but it must also be able to handle an input carry. The same basic rules for addition used in Ch. 4 apply to the full-adder, except the SUM and CARRY output expressions are more complex due to the additional inputs.

Statement of the Problem

English-language statements for both outputs are given here.

SUM *output.* If the augend is zero, the addend is zero, and the input carry is one; or if the augend is zero, the addend is one, and the input carry is zero; or if the augend is one, the addend is zero, and the input carry is zero; or if the augend, addend, and input carry are all one, the sum is one.

OUTPUT CARRY *output.* If the augend is zero, the addend is one, and the input carry is one; or if the augend is one, the addend is zero, and the input carry is one; or if the augend is one, the addend is one, and the input carry is zero; or if the augend, addend, and input carry are all one, the output carry is one.

Substituting letter names for variable names and assigning the logic value of 0 to the numerical value of zero and the logic value of 1 to the numerical value of one, as was done in Ch. 4, puts this problem into a form that can be easily represented in truth tables. In fact, since both the SUM(S) and the OUTPUT CARRY (C_o) outputs are results of the same three inputs, both truth tables can be combined (see Table 5–2). Investigation of the output columns of the truth tables will show that there are four terms that give a 1 (true) output for the S output and four terms that give a 1 (true) output for the C_o output.

The resulting Boolean equations are

$$S = \overline{A} \cdot \overline{B} \cdot C + \overline{A} \cdot B \cdot \overline{C} + A \cdot \overline{B} \cdot \overline{C} + A \cdot B \cdot C$$
$$C_o = \overline{A} \cdot B \cdot C + A \cdot \overline{B} \cdot C + A \cdot B \cdot \overline{C} + A \cdot B \cdot C$$

Note that both equations are in the logical sum of logical products form, which is the form that always results when the *true outputs* are obtained from the truth table. All that remains is to submit each of these equations to algebraic simplification methods to determine if less complex equations may be obtained. The logic diagram for each of the outputs is shown in Fig. 5–12 and is used comparatively when the simplified logic diagram is developed.

Table 5–2   Truth table, full-adder

A	B	C_i	S	C_o
0	0	0	0	0
0	0	1	1	0
0	1	0	1	0
0	1	1	0	1
1	0	0	1	0
1	0	1	0	1
1	1	0	0	1
1	1	1	1	1

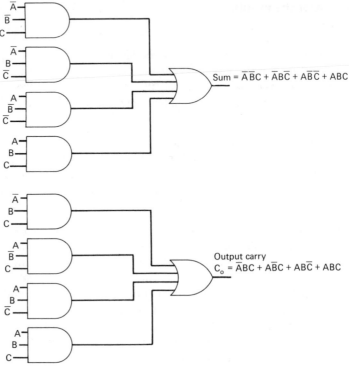

Sum = $\overline{A}\,\overline{B}C + \overline{A}B\overline{C} + A\overline{B}\,\overline{C} + ABC$

Output carry
$C_o = \overline{A}BC + A\overline{B}C + AB\overline{C} + ABC$

Figure 5—12   Logic diagram and equations for a full-adder

Simplification

As with most simplification efforts, numerous methods and approaches are available. In the case of the full adder, an almost *optimum simplification* exists! Due to the widespread use of the full adder in digital equipment many component manufacturers produce integrated circuit (IC) assemblies that are complete full adders (all in one package). In fact, some IC packages contain *more* than one full adder. But such an approach is almost "cheating," and it doesn't provide much experience in the manipulation of Boolean algebra expressions.

Another possibility might be to use two half-adders to make a full adder — if such a method could be devised. Since half-adders are also in common usage in digital equipment, they are available in prepackaged form (two or more to an IC chip). An investigation of the "two half-adder" approach will now follow.

Some additional background concerning the half-adder will be useful while investigating various approaches to the full-adder scheme. The sum equation for the half-adder is commonly found in logic equations and circuit implementation, as pointed out in Sec. 4—3. It may be directly implemented, as shown in Fig. 5—13a, or indirectly implemented, as shown in Figs. 5—13b,

c, and d. Derivation of the indirectly implemented versions of the EXCLUSIVE–
OR, keyed to Fig. 5–13, follows.

$$S = A\overline{B} + \overline{A}B$$

$= A\overline{A} + A\overline{B} + \overline{A}B + B\overline{B}$	Theorems 6, 7, and 8
$= (A\overline{A} + A\overline{B}) + (\overline{A}B + B\overline{B})$	Associative Law
$= A(\overline{A} + \overline{B}) + B(\overline{A} + \overline{B})$	Factoring
$= (A + B)(\overline{A} + \overline{B})$	Distributive Law (see Fig. 5–13b)
$= (A + B)(\overline{AB})$	DeMorgan's Law (see Fig. 5–13c)
$= \overline{\overline{(A + B)(\overline{AB})}}$	Theorem 13
$= \overline{\overline{(A + B)} + \overline{\overline{AB}}}$	DeMorgan's Law
$= \overline{\overline{(A + B)} + AB}$	Theorem 13 (see Fig. 5–13d)

The truth table for the sum output of the half-adder describes the EXCLUSIVE–
OR operation. The output is 1 *only* when either A is 1 or B is 1 – but not
when *both* are 1. Thus the EXCLUSIVE–OR operation excludes the combination of both $A = 1$ and $B = 1$ at the same time, while the normal (inclusive) OR
includes that combination. Many other uses will be discovered for the
EXCLUSIVE–OR operation. The EXCLUSIVE–OR may be constructed with
separate AND, OR, NAND, or NOR gates, or it may be obtained in an IC
package, just like the other logic functions.

The SUM equation for the full adder can be manipulated into many different forms: like terms may be factored; theorems and laws of Boolean algebra
applied; and identities recognized and used. These actions may be to no avail if
the criteria of "logical sum of logical products" form is adhered to. In fact, the
SUM equation as read directly from the truth table is in its simplest form within
the criteria established in this text. Certain deviations from these criteria will be
investigated later, but for our present discussion, the SUM equation is used "as
is." Justification for this assumption is provided in the following chapter, where
graphic simplification methods are demonstrated.

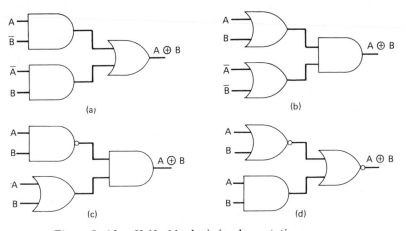

Figure 5–13   Half-adder logic implementations

The CARRY equation, however, may be simplified algebraically with a resultant decrease in not only number of logic elements but also in the number of inputs per logic element. *One* method of simplification is shown.

$$\begin{aligned}
C_o &= \bar{A}BC + A\bar{B}C + AB\bar{C} + ABC \\
&= \bar{A}BC + ABC + A\bar{B}C + AB\bar{C} &&\text{Commutative Law} \\
&= BC(A + \bar{A}) + A(\bar{B}C + B\bar{C}) &&\text{Factoring} \\
&= BC + A(\bar{B}C + B\bar{C}) &&\text{Theorems 12 and 13} \\
&= BC + A(B + C)(\overline{BC}) &&\text{See EXCLUSIVE–OR derivation} \\
&= BC + AB\overline{BC} + AC\overline{BC} &&\text{Distributive Law} \\
&= \underbrace{BC}_{X} + \underbrace{\overline{BC}}_{\bar{X}}(\underbrace{AB + AC}_{Y}) &&\text{Factoring} \\
&\quad X + \bar{X} \cdot Y &&\text{Recognize the form } X + \bar{X}Y \\
C_o &= BC + AB + AC &&\text{Proven identity}
\end{aligned}$$

Thus the original output carry function of the full adder may be implemented with three 2-input AND gates feeding a 3-input OR gate. This represents a reduction of one gate, and the gates used in the simplified form require less inputs per gate.

Now that the groundwork has been laid, the actual investigation of using two half-adders to form one full adder can begin. For reference purposes, the half-adder and the full adder equations and truth tables are reproduced in Fig. 5–14.

Earlier it was stated that the only reason for the full adder configuration was *to include the possible input carry from an addition operation in a previous order of operation.* The half-adder quite adequately performs addition of two single-digit binary numbers. Logically thinking, then, the second half-adder must add the SUM output of the first half-adder to the input CARRY – if any. Neglecting the output carries momentarily and concentrating on the SUM output, the equation for adding the SUM output of the first half-adder (S_1) to the input carry (C_i) is $S_1 \cdot \bar{C}_i + \bar{S}_1 \cdot C_i$.

If A represents the augend and B represents the addend, then $S_1 = A\bar{B} + \bar{A}B$. Substituting, the equation becomes $S_2 = (A\bar{B} + \bar{A}B)(\bar{C}_i) + (\overline{A\bar{B} + \bar{A}B})(C_i)$.

A	B	S	C		A	B	C_1	S	C_0
0	0	0	0		0	0	0	0	0
0	1	1	0		0	0	1	1	0
1	0	1	0		0	1	0	1	0
1	1	0	1		0	1	1	0	1
					1	0	0	1	0
$S = \bar{A}B + A\bar{B}$					1	0	1	0	1
$C = AB$					1	1	0	0	1
					1	1	1	1	1

$$S = \bar{A} \cdot \bar{B} \cdot C + \bar{A} \cdot B \cdot \bar{C} + A \cdot \bar{B} \cdot \bar{C} + A \cdot B \cdot C$$
$$C_0 = \bar{A} \cdot B \cdot C + A \cdot \bar{B} \cdot C + A \cdot B \cdot \bar{C} + A \cdot B \cdot C$$

Figure 5–14   Half-adder and full-adder truth tables and equations

Conversion of this to a logical sum of logical products form results in

$$
\begin{aligned}
S_2 &= (A\overline{B} + \overline{A}B)(\overline{C_i}) + \overline{(A\overline{B} + \overline{A}B)}(C_i) \\
&= A\overline{B}\,\overline{C_i} + \overline{A}B\overline{C_i} + \overline{(A\overline{B} + \overline{A}B)}(C_i) \qquad \text{Distributive Law} \\
&= A\overline{B}\,\overline{C_i} + \overline{A}B\overline{C_i} + (\overline{A}\cdot\overline{B} + AB)(C_i) \qquad \text{Identity} \\
&= A\overline{B}\,\overline{C_i} + \overline{A}B\overline{C_i} + \overline{A}\,\overline{B}C_i + ABC_i \qquad \text{Distributive Law}
\end{aligned}
$$

Note that the equation for S_2 is identical to the equation for the SUM output of the full adder, thus establishing the validity of using two half-adders to obtain the SUM function of the full-adder.

The CARRY output of the full-adder is not as immediately obvious. But a clue or two is available. The CARRY output of the first adder (AB) is also a term in the CARRY output equation of the full adder. The other two terms are to be ORed with AB according to the CARRY output equation. Assuming that the CARRY output is $AB + AC + BC$ implies that an OR gate performs the function if the terms $AC + BC$ or the equivalent can be provided as additional inputs. In other words, the CARRY output of the full adder (C_o) is $AB + (X)$, where X is the equivalent to $AC + BC$. Another way to look at it is that when some input X is ORed with AB, the overall expression is equivalent to $AB + AC + BC$. The only outputs where C is available in combination with A and B is the SUM output of the second half-adder (S_2) and the CARRY output of the second half-adder (C_2). Perhaps C_2 should be checked, since this output has not yet been used.

With the output of the first half-adder ($A\overline{B} + \overline{A}B$) as one input and C_i as the other input to the second half-adder, $C_2 = (A\overline{B} + \overline{A}B)(C_i)$. Combining C_2 with C_1 in an OR gate yields $C_o = AB + C_i(A\overline{B} + \overline{A}B)$. Expanding this expression to $AB + A\overline{B}C_i + \overline{A}BC_i$ yields very little, as far as the required output is concerned. Recognizing that the term ($A\overline{B} + \overline{A}B$) is the EXCLUSIVE–OR function and that sometimes equivalent Boolean expressions work better algebraically than the originals leads to substitution of $(A + B)(\overline{A} + \overline{B})$ for $A\overline{B} + \overline{A}B$. Expansion of this equivalent expression with C_i gives the same answer as the original.

$(A + B)(\overline{AB})$ appears to be more promising. As we mentioned in Sec. 5–5, it is advantageous to be on the lookout for terms in the expression being simplified that are *complements* of each other. This allows the use of the theorems, laws, and identities of Boolean algebra in the simplification process. Expansion and simplification of the CARRY output equation are shown.

$$
\begin{aligned}
C_o &= AB + [(A + B)(\overline{AB})(C_i)] \\
&= AB + [(AC_i + BC_i)(\overline{AB})] \qquad \text{Distributive Law} \\
&= AB + AC_i\overline{AB} + BC_i\overline{AB} \qquad \text{Distributive Law} \\
&= \underbrace{AB} + \underbrace{\overline{AB}}(AC_i + BC_i) \qquad \text{Factoring} \\
&= A + \overline{A} \quad\cdot\quad B \qquad\qquad \text{Recognition of form } A + \overline{A}B \\
C_o &= AB + AC_i + BC_i
\end{aligned}
$$

It has been demonstrated that two half-adders and an OR gate provide the required outputs for the full adder function. The block/logic diagram with Boolean expressions appears in Fig. 5–15. Although very little has been shown in this discussion as far as actual logic elements are concerned, an increasingly popular method of implementing logic functions was used. "Prepackaged" IC logic functions are extremely popular, and where they are cost effective, they are replacing multiple-element implementation of the same function.

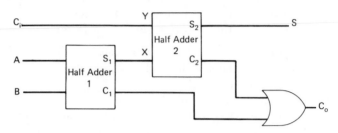

Figure 5–15   Full-adder from two half-adders

PROBLEM APPLICATIONS

5–1.   Define logic simplification.

5–2.   What is another name for the sum of products form of a Boolean expression?

5–3.   What is another name for the product of sums form of a Boolean expression?

5–4.   List the steps necessary to algebraically simplify a logically implemented form of an industrial operation expressed in English-language form.

5–5.   A logic diagram of a Boolean function is supplied. What steps are necessary to algebraically simplify this logic diagram?

5–6.   What algebraic form results from reading a Boolean expression from a truth table?

5–7.   Define single-level, two-level, three-level, and four-level logic circuits.

Algebraically simplify:

5–8. $\overline{AB} + (A\overline{B} + \overline{A}B)$

5–9. $(A + \overline{B})(A\overline{B} + C)(C)$

5–10. $(A + \overline{B})(C) + (A\overline{B} + C)$

5–11. $(AB + C) + (\overline{AB + C})(A + CD)$

5–12. $[(A + C)(A + D)][(B + C)(B + D)]$

5–13. $\overline{A} \cdot \overline{C} + \overline{A}B + BC + \overline{B} \cdot \overline{C} + AB$

5–14. $AB + A\overline{D} + BC + A\overline{B} + BD + \overline{A}B$

5–15. $\overline{A} \cdot \overline{B} + A\overline{D} + B\overline{D} + \overline{B}D + \overline{C} \cdot \overline{D}$

5–16. $[(AB + AC) + (BC + A\overline{B})] + \overline{A}B$

5–17. $A\overline{D} + \overline{A}D + CD + BC + \overline{A}C$

5–18. $AB + BC + \overline{A}C + A\overline{C} + \overline{B}C$

5–19. $(AB)(CD) + \overline{A} + \overline{B} + \overline{C} + \overline{D} + A$

5–20. $AB\overline{C} + \overline{A} \cdot \overline{B} \cdot \overline{C} + \overline{A} + B$

5–21. $\overline{M \cdot L} \cdot H \cdot \overline{J} \cdot K \cdot H + \overline{J}(H + K)$

5–22. $\overline{AB} + \overline{C} + \overline{D}A + ABAC$

5–23. $A\overline{B} + C + DE + \overline{A} + \overline{B} + F + \overline{(\overline{A} + D)(\overline{A} + B + E)\overline{D}}$

5–24. $DE(\overline{A} \cdot \overline{C} + A + \overline{B} + C) + \overline{C} + \overline{D} + E$

5–25–34.   Algebraically simplify Probs. 16–25 in Chapter 6.

6

Graphic
Simplification
Methods

In Chapter 5 we showed how logic expressions* may be simplified by using
the postulates, theorems, laws, and identities of Boolean algebra. It requires
considerable ingenuity, judgment, and experience to perform such logic simplification consistently and satisfactorily. This chapter shows another method of
logic simplification that requires a minimum of Boolean algebra background.

Just as graphic representation of a conventional algebraic expression makes
that expression much easier to visualize, the graphic portrayal of logic expressions reduces the trauma of logic simplification. A minimized expression, requiring the *fewest possible logic elements*, is the goal of both graphic and algebraic
simplification. For those who have studied Ch. 5, it will soon be apparent that
in cases where graphic methods are employed, the simplification solution is more
obvious and easier to obtain. But it may also be apparent that graphic solutions
by themselves may not always work. Combinations of algebraic and graphic
solutions are therefore often required.

Numerous nonalgebraic methods of logic simplification employing charts,
graphs, and maps have been and are being used. The graphic methods of John
Venn, a nineteenth century British mathematician, are still used to visually
demonstrate simple theorems and laws of Boolean algebra.

The staff of the Harvard Computation Laboratory developed a chart
method of simplification in the 1950s. During the same period, W. V. Quine
proposed another chart method. Both the Harvard chart and the Quine method
are somewhat cumbersome and time consuming. However the concepts behind
both are currently being used in computer-aided logic-simplification operations
despite the relative difficulty of their "paper-and-pencil" methods.

In 1952 E. W. Veitch proposed a diagrammatic means for simplifying logic
functions. His method attempted to represent logic functions in a manner that

---

* Justifications and ground rules for simplification of logic expressions appear in
Sec. 5–1, which should be examined at this time if Ch. 5 was not previously read.

would make algebraic factors more evident. The following year, M. Karnaugh suggested a reorganization of the *Veitch diagram.* Identification of factorable terms was more obvious in *Karnaugh's map*, and variations and simplifications have seen widespread use. The similarities of the Veitch diagram and the Karnaugh map have generated much confusion. So as not to perpetuate this confusion, only the term "map" is used in this text.

The purpose of this chapter is to develop the use of the *map method of simplification* as a tool of digital logic analysis. Investigations of computer-based methods are not covered but are left to the individual reader.

6—1 TYPES OF GRAPHS

Venn Diagrams

One graphic method used to display logical expressions of Boolean algebra is the *Venn diagram.* Venn's geometric method of representing logic relationships consists of (1) a rectangle that represents a class of elements with certain characteristics and (2) circles that represent subclasses or more sharply defined groups within the overall class. Since Boolean algebra is considered an algebra of classes and since a *class* is a collection of *all objects having some specific characteristics in common*, Venn diagrams lend themselves well to an investigation of the logic expressions of Boolean algebra.

For example: let A equal all engineering technicians who work on digital equipment and B equal all engineering technicians who are reading this book. The Boolean expression for the combination of both is $A \cdot B$. The Venn diagram representing this condition appears in Fig. 6—1a. Shading is used to indicate the area within the rectangle that represents the special group of engineering technicians who work on digital equipment *and* who are reading this book, i.e., $A \cdot B$. Correlation with the truth table for the AND operation will reveal that the shaded area represents the "true" (1) condition for ANDing two variables. The unshaded area is the "false" (0) condition, which includes $\overline{A} \cdot \overline{B}, \overline{A} \cdot B$, and $A \cdot \overline{B}$. Note the ease with which the expression can be identified and visualized.

Representation of all engineering technicians who work with digital equipment *or* who are reading this book $(A + B)$ is provided in Fig. 6—1b. The shaded

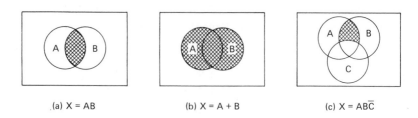

(a) X = AB   (b) X = A + B   (c) X = ABC̄

Figure 6—1   Venn diagrams

area includes all of A, all of B, and the intersection of A and B. Verification of the validity of this graphic representation may be made by noting that the truth table for the OR operation shows a true condition when A is true, B is true, or both A and B are true.

The Venn diagram for the 3-variable term $AB\bar{C}$ is shown in Fig. 6–1c. It is immediately obvious from this diagram that the only condition under which the ANDed combination of A, B, and $\bar{C}$ is true is when both A and B are true and C is not true. Although Venn diagrams are occasionally used with four variables, we recommend that no more than three variables be used. The most common application of Venn diagrams is to demonstrate graphically the simple theorems of Boolean algebra, although simple laws sometimes submit readily to the use of Venn diagrams.

Veitch Diagrams

Veitch diagrams also use rectangles to represent logic relationships. Instead of employing circles to represent variables, however, Veitch divided the rectangle into *squares*. Sufficient squares are provided so that every possible combination of all of the variables in the logic expression is represented. Thus the *Veitch diagram* is just another way of displaying *the same information that the truth table provides*. A 3-variable diagram, as originally presented by Veitch, is shown in conjunction with a 3-variable truth table in Fig. 6–2a. Each square on the diagram represents one line of the truth table. Actually, each square represents the conjunctive combination of the three variables and/or their complements. The row labeled 0 could be written as $\bar{A} \cdot \bar{B} \cdot \bar{C}$; the row labeled 1 as $\bar{A} \cdot \bar{B} \cdot C$; and so on. This labeling scheme, previously used in the manipulation of truth tables, clarifies the configuration of the Veitch diagram. Figure 6–2b shows correlation of truth table entries and Veitch diagram positions.

The intent of the Veitch diagram is to present truth table information in a form that more clearly displays the factorable terms in a Boolean expression.

Label	A	B	C
0	0	0	0
1	0	0	1
2	0	1	0
3	0	1	1
4	1	0	0
5	1	0	1
6	1	1	0
7	1	1	1

		0	1	A
BC	00	0	4	
	01	1	5	
	10	2	6	
	11	3	7	

		0	1	A
BC	00	$\bar{A}\bar{B}\bar{C}$	$A\bar{B}\bar{C}$	
	01	$\bar{A}\bar{B}C$	$A\bar{B}C$	
	10	$\bar{A}B\bar{C}$	$AB\bar{C}$	
	11	$\bar{A}BC$	ABC	

(a) Truth table and original map

(b) Map representation of terms

Figure 6–2   Veitch diagrams

Algebraically factoring a simple Boolean expression such as $\overline{A}BC + ABC$ demonstrates what may be done with the Veitch diagram. By inspection, we see that $\overline{A}BC + ABC = (\overline{A} + A)BC$. Further simplification yields $(1)BC$ or BC. If the two terms are arranged so as to easily identify the existence of common factors such as BC, the algebraic expression is simplified by inspection. (This is what Veitch did when he developed his diagram.)

$\overline{A}BC$ is represented on the diagram by square 3, and ABC by square 7. Marks are placed in each of the identified squares to define the fact that those terms are included in the expression being simplified. All of the squares are then investigated, and when marked squares having a difference of only one variable are recognized, those squares are grouped for future reference. In this case, squares 3 and 7 can be grouped together, since B and C remain the same for both squares and since A is 0 for square 3 and 1 for square 7. When a variable and its complement are both present in an AND–OR expression, that variable may be discarded, as demonstrated in the algebraic simplification of the expression being used. Thus in a 3-variable Veitch diagram, when two squares are properly grouped, the new term described by the grouped squares has one less variable than the original.

A bit of difficulty exists, however, in identifying those squares that can be grouped together. To take advantage of the factoring idea, recall that squares may differ by only *one* variable. No difficulty exists in terms of the squares making up the $A = 0$ column and the $A = 1$ column. Obviously, squares 0, 1, 2, and 3 are opposite in A value from squares 4, 5, 6, and 7. But what about squares 0 and 2 in terms of the variable B? Or squares 1 and 3 in terms of B? Or 4 and 6, or 5 and 7? These combinations all meet the criterion of squares differing by only one variable, although from the circling or grouping operation it may not be immediately apparent. Considerable familiarity with the use of Veitch diagrams is required before such relationships are automatically recognized. Since other methods of graphic simplification not requiring the added effort of mentally grouping squares are available, the *original* Veitch diagram is rarely used.

Many additional capabilities of the Veitch diagram (along with the intricacies of its use) have not been covered here. But the contributions of Veitch should not be overlooked. Without his early efforts in graphically portraying logic simplification, the revised diagrams and maps now in use might have taken much longer to develop.

Karnaugh Maps

The same general concepts exist for both Veitch diagrams and Karnaugh maps — *rectangles* represent *logic relationships* and *squares* represent *conjunctive combinations* of all variables in the expression being evaluated. The basic difference between the two types of graphic representation lies in the *arrangement of the squares* inside the rectangle. By rearranging the squares, it is possible to have any square that is adjacent to any other square differ by only one variable.

Figure 6—3a shows the 3-variable map as originally presented by Karnaugh. The commonly used simplified version is reproduced in Fig. 6—3b. By filling in the actual terms for each square, it is obvious that no square differs from any adjacent square by more than one variable, thereby facilitating the factoring process (see Fig. 6—3c).

Placing *identification numbers* in each square reveals that there may not be an easy method for determining how Karnaugh maps are constructed (Fig. 6—3d). So it may be interesting to consider some of the ideas behind the construction. Random tries could be initiated — terms tried in various locations until the proper combination is found. But trial and error is not really efficient! What really needs to be done is to find a way of identifying the squares so that there is no doubt of the outcome and so that a minimum of time is spent in determining how to construct a map for a situation not previously encountered. A recognizable means of identifying the squares is also desirable. It is apparent that sequentially using the binary-coded-decimal (BCD) equivalent of each row on the truth table is of no avail. Perhaps some version of the BCD code other than the 8—4—2—1 version could be used.

One such version of the BCD code, called the *Gray code*, has only a one-variable difference as counting progresses from one number to the next. Straight BCD code (4—2—1 BCD) and the Gray code are compared in Table 6—1. But can this code help build a Karnaugh map? Taking the original arrangement that Karnaugh proposed for the physical layout of his map and placing the appropriate

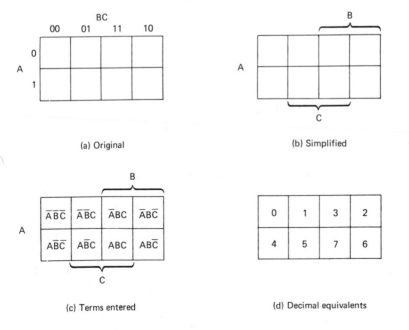

(a) Original

(b) Simplified

(c) Terms entered

(d) Decimal equivalents

Figure 6—3   Karnaugh maps

Table 6–1   4–2–1 versus Gray code

| | 4–2–1 | | | | Gray | |
A	B	C	Decimal	A	B	C
0	0	0	0	0	0	0
0	0	1	1	0	0	1
0	1	0	2	0	1	1
0	1	1	3	0	1	0
1	0	0	4	1	1	0
1	0	1	5	1	1	1
1	1	0	6	1	0	1
1	1	1	7	1	0	0

Boolean terms in each square (Fig. 6–3c) again leads to an interesting observation. Note that each square differs from each adjacent square by *only one* variable. If the decimal equivalent of the Gray code version of each term is entered (Fig. 6–4), a numbering pattern emerges. Thus the Karnaugh map is *not* a random development. It resulted from a recognition of the basic characteristics of Boolean expressions, number systems, and truth tables.

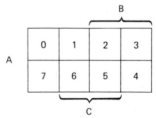

Figure 6–4   Three-variable Karnaugh map using Gray Code numbering

6–2 PREPARATION FOR THE USE OF MAPS

Map Construction

Maps used in graphic simplification of Boolean expressions are actually truth tables in another form. As such, the same number of possible combinations presented in the truth table must be accounted for in the map. Thus for *n* variables, 2^n combinations exist, so a 2-variable truth table must have 2^2 or 4 entries, a 3-variable truth table must have 2^3 or 8 entries, and so on. Two-variable maps must then have 4 squares, 3-variable maps 8 squares, and so on. The actual physical placement of the squares on the map depends on individual variations of the common map methods. As long as the criterion of each square varying from each adjacent square by only one variable is met, the physical layout is unimportant. Figure 6–5 shows the map layout used in this text for 2, 3, 4, and 5-variable expressions.

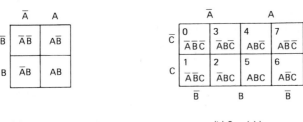

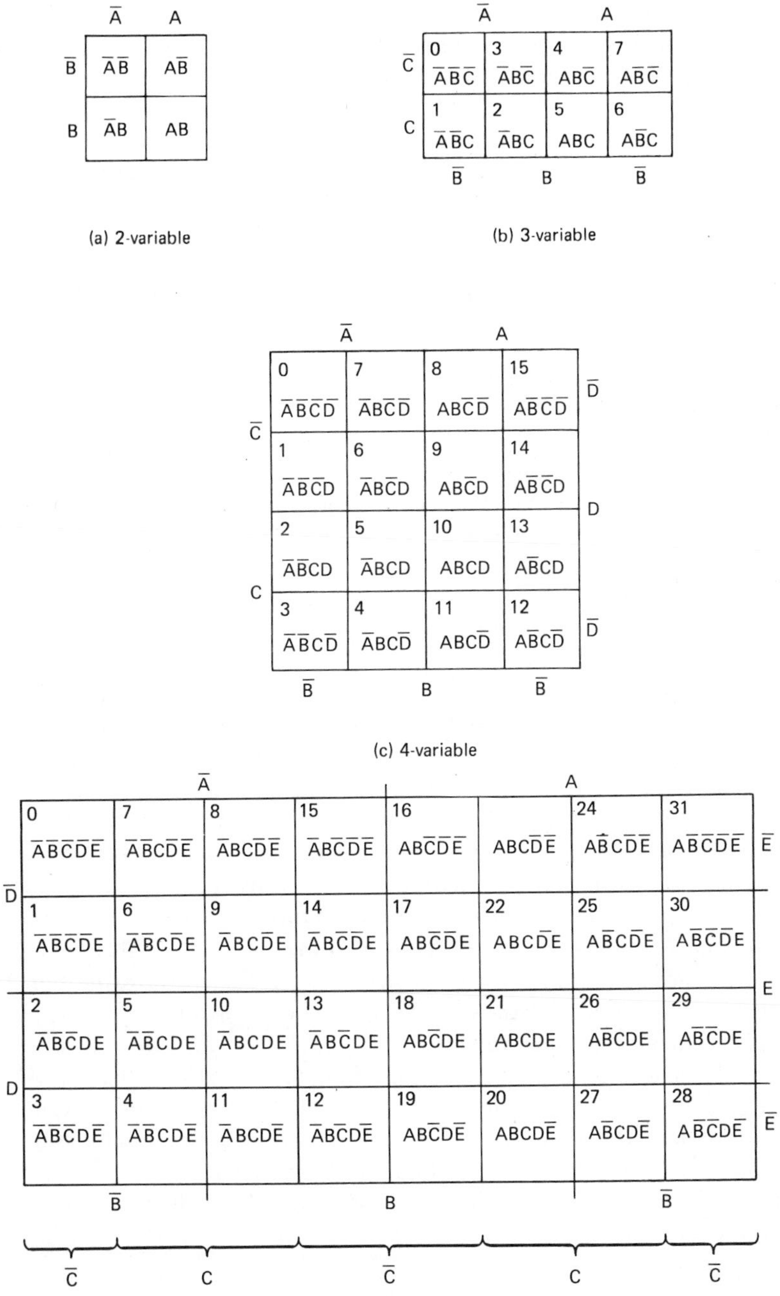

(a) 2-variable (b) 3-variable

(c) 4-variable map:

	$\bar{A}$		A	
$\bar{C}$ $\bar{D}$	0 $\bar{A}\bar{B}\bar{C}\bar{D}$	7 $\bar{A}B\bar{C}\bar{D}$	8 $AB\bar{C}\bar{D}$	15 $A\bar{B}\bar{C}\bar{D}$
$\bar{C}$ D	1 $\bar{A}\bar{B}\bar{C}D$	6 $\bar{A}B\bar{C}D$	9 $AB\bar{C}D$	14 $A\bar{B}\bar{C}D$
C D	2 $\bar{A}\bar{B}CD$	5 $\bar{A}BCD$	10 $ABCD$	13 $A\bar{B}CD$
C $\bar{D}$	3 $\bar{A}\bar{B}C\bar{D}$	4 $\bar{A}BC\bar{D}$	11 $ABC\bar{D}$	12 $A\bar{B}C\bar{D}$

Bottom column labels: $\bar{B}$ B $\bar{B}$; right-side labels: $\bar{D}$, D, D, $\bar{D}$

(c) 4-variable

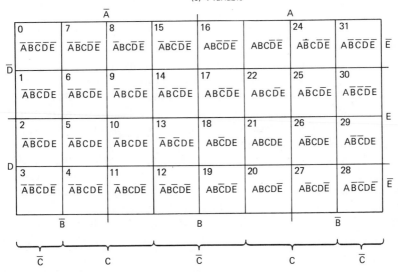

(d) 5-variable

Figure 6–5 Construction of maps

In actual use, maps do not have the number identification (terms) indicated within each square. These are provided initially so that the pattern of square identification can be established. Careful study of the four map forms shows that (1) each square varies from each adjacent square by only one variable and (2) an identification pattern does exist. Note that the identification numbers are based on the Gray code* representation of decimal numbers as are Karnaugh maps.

Forms of Boolean Expressions

To use map methods of simplification, the Boolean expression must be in a form that can be represented on the map. Two basic forms of Boolean expressions exist (neglecting negated expressions).

1. The logical sum of logical products form, previously mentioned, contains terms ORed together, and each term consists of one or more variables ANDed together.
2. The logical product of logical sums form contains factors ANDed together, and each factor consists of one or more variables ORed with other variables.

Each of these is discussed below.

Logical sum of logical products expressions are known by many different names: *AND—OR*, *sum-of-products*, *standard sum*, and *minterm form*. Actually, the last two names require a definition more explicit than merely "logical sum of logical products." Not only must the expression be in the logical sum of logical products form, but each term must contain all variables in either the normal or complemented (negated) form. Rather than confuse the situation with numerous definitions, *minterm* is used in this text to define a Boolean expression in which *each term contains all of the variables* (in either the normal or complemented form) AND*ed together and all of the terms* OR*ed together.* Minterm comes from a mathematical definition of a particular type of expression called a "minimal polynomial term."

The logical product of logical sums form is also known as the *OR—AND*, the *products of sums*, the *standard product*, and the *maxterm form*. Again, the standard product and the maxterm forms have more explicit definitions than the product of sums form definition. In the standard product and the maxterm forms, *all variables* (in either the normal or complemented form) *are* OR*ed together in each factor and all of the factors are* ANDed. *Maxterm* is used exclusively in this text.

Conversion of Boolean expressions to minterm form (the form required for entry of the expression into the logic map) requires use of the theorems, laws, and common identities of Boolean algebra. For example, $\overline{A + B}$ is not in minterm

---

* See Chapter 14 for the conversion table from BCD to Gray to decimal codes.

form. However the vincula may be split, using DeMorgan's Law, and $\overline{A + B}$ becomes $\overline{A} \cdot \overline{B}$. Individual terms should also be simplified, like a variable that appears more than once within the term ($ABCA = ABC$). When a variable and its complement appear within the same ANDed term, simplification should also be accomplished ($ABC\overline{A} = 0$). Where appropriate, the Distributive Law should be applied to remove the parentheses $AB(C + D) = ABC + ABD$. The reader will be provided with considerable practice examples and problems.

Another way to obtain the minterm form for *any* expression is to use truth tables. Construct a truth table for the expression being used; then examine the output column for combinations of inputs that provide an output equal to 1. Each of these terms may then be combined disjunctively to give the minterm form. Thus when a Boolean expression is read from a truth table, it is in the logical sum of logical products or minterm form.

Truth Table Forms

Since the map is merely another way of presenting truth table information, the actual *form of the information* in the truth table must be realized. As originally defined, a truth table is a tabular listing of all possible combinations of input conditions to a logic device with a resultant output for each case. This definition remains valid in studying map methods of logic simplification. Right now, the important consideration is the form in which the *output conditions* are read from the truth table.

To take advantage of the *factorability* of logical expressions, the expressions should be in the logical sum of logical products form. When in this form, factorable terms are more easily identified, as demonstrated during the Veitch diagram discussion. *The major premise of the map method of logic simplification is based on having the Boolean expression in the logical sum of logical products form.* If the expression is not in this form, algebraic methods must be used to convert it. In many cases, the algebraic conversion results in the simplest form, but the converted expression should be examined with maps to assure that such is the case. When the expression is converted to the logical sum of logical products form, the truth table is constructed, and the output is determined in this form. Once again, the Boolean expression must be in the logical sum of logical products form to apply map simplification operations. This is why one of the criteria of "simplest circuit" in Chapter 5 was that the final expression be in logical sum of logical products form.

Each term contains *all* of the variables in either the inverted or noninverted form when the expression is read from the truth table. Thus it may be entered directly into the simplification map. An expression that is not in the standard sum form must either be subjected to truth table manipulations or be converted to the proper form. The Boolean expression $X\overline{Y}Z + XY + \overline{X}Y\overline{Z}$ is not in standard sum form because it does not contain all of the variables in the second term. Expansion to standard sum form by both truth table and algebraic methods is shown in Fig. 6–6.

X	$\overline{X}$	Y	$\overline{Y}$	Z	$\overline{Z}$	$X\overline{Y}Z$	XY	$\overline{X}Y\overline{Z}$	$X\overline{Y}Z + XY + \overline{X}Y\overline{Z}$	
0	1	0	1	0	1	0	0	0	0	
0	1	0	1	1	0	0	0	0	0	
0	1	1	0	0	1	0	0	1	1	$\overline{X}Y\overline{Z}$
0	1	1	0	1	0	0	0	0	0	
1	0	0	1	0	1	0	0	0	0	
1	0	0	1	1	0	1	0	0	1	$X\overline{Y}Z$
1	0	1	0	0	1	0	1	0	1	$XY\overline{Z}$
1	0	1	0	1	0	0	1	0	1	XYZ

As shown in the truth table $X\overline{Y}Z + XY + \overline{X}Y\overline{Z} = \overline{X}Y\overline{Z} + X\overline{Y}Z + XY\overline{Z} + XYZ$

(a)

$$\begin{aligned} X\overline{Y}Z + XY + \overline{X}Y\overline{Z} &= X\overline{Y}Z + XY(1) + \overline{X}Y\overline{Z} \\ &= X\overline{Y}Z + XY(Z+\overline{Z}) + \overline{X}Y\overline{Z} \\ &= X\overline{Y}Z + XYZ + XY\overline{Z} + \overline{X}Y\overline{Z} \end{aligned}$$

$XY(1) = XY$
$Z + \overline{Z} = 1$
Distributive laws

(b)

Figure 6–6   Expansion of an expression to standard sum
form; (a) using the truth table, (b) algebraic expansion

Maxterms are less commonly used than minterms. The maxterm form of
a given expression can be obtained (1) by writing the inverse of the desired
expression in minterm form and (2) by inverting the expression obtained in
step (1). For example, the maxterm is derived for the expression described by
the truth table, Table 6–2. .

The minterm form is read directly from the truth table as

$$W = \overline{X}\cdot\overline{Y}\cdot\overline{Z} + \overline{X}\cdot\overline{Y}\cdot Z + \overline{X}\cdot Y\cdot\overline{Z} + X\cdot\overline{Y}\cdot\overline{Z} + X\cdot\overline{Y}\cdot Z$$

Writing the inverse of W in minterm form from the truth table yields

$$\overline{W} = \overline{X}\cdot Y\cdot Z + X\cdot Y\cdot\overline{Z} + X\cdot Y\cdot Z$$

Inverting $\overline{W}$ provides the required maxterm form

$$W = (X + \overline{Y} + \overline{Z})(\overline{X} + \overline{Y} + Z)(\overline{X} + \overline{Y} + \overline{Z})$$

The use of maxterm forms in map simplification will be demonstrated later.

Table 6–2   Truth table $W = \overline{X}\,\overline{Y}\,\overline{Z} + \overline{X}\,\overline{Y}Z + \overline{X}Y\overline{Z} + X\overline{Y}\,\overline{Z} + X\overline{Y}Z$

X	Y	Z	W	$\overline{W}$
0	0	0	1	0
0	0	1	1	0
0	1	0	1	0
0	1	1	0	1
1	0	0	1	0
1	0	1	1	0
1	1	0	0	1
1	1	1	0	1

Entering Information on Maps

Each square on a map represents only *one unique combination* of the variables in the expression being entered. It is assumed that the expression has been converted or expanded, as necessary, to the form required for entry. Each term of the expression is examined, and a 1 is placed in the square representing the true condition for that term. Zeros may be inserted in all of the remaining squares to represent the false terms, but common practice dictates omitting the zeros for clarity. 2-variable through 5-variable maps are shown in Fig. 6–7. Each map has a Boolean expression, demonstrating the information entry process.

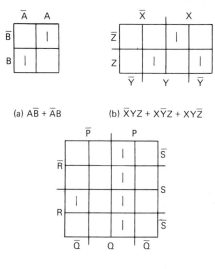

(a) $A\overline{B} + \overline{A}B$ (b) $\overline{X}YZ + X\overline{Y}Z + XY\overline{Z}$

(c) $PQRS + PQ\overline{R}S + PQ\overline{R}\overline{S} + PQ\overline{R}\overline{S} + \overline{P}QRS$

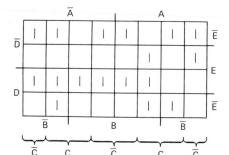

(d) $ABCDE + AB\overline{C}DE + ABC\overline{D}E + ABC\overline{D}E + \overline{A}BCDE + \overline{A}BC\overline{D}E + \overline{A}B\overline{C}DE + \overline{A}BCDE + \overline{A}\overline{B}CD\overline{E} +$
$A\overline{B}CD\overline{E} + A\overline{B}C\overline{D}E + A\overline{B}C\overline{D}E + \overline{A}\overline{B}C\overline{D}E + \overline{A}\overline{B}CD\overline{E} + \overline{A}BC\overline{D}\overline{E} + ABCD\overline{E}$

Figure 6–7   Completed maps

The meaning of the words minterm and maxterm may become more apparent by examining the 3-variable map. Construction of the map is such that eight separate squares are contained within the rectangle defining the 3-variable expression. The *minimum* distinguishable area capable of being individually identified is *one* of the squares. Identification of a single square is made by conjunctively combining all three of the variables in either their normal or complemented form. A term composed of all of the variables conjunctively combined in either their normal or complemented form has been defined as a minterm. Thus *a minterm is characterized by the minimum distinguishable area on a logic map.* Since the maxterm is the complement of the minterm, all of the rest of the squares on the map represent the maxterm. *The maxterm*, then, *is characterized by the maximum distinguishable area on a logic map.* This relationship may assist the reader in remembering and identifying the minterm and maxterm forms when they are encountered during map simplification exercises.

6–3   2-VARIABLE MAP SIMPLIFICATION

A Simple Two-variable Expression $AB + \overline{A}B$

An easy way to get started with the map method of simplification is to compare *map simplification* (of a simple 2-variable expression) with *algebraic simplification* (of the same expression). The many advantages of map simplification may not be immediately obvious with simple 2-variable expressions, but the use of maps can be demonstrated without too much difficulty. Consider the expression $AB + \overline{A}B$. Since the variable B is common to both terms, it may be factored, resulting in $B(A + \overline{A})$. $(A + \overline{A}) = 1$ by Theorem 12, thus simplifying the expression to $B(1)$. Theorem 3 states that $B(1) = B$, so the original expression reduces simply to B. The truth table and comparative logic diagram appear in Fig. 6–8.

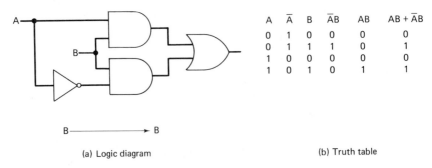

A	$\overline{A}$	B	$\overline{A}B$	AB	$AB + \overline{A}B$
0	1	0	0	0	0
0	1	1	1	0	1
1	0	0	0	0	0
1	0	1	0	1	1

(a) Logic diagram   (b) Truth table

Figure 6–8   Logic diagram and truth table, $AB + \overline{A}B$

Figure 6—5 showed the physical layout of a map for simplification of 2-variable expressions. Section 6—2 defined the form of the Boolean expression required for entry on the map and demonstrated such entry. Examination of the expression to be simplified discloses that it is in the required minterm form — all we must do is insert 1s into the squares corresponding to each of the minterms. The completely filled-in map appears in Fig. 6—9a.

The next step is to learn how to use the map. Since the physical layout of the map is such that each square varies from each adjacent square by only one variable, examination of adjacent squares can yield valuable information. For example, the 1s in Fig. 6—9a are adjacent. While each 1 occupies a square that is equivalent to the variable B, 1 of the squares shares B with $\overline{A}$ and the other square shares B with A. It should be evident that the map is displaying graphically the same thing as the algebraic factoring $B(A + \overline{A})$. And it has been shown that when a variable and its complement are ORed together, the result is logic 1. It is also known that a variable ANDed with 1 is equivalent to that variable. Thus the map performs factoring in much the same manner as algebraic factoring.

Identification of patterns of filled squares is usually made by circling groups of squares. When using the 2-variable map, the following patterns (shown in order of preferences) are allowed.

1. All four squares filled represent a constant value of logic 1 for the expression entered.
2. Any two adjacent squares filled represents a single-variable term.
3. A single square filled represents a 2-variable term.

In the example being simplified, only the two adjacent squares shown in Fig. 6—9b can be grouped. By the given rule, this must represent a single-variable. term. When the grouping operation encompasses both a variable and its complement, that variable and its complement are cancelled. Only the variable that is included in the grouping operation by itself (i.e., with no complement) is left. In this case, $A + \overline{A}$ cancels, leaving only B. As stated earlier, when two squares are properly grouped, the new term described by the grouped squares has *one less* variable than the original.

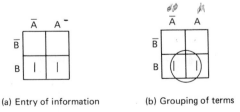

(a) Entry of information   (b) Grouping of terms

Figure 6—9   Map simplification of $AB + \overline{A}B$

A Complex Two-variable Expression $A(A + B) + B(\overline{A} + B)$

A more complex 2-variable logic simplification problem is now demonstrated. To enter this expression in the map, it must be in minterm form. The steps required to perform the conversion to this standard form are presented here.

$$A(A + B) + B(\overline{A} + B) = AA + AB + \overline{A}B + BB \qquad \text{Distributive Law}$$
$$= A + B + AB + \overline{A}B \qquad \text{Theorem 5,}$$
$$\text{Commutative Law}$$
$$= A(1) + B(1) + AB + \overline{A}B \qquad \text{Theorem 3}$$
$$= A(B + \overline{B}) + B(A + \overline{A}) + AB + \overline{A}B \qquad \text{Theorem 12}$$
$$= AB + A\overline{B} + AB + \overline{A}B + AB + \overline{A}B \qquad \text{Distributive Law,}$$
$$\text{Theorem 5}$$
$$\therefore \ A(A + B) + B(\overline{A} + B) = AB + A\overline{B} + \overline{A}B \qquad \text{Theorem 11}$$

Truth tables and logic diagrams for this expression are shown in Fig. 5–6. The minterm form entered in a 2-variable map is shown in Fig. 6–10a.

The same types of patterns used in the previous example are used in this map. Considering the map as two separate maps, we see that two 1s can be grouped (*looped*) at the bottom of the map (Fig. 6–10b) and that two 1s can also be looped at the right-hand side of the map (Fig. 6–10c). The two 1s at the bottom of the map represent the variable B. Actually, one square represents $\overline{A}B$, and the other square represents AB. It has been shown previously, however, that the map performs the task of algebraically simplifying expressions of the form $AB + \overline{A}B$. Therefore A and $\overline{A}$ cancel, allowing the two looped 1s at the bottom of the map to represent the single variable B. Similar reasoning can be applied to the two 1s at the right-hand side of the map to obtain the single variable A. Placing the two single maps together (Fig. 6–10d) discloses that the

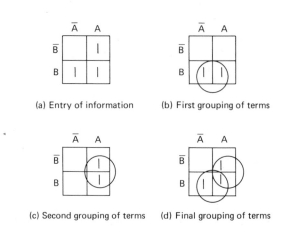

(a) Entry of information   (b) First grouping of terms

(c) Second grouping of terms   (d) Final grouping of terms

Figure 6–10   Map simplification of $AB + A\overline{B} + \overline{A}B$

square equivalent to AB is shared by both maps. When a 1 is placed in a square, that square is considered filled. If mapping another term requires a 1 in a filled square, no additional action is necessary. When such a situation occurs, it merely means that a redundancy exists for the term indicated by the filled square. All additional terms falling within the filled square will be ignored during the map-reading operations. Sharing squares is equivalent to Postulate 8 or $(1 + 1 = 1)$.

When more than one term is required to describe the patterns of 1s on the map, the terms are joined by the OR operator. Since the map is another way of presenting truth table information, the *joining of terms by the* OR *operator* is analogous to reading the output column of the truth table in logical sum of logical products form. Interpretation of map simplification of the Boolean expression $A(A + B) + B(\overline{A} + B)$, using the above information, results in the simplified expression $A + B$.

6–4 3-VARIABLE MAPS

A Complex Three-variable Expression

$$XY\overline{Z} + \overline{X}Y\overline{Z} + \overline{X + \overline{Y} + \overline{Z}} + \overline{X + Y + Z} + Z(\overline{X} + Y)$$

The real advantages of map simplification begin to appear when three or more variables are encountered. Entry of the expression into the map is not feasible in the original form. Algebraic manipulation is necessary to obtain the minterm form. Thus

$XY\overline{Z} + \overline{X}Y\overline{Z} + \overline{X + \overline{Y} + \overline{Z}} + \overline{X + Y + Z} + Z(\overline{X} + Y)$	Original expression
$XY\overline{Z} + \overline{X}Y\overline{Z} + \overline{X}Y Z + \overline{X}\cdot\overline{Y}\cdot\overline{Z} + Z(\overline{X} + Y)$	DeMorgan's Law
$XY\overline{Z} + \overline{X}Y\overline{Z} + \overline{X}YZ + \overline{X}\cdot\overline{Y}\cdot\overline{Z} + \overline{X}Z + YZ$	Distributive Law
$XY\overline{Z} + \overline{X}Y\overline{Z} + \overline{X}YZ + \overline{X}\cdot\overline{Y}\cdot\overline{Z} + \overline{X}YZ + \overline{X}\cdot\overline{Y}\cdot Z + XYZ + \overline{X}YZ$	Expansion
$XY\overline{Z} + \overline{X}Y\overline{Z} + \overline{X}YZ + \overline{X}YZ + \overline{X}YZ + \overline{X}\cdot\overline{Y}\cdot\overline{Z} + \overline{X}\cdot\overline{Y}\cdot Z + XYZ$	Commutative Law
$XY\overline{Z} + \overline{X}Y\overline{Z} + \overline{X}YZ + \overline{X}\cdot\overline{Y}\cdot\overline{Z} + \overline{X}\cdot\overline{Y}\cdot Z + XYZ$	$(A + A = A)$

Since the expression contains three variables (either in normal or comple-mented form), a total of eight possible combinations exists; hence a map with eight squares is required. The squares are arranged as we discussed in Sec. 6–2 so that each square differs from each adjacent square by only one variable. But now a problem arises. What about the squares on the left side and on the right side of the rectangle that have no adjacent squares on one side? Reference to Fig. 6–5 may help solve the dilemma. Note that the square numbered 0 $(\overline{A}\cdot\overline{B}\cdot\overline{C})$ differs from the square numbered 7 $(A\cdot\overline{B}\cdot\overline{C})$ and that the square numbered 1 $(\overline{A}\cdot\overline{B}\cdot C)$ differs from the square numbered 6 $(A\overline{B}C)$ by only one variable. Thus square 0 is adjacent to square 7, and square 1 is adjacent to square 6, as though the map were inscribed on a *cylinder*. If the 3-variable map is visualized in this manner, no difficulty should be encountered in simplifying 3-variable Boolean expressions.

As in the 2-variable map, a 1 is placed in the square representing each of the terms in the expression. Figure 6–11a shows the completed map. Rules for simplifying the expression by use of the map are similar to the 2-variable map rules. Certain expansions are necessary, however, to compensate for the additional squares. The following patterns of 1s (once again shown in order of preference) are allowed.

1. All eight squares filled represents a constant value of logic 1 for the expression entered.
2. Any four adjacent squares filled represents a single-variable term. Remember, squares at opposite ends of rows are adjacent.
3. Any two adjacent squares filled represents a 2-variable term.
4. A single square filled represents a 3-variable term.

The preference in all cases is to detect *patterns having the maximum number of* 1*s*, as long as the number of 1s is a power of 2 (i.e., 1, 2, 4, 8, 16, etc.). In the 3-variable map, the first order of preference is to have *all* eight squares

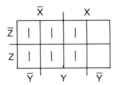

(a) All information entered

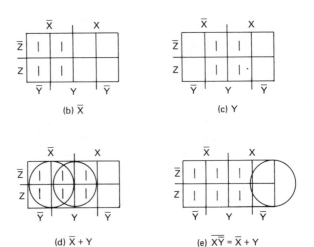

(b) $\bar{X}$

(c) Y

(d) $\bar{X} + Y$

(e) $X\bar{Y} = \bar{X} + Y$

Figure 6–11   Map simplification of $X Y \bar{Z} + \bar{X} Y \bar{Z} + \overline{X + \bar{Y} + \bar{Z}} + \overline{X + Y + Z} + Z(\bar{X} + Y)$

filled. This would represent a logic value of 1 for the complete expression, and all of the gates required to implement the expression could be replaced with a direct connection. In this example, squares without 1s exist, so the first order of preference cannot be met. The second order of preference is a grouping of four 1s in adjacent squares. Figure 6–11b shows one possible grouping, and Fig. 6–11c shows another possible grouping. As in the 2-variable map, 1s may be shared in more than one loop. Thus two loops of four 1s exist, and all 1s are within loops (Fig. 6–11d). The resulting equation should be a 2-term equation, consisting of one variable or its complement in each term. The grouping in Fig. 6–11b represents $\bar{X}$. Actually, it represents $\bar{X}YZ + \bar{X}\cdot\bar{Y}\cdot Z + \bar{X}Y\bar{Z} + \bar{X}\cdot\bar{Y}\cdot\bar{Z}$, but the factoring properties of the map allow *direct* interpretation of the display without any algebraic manipulation. Similarly, the grouping in Fig. 6–11c represents Y. Thus the simplified version of the complex expression is $\bar{X} + Y$.

Truth-table proof is shown in Table 6–3 to verify this simplification.

Another very important characteristic of map simplification is demonstrated by the Boolean expression currently being examined. If the expression is viewed as an equation such that $F = XY\bar{Z} + \bar{X}Y\bar{Z} + X + \bar{Y}+\bar{Z} + \overline{X+Y+Z} + Z(\bar{X} + Y)$, then the completed map can be considered to represent F. The portion of the map that is *not* marked represents all that is not F, or $\bar{F}$. Circling of the unmarked squares (Fig. 6–11e) and interpreting their meaning results in the Boolean equation $\bar{F} = X\bar{Y}$. Negating both sides of the equation and applying DeMorgan's Law provide the equation $F = X + \bar{Y}$, which is identical to the result obtained by interpreting the marked squares. In cases where there are relatively few unmarked squares, this approach may prove effective in quickly recognizing the simplified expression.

Graphic Simplification Starting with a Logic Diagram

Often, the only information available from which to perform logic simplification is *a portion of a logic diagram*. It then becomes necessary to first develop the basic algebraic expression that describes the diagram's operation. Starting at the input side of the diagram, each logic element is examined and its output recorded. The logic diagram in Fig. 6–12 is evaluated in this manner.

OR 1	$A + \bar{C}$
OR 2	$\bar{A} + C$
AND 1	$(A + \bar{C})(\bar{A} + C)$
NOR 1	$\overline{B + (A + \bar{C})(\bar{A} + C)}$
NOR 2	$\overline{A + C}$
AND 2	$AB\bar{C}$
OR 3	$\overline{B + (A + \bar{C})(\bar{A} + C)} + \overline{A + C} + AB\bar{C}$

Table 6–3   Truth table proof $XY\overline{Z} + \overline{X}Y\overline{Z} + \overline{X+\overline{Y}+\overline{Z}} + \overline{X+Y+Z} + Z(\overline{X}+Y) = \overline{X}+Y$

X	$\overline{X}$	Y	$\overline{Y}$	Z	$\overline{Z}$	XYZ	$XY\overline{Z}$	$\overline{X}Y\overline{Z}$	$X+\overline{Y}+\overline{Z}$	$\overline{X+\overline{Y}+\overline{Z}}$	$X+Y+Z$	$\overline{X+Y+Z}$	$(\overline{X}+Y)$	$Z(\overline{X}+Y)$	W^*	$\overline{X}+Y$
0	1	0	1	0	1	0	0	0	1	0	0	1	1	0	1	1
0	1	0	1	1	0	0	0	0	1	0	1	0	1	1	1	1
0	1	1	0	0	1	0	0	1	1	0	1	0	1	0	1	1
0	1	1	0	1	0	0	0	0	0	1	1	0	1	1	1	1
1	0	0	1	0	1	0	0	0	1	0	1	0	0	0	0	0
1	0	0	1	1	0	0	0	0	1	0	1	0	0	0	0	0
1	0	1	0	0	1	0	1	0	1	0	1	0	1	0	1	1
1	0	1	0	1	0	1	0	0	1	0	1	0	1	1	1	1

(The last two columns W^* and $\overline{X}+Y$ are marked **Equal**.)

* $W = XY\overline{Z} + \overline{X}Y\overline{Z} + \overline{X+\overline{Y}+\overline{Z}} + \overline{X+Y+Z} + Z(\overline{X}+Y)$

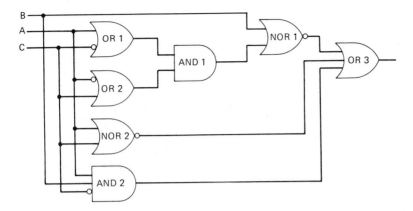

Figure 6—12   Sample logic diagram

Conversion to sum of products form follows.

$$\overline{B + (A + \overline{C})(\overline{A} + C)} + \overline{A} + C + AB\overline{C}$$

$$\overline{B(A + \overline{C})} + \overline{(\overline{A} + C)} + \overline{A} \cdot \overline{C} + AB\overline{C} \qquad \text{DeMorgan's Law}$$

$$\overline{B}(\overline{A}C) + A\overline{C} + \overline{A} \cdot \overline{C} + AB\overline{C} \qquad \text{DeMorgan's Law}$$

$$\overline{B} \cdot \overline{A} \cdot C + A\overline{C} + \overline{A} \cdot \overline{C} + AB\overline{C} \qquad \text{Distributive Law}$$

Then to the minterm form.

$$\overline{B} \cdot \overline{A} \cdot C + A\overline{C}(B + \overline{B}) + \overline{A} \cdot \overline{C}(B + \overline{B}) + AB\overline{C} \qquad \text{Theorem 3}$$

$$\overline{B} \cdot \overline{A} \cdot C + AB\overline{C} + A \cdot \overline{B} \cdot \overline{C} + \overline{A}B\overline{C} + \overline{A} \cdot \overline{B} \cdot \overline{C} + AB\overline{C} \qquad \text{Distributive Law}$$

$$\overline{A} \cdot \overline{B} \cdot C + A \cdot \overline{B} \cdot \overline{C} + \overline{A}B\overline{C} + \overline{A} \cdot \overline{B} \cdot \overline{C} + AB\overline{C} \qquad \begin{array}{l}\text{Theorem 11,}\\ \text{Commutative}\\ \text{Law}\end{array}$$

The minterm form expression is entered on the 3-variable map, and the grouping of 1s performed (Fig. 6—13b) according to rules previously defined. $\overline{C}$ is interpreted from the "4-loop," and $\overline{A} \cdot \overline{B}$ is interpreted from the "2-loop." Both terms are joined by the OR operator to form the expression $\overline{C} + \overline{A} \cdot \overline{B}$. The simplified logic diagram appears in Fig. 6—13a.

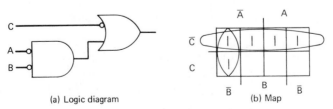

(a) Logic diagram   (b) Map

Figure 6—13   Simplified logic diagram and map for Figure 6—12

Using Logical Sum of Logical Products Expressions

Now that a number of simplifications have been accomplished using the map method, it is time to investigate some "shortcuts." Note that much of the effort expended with the map method is in *preparing the expression* so that it can be entered on the map. Getting from the original expression to the minterm expression may be difficult at times, to say the least. The minterm form, of course, is a special case of the logical sum of logical products form, where all variables in the expression are represented in each term in either their normal or complemented form. As you may have noted, many terms in AND–OR form expressions do not contain all of the variables. It has been necessary to expand such terms to include the missing variable(s) to reach the minterm form. In each case, additional terms have been generated, and the complexity of the expression has grown.

Recognition of the operations performed allows these operations to be performed mentally without the attendant increase in complexity. An example may prove helpful. Consider the expression just simplified. Application of DeMorgan's Law and the Distributive Law to the expression $\overline{B + (A + \overline{C})(\overline{A} + C)} + \overline{A} + C + AB\overline{C}$ resulted in the 4-term AND–OR expression $\overline{A} \cdot \overline{B} \cdot C + A\overline{C} + \overline{A} \cdot \overline{C} + AB\overline{C}$. Expansion to minterm form resulted in the 6-term expression $\overline{A} \cdot \overline{B} \cdot C + AB\overline{C} + A \cdot \overline{B} \cdot \overline{C} + \overline{A}B\overline{C} + \overline{A} \cdot \overline{B} \cdot \overline{C} + ABC$. The 6-term expression fulfilled the requirements for the minterm form.

The operation that generated the additional terms in the minterm form was the expansion of the 2-variable terms $A\overline{C}$ and $\overline{A} \cdot \overline{C}$ to include both the normal and the complemented form of the missing variable in each case. Obviously, a relationship exists between the *number of variables* in a term and the *number of squares* occupied by that term on the map. In a 3-variable map, for example, a term containing all three variables occupies only one square. That one square is completely defined by the intersection of the rows and columns on the map used to describe each of the variables. When the number of variables in the term of a 3-variable map drops to only two, then two squares must be used to describe that term. This occurs because the 2-variable term may occupy the square that it shares with the missing variable and also the square that it shares with the complement of the missing variable. Reduction of the number of variables in a term to one results in four squares being required to describe that term. Now the variable in the term must share squares with the normal and complemented forms of *both* of the missing variables. The implied relationship between number of variables in a term and the size of the map can be expressed in the following manner.

1-variable term	1/2 of the squares
2-variable term	1/4 of the squares
3-variable term	1/8 of the squares
4-variable term	1/16 of the squares
5-variable term	1/32 of the squares

Note that as the number of variables in the term increases, the number of squares occupied decreases by a power of 2 (or is halved). In the 3-variable map, a 1-variable term occupies half (four) of the squares. A 2-variable term occupies one-fourth (two) of the squares. Note the reduction by a power of 2. A 3-variable term occupies one-eighth (one) of the squares. Again, note the reduction by a power of 2. Thus an n-variable term occupies $1/2^n$ squares. This relationship holds for all maps to be discussed in this chapter.

Using this relationship, it can be seen that the two terms $A\overline{C}$ and $\overline{A} \cdot \overline{C}$ each occupy two squares, since a 2-variable term must occupy one-fourth of the total squares on the map. The algebraic expansion to minterm form has shown that in each case, the two squares are those squares shared with the missing variable and its complement. With a little practice, placement of logical sum of logical product form expressions in less than full minterm form on maps becomes a real timesaver.

Some of the examples that follow show entry of logical sum of logical products form expressions directly on the map without conversion to minterm form so that the techniques of such operations can be demonstrated. When in doubt, however, always expand to the full minterm form. A little time may be lost, but it ensures that the complete expression is entered.

Logic Operations with Maps

Direct ANDing or ORing of maps is another valuable simplification operation. When a complex expression such as

$$[\overline{D}(E + \overline{F}) + E\overline{F}]\,[\overline{DE\overline{F} + \overline{D + \overline{E}}}]\,[\overline{D} + \overline{F}(D + \overline{E})]$$

is to be simplified, something other than pure algebraic simplification appears advisable. Conversion of the entire expression to AND–OR form is a laborious task, requiring many steps. It is much quicker to convert each of the three separate expressions within the complete expression to logical sum of logical products form and then AND the maps to simplify. The AND–OR form derivation for each of the separate expressions is shown in preparation for map simplification.

$$[\overline{D}(E + \overline{F}) + E\overline{F}] = [(\overline{D}E + \overline{D}\overline{F}) + E\overline{F}] \qquad \text{Distributive Law}$$
$$= (\overline{D}E + \overline{D}\overline{F} + E\overline{F}) \qquad \text{Associative Law}$$
$$[\overline{DE\overline{F} + \overline{D + \overline{E}}}] = (\overline{DE\overline{F}} + \overline{DE}) \qquad \text{DeMorgan's Law}$$
$$[\overline{D} + \overline{F}(D + \overline{E})] = [\overline{D} + (D\overline{F} + \overline{E}\overline{F})] \qquad \text{Distributive Law}$$
$$= (\overline{D} + D\overline{F} + \overline{E}\overline{F}) \qquad \text{Associative Law}$$

In its basic form, ANDing of maps requires a *separate map* for each of the logical sum of logical product form expressions. (Conversion to minterm form is not accomplished for this example, but the reader may perform the conversions to verify the adequacy of the logical sum of logical products form.) None of the three terms in the first expression contains all three variables, so it is necessary to apply the concepts mentioned previously to enter the expression on the map. The first term, $\overline{D}E$, contains only two of the three variables, so it must occupy one-fourth (two) of the eight squares on the map. F is the missing variable that must be combined, in its normal and complemented form, with $\overline{D}E$. The resulting two terms, $\overline{D}EF$ and $\overline{D}E\overline{F}$, are then plotted on the map. Actually, the expansion is performed mentally, and the required two squares are plotted as though two separate 3-variable terms were being plotted. Figure 6–14a is the map of the term $\overline{D}E$. Similarly, $\overline{D}F$ is combined with missing variable E and its complement and mapped in Fig. 6–14b. $E\overline{F}$ is shown in Fig. 6–14c.

$(\overline{D}E + \overline{D}\cdot\overline{F} + E\overline{F})$ is a 3-term logical sum of logical products expression. It consists of two variables ANDed together in each term with all terms ORed together. Since a map has been developed to represent each term, it stands to reason that a map representing the whole expression can be developed. In the algebraic expression, the terms were ORed together. Maps representing terms may also be ORed together by following the basic rules for disjunction.

$$0 + 0 = 0, \quad 0 + 1 = 1, \quad 1 + 0 = 1, \quad \text{and} \quad 1 + 1 = 1$$

Superimposing the three maps and applying the disjunctive rules is equivalent to ORing the three terms of the expression. When a 1 on one of the maps coincides with a 1 on another of the maps, this is equivalent to $1 + 1 = 1$, and the map for the complete expression has a 1 in that square. $0 + 1 = 1$ and $1 + 0 = 1$ are represented when one of the maps has a 1 in a square corresponding to a square on another map that has a 0 or no mark. Thus a 1 in any square in any of the maps being combined into a single map under disjunctive rules will cause a 1 to be in a like square on the single map. Of course, $0 + 0 = 0$, and when like squares in maps all have 0s, the single map will also have a 0 in that square. Algebraic terms, therefore, may be ORed easily by plotting maps of each of the terms and then combining the maps disjunctively to form that total expression (see Fig. 6–14d).

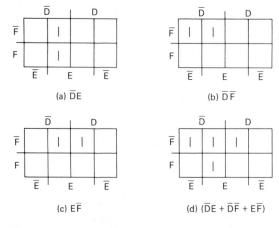

Figure 6–14   Map entries $(\overline{D}E + \overline{D}\overline{F} + E\overline{F})$

The second expression, $(\overline{DEF} + \overline{DE})$, is a bit easier. No difficulty should be encountered with the first term $\overline{DEF}$, since it describes only one square (Fig. 6−15a). The second term, $\overline{DE}$, requires two squares − one of which is the same as the first term (Fig. 6−15b). Combining the two maps in an OR relationship results in Fig. 6−15c, the map for the expression $(\overline{DEF} + \overline{DE})$.

$(\overline{D} + D\overline{F} + \overline{E}\cdot\overline{F})$, the third expression, demonstrates the entry of a single variable on a 3-variable map. As we pointed out earlier, the single variable must share squares with all possible combinations of *both* missing variables. With two variables missing, E and F, four combinations exist. Four squares are required to plot $\overline{D}$ (Fig. 6−16a). $D\overline{F}$ shares two squares with E and $\overline{E}$ (Fig. 6−16b), and $\overline{E}\cdot\overline{F}$ shares two squares with D and $\overline{D}$ (Fig. 6−16c). The map of the complete expression is shown in Fig. 6−16d.

Completion of the maps for each of the three separate expressions is all that is required to prepare for the ANDing operation. The separate maps are superimposed, and the resulting combinations of 1s and 0s in each square are combined according to the four rules of conjunction;

$$0\cdot0 = 0, \qquad 0\cdot1 = 0, \qquad 1\cdot0 = 0, \quad \text{and} \quad 1\cdot1 = 1$$

In other words, unless corresponding squares in *all* of the maps being used have 1s entered, the logic value for that square in the single map is 0. Only in squares that have 1s in each corresponding position in each map is the logic value 1. The maps for each separate expression are shown in Figs. 6−17a, b, and c. The single map that is the result of ANDing the three separate maps is shown in Fig. 6−17d. In situations where it is relatively easy to obtain logical sum of logical product form expressions from the expressions that are initially provided, logic operations with maps prove to be valuable for combinational logic-circuit analysis.

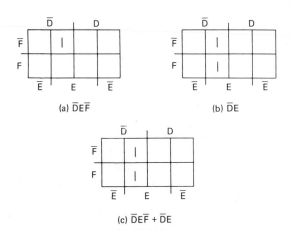

(a) $\overline{D}E\overline{F}$ (b) $\overline{D}E$

(c) $\overline{D}E\overline{F} + \overline{D}E$

Figure 6−15   Map entries $(\overline{DEF} + \overline{DE})$

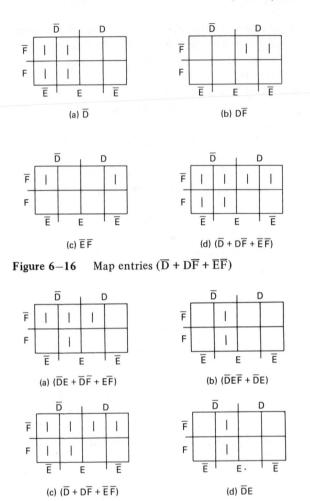

Figure 6—16   Map entries $(\overline{D} + D\overline{F} + \overline{E}\,\overline{F})$

Figure 6—17   ANDing of maps

6—5 4-VARIABLE MAPS

Advancing to 4-variable maps does not appreciably complicate matters. With four variables, sixteen possible combinations exist, so the map must have sixteen squares on which to plot variable combinations. The map is constructed (Fig. 6—5c) so that all adjacent squares differ by only one variable to take advantage of factoring possibilities. The techniques already discussed also apply to 4-variable maps. Following entry of the algebraic expressions on the map, the only difference in map analysis is the *meaning of identified adjacencies.* The general rule in Sec. 6—4 can be followed, and very little difficulty should be encountered. As stated, a 1-variable term is identified by one-half of the squares being marked. In the case of the 4-variable map, any allowable grouping of eight squares represents a 1-variable term. A 2-variable term is represented by a grouping of four squares, a 3-variable term by two squares, and a 4-variable term by one square. However, identification of allowable groups varies somewhat from that of the 2-variable and 3-variable configuration.

Adjacency Identification

Adjacency identification can be demonstrated by map simplification of the expression

$$AB + \overline{A} \cdot \overline{B} \cdot \overline{C} \cdot \overline{D} + \overline{A}BCD + \overline{B}C\overline{D} + \overline{B} \cdot \overline{C} \cdot D + \overline{A} \cdot \overline{D} + A\overline{B} + \overline{A}B\overline{C}D.$$

Once again, the logical sum of logical products form of the expression is entered on the map rather than conversion to full minterm form. The reader should verify each of the entries for the expression on the map (Fig. 6—18e). Should any difficulty arise in verifying the entries, the expression should be expanded to minterm form and separate maps used for each term.

Reading the map progresses in almost a straightforward manner. It is apparent that not all squares are filled, so the value of the expression is not logic 1. Since sixteen squares cannot be circled, eight (or one-half of the squares) should be examined. Figure 6—18a shows an identifiable group (A), Fig. 6—18b another (B), and Fig. 6—18c a third ($\overline{C}$). But another group of eight squares can be circled. Just as the 3-variable map was visualized as being inscribed on the surface of a cylinder, so the 4-variable map should be visualized as being inscribed on the surface of a *sphere*. Under these circumstances, the left-hand column of squares is adjacent to the right-hand column of squares, and the top row of squares is adjacent to the bottom row of squares. Reference to Fig. 6—5c verifies that each square in the left-hand column differs from its adjacent square in the right-hand column by only one variable. The same is true for the top and bottom rows. Thus the eight squares comprising the top and bottom rows can be circled (Fig. 6—18d). The simplified expression then becomes A OR B OR $\overline{C}$ OR $\overline{D}$ and is shown in complete map form in Figs. 6—18e and f.

This expression also lends itself to simplification by using the *unplotted squares method* discussed in Sec. 6—4. The unplotted square is equivalent to $\overline{A} \cdot \overline{B} \cdot C \cdot D$ and is the *complement* of the mapped expression. If the mapped expression is equated to X, then the unplotted square is equivalent to $\overline{X}$. Complementing both sides to obtain X and applying DeMorgan's Law result in $X = \overline{\overline{A} \cdot \overline{B} \cdot C \cdot D}$ or $A + B + \overline{C} + \overline{D}$, which was the simplified expression obtained by circling all of the marked squares. Either method of map reading is equally valid. Just as in the 3-variable example, this operation is the same as taking the maxterm form directly from the map.

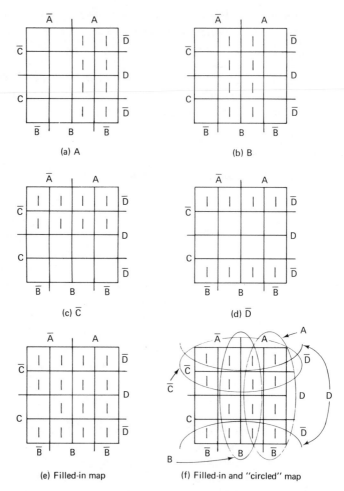

Figure 6—18　　Map entries — $AB + \overline{A}\,\overline{B}\,\overline{C}\,\overline{D} + \overline{A}BCD + \overline{B}C\overline{D} + \overline{B}C\overline{D} + \overline{A}\overline{D} + A\overline{B} + \overline{A}B\overline{C}\overline{D} = (A + B + \overline{C} + \overline{D})$

Simplification Starting with a Truth Table

Consider now a situation where the only information given is a truth table of an unknown logic function. As demonstrated in the previous chapter, the Boolean algebra expression may be read directly from the output column. For example, Table 6—4 shows all possible combinations of four variables with an output called E, which has some logical relationship to those four variables. Remember, the truth table output column is read to show all of the conditions of the variables that provide a true (1) output. Note that the expression is recovered from the truth table in logical sum of logical products form.

Table 6–4   Truth table of a 4-variable complex function

A	$\bar{A}$	B	$\bar{B}$	C	$\bar{C}$	D	$\bar{D}$	$\bar{A}\bar{B}\bar{C}\bar{D}$	$\bar{A}B\bar{C}\bar{D}$	$AB\bar{C}\bar{D}$	$\bar{A}B\bar{C}D$	$\bar{A}BC\bar{D}$	$ABCD$	E^{*}
0	1	0	1	0	1	0	1	1	0	0	0	0	0	1
0	1	0	1	0	1	1	0	0	0	0	0	0	0	0
0	1	0	1	1	0	0	1	0	0	0	0	0	0	0
0	1	0	1	1	0	1	0	0	0	0	0	0	0	0
0	1	1	0	0	1	0	1	0	1	0	0	0	0	1
0	1	1	0	0	1	1	0	0	0	0	1	0	0	1
0	1	1	0	1	0	0	1	0	0	0	0	1	0	1
0	1	1	0	1	0	1	0	0	0	0	0	0	0	0
1	0	0	1	0	1	0	1	0	0	0	0	0	0	0
1	0	0	1	0	1	1	0	0	0	0	0	0	0	0
1	0	0	1	1	0	0	1	0	0	0	0	0	0	0
1	0	0	1	1	0	1	0	0	0	0	0	0	0	0
1	0	1	0	0	1	0	1	0	0	1	0	0	0	1
1	0	1	0	0	1	1	0	0	0	0	0	0	0	0
1	0	1	0	1	0	0	1	0	0	0	0	0	0	0
1	0	1	0	1	0	1	0	0	0	0	0	0	1	1

* $E = \bar{A}\bar{B}\bar{C}\bar{D} + \bar{A}B\bar{C}D + \bar{A}B\bar{C}\bar{D} + \bar{A}BC\bar{D} + AB\bar{C}\bar{D} + ABCD$

With the algebraic expression available, simplification can proceed in the usual manner, using either algebraic simplification or map simplification. This expression yields readily to map simplification. The 4-variable map is used to plot the minterms of the derived algebraic expression. Since minterms have been used, each term is plotted in only one square on the map. The seven terms in the expression thus occupy seven squares, as shown in Fig. 6—19a. Applying the concepts discussed earlier results in identification of two groups of four filled squares: one group represents the term $B\overline{C}$ (Fig. 6—19b), and the other group represents $\overline{A}\cdot\overline{D}$ (Fig. 6—19c). Note that the identification of the latter group depends on the visualization of the map inscribed on a sphere to obtain adjacent squares. The completed map appears in Fig. 6—19d.

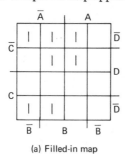

(a) Filled-in map

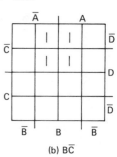

(b) B$\overline{C}$

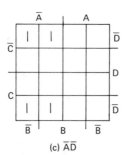

(c) $\overline{A}\overline{D}$

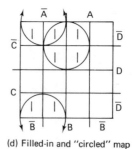

(d) Filled-in and "circled" map

Figure 6—19   Map simplification of a 4-variable complex function (Table 6—4)

6—6 5-VARIABLE MAPS

Mapping of expressions with more than four variables becomes increasingly difficult. Physical visualization of adjacent squares varying by no more than one variable may not be immediately obvious. Construction of the map is complex, and entering information on and extracting information from the map presents a certain amount of difficulty. Thus information concerning 5-variable maps will not be as extensive as for the preceding four maps.

Concepts of 5-variable Maps

The 5-variable map may be more easily understood if it is viewed as *two* 4-variable maps *back-to-back*. One map combines all possible combinations of four of the five variables with the fifth variable, while the other map combines all possible combinations of the four variables with the complement of the fifth variable. The basic 5-variable map of Fig. 6–7d is effectively split in half – the segment representing the complement of variable A is left in place, while the segment representing A is folded back behind the stationary segment. Rules for determining adjacencies of squares for 5-variable maps are stated here.

1. Squares adjacent in the same row or column may be grouped.
2. Squares at opposite ends of rows or columns may be grouped.
3. Squares that occupy identical positions in the adjacent maps may be grouped.

Figure 6–20 represents the concept of "back-to-back" logic maps. Rule 1 was easily demonstrated on 2, 3, and 4-variable maps. 4-variable maps were visualized as inscribed on the surface of a sphere to help explain how Rule 2 was developed. By now, enough examples have been worked out so that it should no longer be necessary to resort to the spherical visualization of 4-variable maps to determine which squares are adjacent. In other words, the reader should be able to determine which squares can be grouped without the visual representation of the map on the spherical surface. The 2, 3, and 4-variable maps may therefore be used on a flat surface to effectively represent Boolean expressions.

When the "back-to-back" 4-variable map representation of a 5-variable expression is used, it is necessary to employ a 2-dimensional picture. As the portion of the map representing A is rotated around behind the portion representing $\overline{A}$, we see that 1 column becomes adjacent to column 8, column 2 adjacent to column 7, column 3 adjacent to column 6, while column 4 remains adjacent to column 5. Identical positions in columns 1 and 8 are then adjacent. The same is true of columns 2 and 7, 3 and 6, and 4 and 5. Thus Rule 3 is demonstrated. The reader is invited to assign variable names to each of the squares as in Fig. 6–7d and compare square for square in the newly defined adjacenies. Again, it becomes apparent that Rule 3 is valid. Breaking off the rear map after rotating it behind the front map and moving the rear map away results in the 2-dimensional representation of Fig. 6–20b. Variables can now be grouped on the front map, on the rear map, and between the front and rear maps.

Since it is inconvenient to draw the maps in the back-to-back configuration, the 5-variable map is used as shown in Fig. 6–7d, keeping the three rules for determining adjacencies of squares in mind. The only really new concept that has to be remembered is that columns 2 and 7, and 3 and 6 are adjacent; 1 and 8, and 4 and 5 are already adjacent merely by construction of the map and the previously defined rules. In the example shown, the new concepts of adjacency are stressed so that the reader can become familiar with grouping variables that fit into this category.

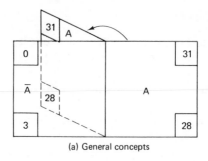

(a) General concepts

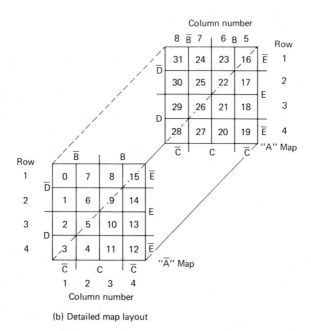

(b) Detailed map layout

Figure 6—20   Five-variable map concepts

A Complex 5-variable Expression $[AB(C+D+E)+DBA][\overline{A}+\overline{B}+\overline{E}]$

Entering information on and reading information from 5-variable maps is demonstrated by attempting to simplify this expression. A number of different approaches may be used, but the easiest is to convert the first portion to logical sum of logical products form, plot both of the expressions separately, and then AND the maps. A minimum amount of new rule usage is required, and the rather "different" physical layout of the map is easily demonstrated.

The first portion of the expression is quickly converted to $ABC+ABD+ABE$. Entering the converted expression on the 5-variable map can be accomplished with the aid of the relationships defined in Sec. 6—4. Since the 5-variable map contains 32 squares, a 3-variable term will occupy one-eighth (four) of the squares. ABC is plotted in Fig. 6—21a, ABD in Fig. 6—21b, and ABE in Fig. 6—21c. Combining the three maps disjunctively (ORing the maps) results in the map in Fig. 6—21d, which represents the complete expression $ABC + ABD + ABE$.

The second portion of the expression proceeds in a similar manner. A 1-variable term occupies one-half (sixteen) of the squares. A is plotted in Fig. 6—22a, $\overline{B}$ is plotted in Fig. 6—22b, and $\overline{E}$ is plotted in Fig. 6—22c. Once again, ORing the three maps results in the map in Fig. 6—22d, which represents the complete expression $\overline{A} + \overline{B} + \overline{E}$.

As previously demonstrated, superimposing the maps representing the terms to be ANDed and keeping only those squares common to both maps will result in a map of the overall expression. Figure 6—23 is the result of ANDing Figs. 6—21d and 6—22d. Only three squares are common to both maps. the adjacencies are easily identified without using any new concepts. Grouping the two marked squares at the bottom of row 4 provides the term $ABD\overline{E}$, while grouping the two squares at the top and bottom of column 6 results in the term $ABC\overline{E}$. The simplified expression then becomes $ABD\overline{E} + ABC\overline{E}$. Proof of the equality of the original expression and the simplified expression is left to the reader.

6—7 SIMPLIFYING THE BINARY ADDER GRAPHICALLY

One practical application of simplification by map methods may be illustrated by simplifying the problem used in Ch. 5 to demonstrate arithmetic addition in the binary number system. Most of the concepts of map usage developed in this chapter will be included in this demonstration. The English-language statements for both outputs of the binary adder, plus the descriptive truth table and Boolean equations, were developed in Sec. 5—7. Final simplification is compared with Fig. 5—12 to determine the actual reduction in logic component requirements.

Simplification

The conventional approach to binary-adder map simplification is to treat each of the output expressions *individually*, i.e., develop two separate simplified logic circuits for each expression. Both the SUM and the OUTPUT CARRY expressions are in minterm form, as read from the truth table. Entering the

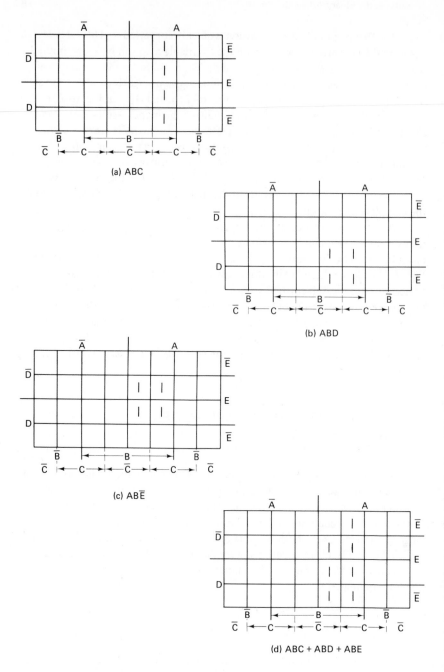

(a) ABC

(b) ABD

(c) ABĒ

(d) ABC + ABD + ABE

Figure 6−21   Map simplification ABC + ABD + ABE

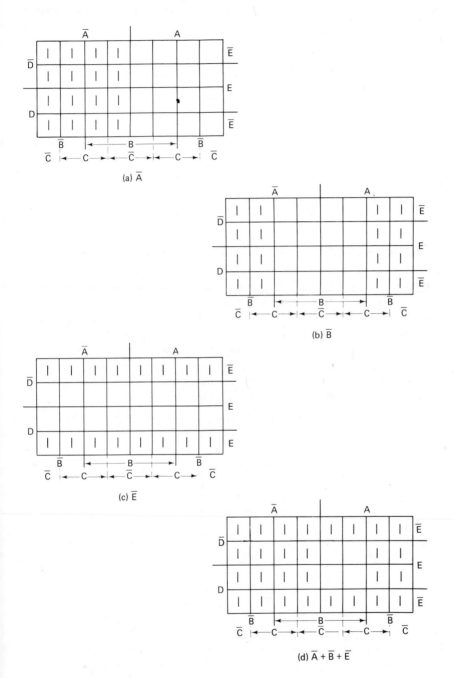

Figure 6–22   Map simplification $\overline{A} + \overline{B} + \overline{E}$

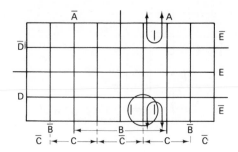

Figure 6–23   Combined maps, Figures 6–21 and 6–22
($\overline{ABCE}$ + $\overline{ABDE}$)

SUM expression into the required 3-variable map (Fig. 6–24a) and applying the rules for looping adjacent squares reveals that no real simplification is attained by using the map method. No two marked squares are adjacent, which means that none of the terms may be simplified. Thus the original expression for the SUM output of the binary adder does not yield to simplification using the map method in the conventional manner.

Entering the OUTPUT CARRY expression into the 3-variable map, however, shows some promise. The four marked squares in Fig. 6–24b can be looped (Fig. 6–24c), resulting in three 2-variable terms instead of four 3-variable terms. The two equations for the full adder are now

$$S = \overline{A}\cdot\overline{B}\cdot C + \overline{A}\cdot B\cdot\overline{C} + A\cdot\overline{B}\cdot\overline{C} + A\cdot B\cdot C$$
$$C_o = A\cdot B + A\cdot C + B\cdot C$$

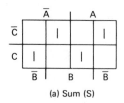

(a) Sum (S)

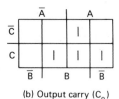

(b) Output carry (C_o)

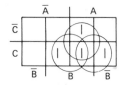

(c) Simplified C_o (AB + AC + BC)

Figure 6–24   Full adder-map simplification

The logic diagram resulting from direct map simplification shows a reduction of one 3-input AND gate and the replacement of three 3-input AND gates with three 2-input AND gates. From a practical standpoint, the change from 3-input AND gates to 2-input AND gates would probably be disregarded (unless 2-input AND gates were commonly available in the logic system being used), since 3-input AND gates are fairly standard.

Another approach that sometimes yields a reduction in logic-element count is to look for *common squares* on the maps of the separate outputs. The binary adder is typical of many combinational logic circuits — it has common inputs but separate outputs. By placing the maps of the separate outputs in proximity to each other, common squares may be identified. If the same square is marked on each map, it means that the term represented by those squares is required in both output expressions. One gate can be used to form the common term, and the rest of the terms developed by additional gates can form the separate outputs. Although significant simplification cannot be accomplished this way by the binary adder, the small amount of simplification that does result is a good demonstration of the technique. Comparison of the SUM expression map of Fig. 6–25a and the CARRY OUTPUT expression map of Fig. 6–25b reveals that one square is common to both maps. The term ABC is represented by the common square, and this is required in both of the output expressions. Thus logic-element count is reduced by one AND gate by this method.

From a strict logic-element count, hardware implementation of this simplified binary adder does not appear any more advantageous than the previously discussed simplification. Where many binary adders are used, however, some reduction in physical wiring and interconnection requirements may be realized. This is really a design problem, but it does point out one reason why there may be different implementations of the simple binary adder function.

Numerous other configurations of binary adders exist. Both algebraic and map methods of simplification are used to determine these configurations, but the techniques are generally beyond the scope of this text. Further discussion tends to complicate and detract from the real purpose of discussing map methods of simplification — applying the concepts of maps to analysis of digital logic circuits. As more complex logic circuits are encountered, the funda-

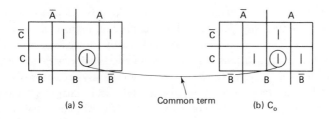

Figure 6–25   Common square in the SUM and CARRY outputs

mentals discussed in this chapter will become increasingly important. The ability to manipulate Boolean algebra expressions into different forms either algebraically or by the use of maps is of great value in determining operation of logic circuits and circuit equivalency.

6–8 COMPARISON OF SIMPLIFICATION METHODS

Many of the commonly encountered combinational logic simplification situations have been discussed in this chapter. Of course, variations of all these methods, plus some expansion of most of the methods to more complex problems, will be encountered in logic engineering tasks. For the digital logic analyst, though, a sufficient variety has been presented. The map methods, in conjunction with the algebraic simplification methods of earlier chapters, should handle practically all the logic analysis problems that he will encounter.

The reader has undoubtedly noted that work in both map and algebraic simplification has much in common. A good percentage of the expressions requiring simplification will not be in the required logical sum of logical products form needed for map simplification. Algebraic methods are required to obtain the required form. Until considerable experience is gained, it is good practice to convert all expressions to be simplified to the logical sum of logical products form before making the decision about which simplification method to employ. If the logical sum of logical products form of the expression contains many more than four terms and two variables, the map method will usually yield a faster answer. (We should also mention that the map method may yield more than one simplified answer.) At this point, the logic analyst will have to depend on his ingenuity and insight to determine which of the multiple answers (if any) is the one that will best simplify in accordance with the definitions used for the system being analyzed.

If more than five variables are contained in the expression, both the map method and the algebraic method become exceedingly complex. The expression should be separated into two or more smaller expressions; each of those expressions should be simplified; and the results should then be combined either algebraically or with maps to yield an overall simplification.

The logic analyst should also be on the lookout for expressions that do not have to be converted to logical sum of logical product form to provide a simplified expression. This type of expression will usually be encountered when working with multiple applications of DeMorgan's Law, as demonstrated in Sec. 5–4.

Devising a fixed set of rules to cover all logic analysis cases is out of the question. The general guidelines and examples shown in Chapters 5 and 6 merely prepare the reader for the more complex work to follow. Each analyst must choose the method that best suits his application needs. Hopefully, sufficient information has now been provided to enable the reader to make such a decision.

6–9 SUMMARY

Entering Information on Logic Maps*

1. Convert the Boolean algebra expression to AND–OR (logical sum of logical products) or minterm form.
2. Plot one term at a time, according to the following rules.
   (a) Determine the column(s) according to the column headings.
   (b) Determine the row(s) according to the row headings.
   (c) Place 1s in all squares covered by the intersection of the selected row(s) and column(s). When more than one term generates a 1 in the same square, a single 1 is sufficient. This is equivalent to the logic operation $1 + 1 = 1$.
3. The number of squares to be marked for any term is determined by the relationship $X = 1/2^n$, where X is the number of squares to be marked and n is the number of variables in the term. For example,

1-variable term	1/2 of the squares
2-variable term	1/4 of the squares
3-variable term	1/8 of the squares

Extracting Information from Logic Maps

1. Look for groups of squares that are adjacent to each other.
   (a) An adjacent square is one which differs from another square by only one variable.
   (b) Squares may be considered adjacent under the following conditions.
      (1) They are physically next to each other in the same row or column.
      (2) They are at opposite ends of rows or columns.
      (3) They occupy identical positions in adjacent maps.
2. Adjacent squares may be grouped in integral powers of 2, i.e., 1, 2, 4, 8, etc. Groups should generally have as many squares included together as possible, as long as they are in integral powers of 2. Circles or loops are usually drawn around the group for identification purposes.
3. The number of variables in each term can be determined by the relationship shown in the rules for entering information on logic maps.
4. A marked square may be used in as many loops as needed.

_____

* Figure 6–5 shows how to construct 2 through 5-variable maps.

5. Combine all of the extracted terms disjunctively (logic OR) to form the simplified expression.

6. When unmarked squares consist of only one term, those squares may be grouped and considered as the complement of the expression.

7. Logic maps can be used to simplify multioutput logic expressions where more than one output is developed from common inputs. The maps for each of the output expressions are compared. If the same squares are marked, the terms are common and can be used by all output expressions.

PROBLEM APPLICATIONS

6–1. What is the basic reason for performing logic simplification?
6–2. How does a Venn diagram differ from a Veitch diagram?
6–3. What is the major application of Venn diagrams?
6–4. Construct a Venn diagram representing $\overline{A}B$.
6–5. What was the original intent of the Veitch diagram?
6–6. Construct a Veitch diagram representing $\overline{A} + BC$.
6–7. How does a Veitch diagram differ from a Karnaugh map?
6–8. What was the original intent of the Karnaugh map?
6–9. Construct a Karnaugh map representing $\overline{A} + BC$.
6–10. What number system code is used to label divisions of a Karnaugh map? Discuss the characteristics of the code.
6–11. Define "sum of products" form as applied to Boolean expressions.
6–12. Define "minterm" form as applied to Boolean expressions.
6–13. Define "product of sums" form as applied to Boolean expressions.
6–14. Define "maxterm" form as applied to Boolean expressions.
6–15. What form must a Boolean expression possess to be entered on a logic map?

Graphically simplify the following Boolean expressions.

6–16. $B\overline{C} + ABC$
6–17. $AB + \overline{A}\overline{B} + B\overline{C} + \overline{B}C$
6–18. $ABC + \overline{A}BC + \overline{B}\cdot\overline{C}$
6–19. $\overline{A}C + B\overline{C} + \overline{B}\cdot\overline{C}$
6–20. $AB + \overline{A}\cdot\overline{B} + B\overline{C} + AC + \overline{B}\cdot\overline{C}$
6–21. $ABCD + ACD + AD$
6–22. $\overline{A}B\overline{C}D + \overline{A}BCD + AB\overline{C}D + ABCD$
6–23. $W\overline{Z} + W\overline{Y}Z + WXY + \overline{W}\cdot\overline{X}\cdot\overline{Y}\cdot Z$
6–24. $[W(X\overline{Z} + YZ) + (\overline{W + X})YZ][\overline{X + Y + W\overline{Z}} + \overline{Z}(\overline{X}Y + \overline{W}X) + \underline{WX\overline{Y}Z}]$
6–25. $E(AC + B\overline{C}) + \overline{B + C + D + E} + C(BE + A\overline{B} + \overline{A + B + D + E}) + \overline{\overline{A} + \overline{B}} + C$

7

Analysis and Troubleshooting of Combinational Logic

Analysis and troubleshooting of combinational logic circuits follows a relatively *standard sequence of events*. This chapter provides a proposed procedure, listing and explaining each preparatory and procedural step that leads to a complete understanding of combinational logic-circuit operation.

Troubleshooting is also organized into a procedural operation. Following discussion of the systems concept and troubleshooting philosophy, the flowchart concept is used to develop a step-by-step method of troubleshooting combinational logic circuits.

7–1 A PROPOSED ANALYSIS PROCEDURE

Preparatory Steps

Any investigation of combinational logic-circuit operation must begin with an overall description of the logic function(s) of the circuit being analyzed. The description may be in any of the commonly encountered forms: English-language statements, Boolean algebra, or waveforms. From a practical standpoint, either the Boolean algebra or the waveform description is the most desirable, since analysis is usually performed either algebraically or with waveforms. If English-language operational descriptions are provided, conversion to the Boolean algebra form *must* be performed.

A normal "fallout" of the logic description of a combinational logic circuit is *a definition of the inputs and outputs*. In other words, if the operation of a logic circuit has been described from an overall viewpoint, it will usually have been described in terms of outputs for various inputs. If such is not the case, all inputs should be provided with arbitrary names so that the rules of Boolean algebra can be applied during investigation. In any case, it should be possible with either the information at hand or that arbitrarily provided to describe all outputs in terms of inputs following logic-circuit analysis. Some logic circuits also operate with *timed inputs*, i.e., they are enabled to perform their function for a given period of time and then disabled until needed again. If such timing is used in the circuit being evaluated, it should be described either in terms of Boolean algebra or waveforms, depending on the method(s) of analysis being used.

Finally, any nonstandard logic symbols or operations being performed should be carefully defined.* Most manufacturers provide Boolean algebra descriptions of their logic packages; so as new logic operations become available, it will be necessary to apply only the provided descriptions to understand the operation of the logic element.

In summary: *certain basic steps must be performed prior to the start of any combinational logic-circuit investigation.* The logic description of the over- all function should be obtained, along with definitions of inputs and outputs if possible. Where inputs and outputs are not adequately defined, arbitrary naming is allowed. Any necessary timing should be established, even if it is necessary to once again arbitrarily assign values for the purpose of analysis. Finally, any nonstandard logic symbols must be defined in terms of MIL-STD- 806B symbols to facilitate understanding.

Procedural Steps

Combinational logic analysis usually begins at the *input portion of the logic circuit* and proceeds to the output(s). If the preparatory steps have been adequately performed, all inputs and timing will have been identified or synthesized, and investigation can begin with the selection of *one arbitrary path of signal flow* from input toward the output. The selected path should be followed until a level of logic is reached that requires other inputs. All of the tools for analysis that have been discussed in previous chapters should be applied where needed during the evaluation of the combinational logic circuit. Boolean algebra descriptions are commonly used in the case of simple combinational logic circuits, but other methods are equally applicable. Waveforms often prove invaluable by making the actual circuit operation more "visible." Truth tables and maps may also be used to help understand circuit operation.

When the level of logic is reached where additional inputs are needed, we must return to the beginning and develop the required inputs. This process con- tinues — new inputs being developed as required — until *all* inputs and outputs are accounted for. As might be imagined, some combinational logic-circuit analysis problems can become quite complex. It is therefore quite necessary to adequately document each step being performed to reduce errors caused by complex circuit configurations.

---

* Unless otherwise stated, it will be assumed that all symbols used in logic diagrams are from MIL-STD-806B. Nonstandard symbols are described in Appendix C, where they are correlated with MIL-STD-806B.

The complete procedure is now outlined.

A Proposed Analysis Procedure

    I. *Preparatory steps*
        A. Obtain description of overall logic function.
        B. Define or synthesize inputs and outputs.
        C. Establish or synthesize timing if necessary.
        D. Define nonstandard rules of logic.

   II. *Procedural steps*
        A. Perform preparatory steps.
        B. Select an arbitrary path from input to output.
        C. Using the tools of logic analysis, proceed from the input toward the outputs until a level of logic is reached that requires other inputs.
        D. Return to the beginning and develop those inputs.
        E. Continue, developing new inputs as required until all outputs are obtained.
        F. Document results of each step as developed.

7–2 A TYPICAL ANALYSIS

Our proposed procedure can be demonstrated by a typical problem. Figure 7–1a is the logic diagram of a device that is to be analyzed. Note that the diagram consists of standard logic symbols and is made up completely of NAND gates. The only other information available is that the logic circuit performs the full-adder function.

Preparatory Steps

With the logic diagram and its functional description available, the preparatory steps of the procedure may be accomplished. The knowledge that the logic circuit is a *full adder* is of considerable aid in developing the complete description of the overall logic function. A small amount of analysis has already been performed on the full adder, and advantage should be taken of the information gained in earlier chapters. Both the Boolean algebra description and the truth table form of the full adder are reproduced in Fig. 7–1b. Derivation of the truth table and Boolean expressions from the basic English-language description were previously discussed.

Waveform analysis is attempted following algebraic and operational analysis using the binary inputs.

Logic circuit inputs and outputs are defined by the Boolean algebra description. However it may not be immediately obvious which output is the SUM and

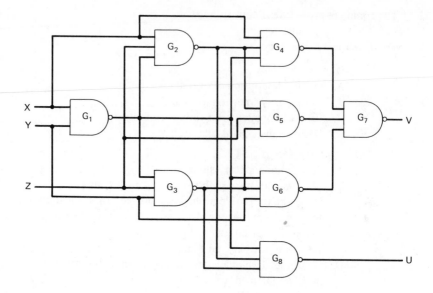

$$\text{Sum} = \overline{A} \cdot \overline{B} \cdot C + \overline{A} \cdot B \cdot \overline{C} + A \cdot \overline{B} \cdot \overline{C} + A \cdot B \cdot C$$

$$\text{Carry out} = \overline{A} \cdot B \cdot C + A \cdot \overline{B} \cdot C + A \cdot B \cdot \overline{C} + A \cdot B \cdot C = A \cdot C + A \cdot B + B \cdot C$$

A	B	C_i	S	C_o
0	0	0	0	0
0	0	1	1	0
0	1	0	1	0
0	1	1	0	1
1	0	0	1	0
1	0	1	0	1
1	1	0	0	1
1	1	1	1	1

Figure 7—1   Boolean algebra description and truth table
(full-adder)

which is the CARRY OUT. Additionally, the proper labels to be applied to each
input may be a mystery. Temporarily, the inputs are arbitrarily named X, Y,
and Z, and the outputs are named U and V. Upon completion of the procedural
steps, the proper assignment of variables will be apparent. No timing requirements are indicated on the logic diagram, since the required number of inputs for
a full adder (3) are the only inputs shown. Therefore it is not necessary to establish or synthesize circuit timing.

   Since standard logic symbology prevails in this logic diagram, no additional
rules of logic are required. The logic circuit is now ready for the procedural
steps.

Procedural Steps

Starting at the left side of the diagram (where the inputs are *usually* located), the first gate to be encountered is G_1. The inputs to G_1 are X and Y, and since G_1 is a NAND gate, the output is $\overline{XY}$. The output of G_1 is routed to both G_2 and G_3, so a purely arbitrary direction of analysis is assumed. If the output U is to be developed first, then G_2 may be examined next. With inputs X, Z, and $\overline{XY}$ applied, the output of NAND gate G_2 is $\overline{(X)(Z)(\overline{XY})}$. G_8 is the next gate encountered. Two inputs for G_8 have already been determined (from G_1 and G_2), so it is necessary to determine only the third input, furnished by G_3.

A level of logic has now been reached that requires returning to a lower (previous) level to obtain other inputs. Evaluation of G_3 shows that the three required inputs are Y, Z, and the output of G_1, which is $\overline{XY}$. G_3's output, then, is $\overline{(Y)(Z)(\overline{XY})}$. The three inputs to G_8 are therefore $\overline{XY}$, $\overline{(X)(Z)(\overline{XY})}$, and $\overline{(Y)(Z)(\overline{XY})}$. Combining these three inputs in a NAND operation results in $\overline{[\overline{XY}]\ [\overline{(X)(Z)(\overline{XY})}]\ [\overline{(Y)(Z)(\overline{XY})}]}$, which is the output of G_8 (labeled U in the logic diagram). But the expression for U does not look anything like *either* of the full-adder outputs shown in Fig. 7–1. The time has come to apply some of the tools of logic analysis to determine which of the full-adder outputs is being represented by U.

The expression for the output of G_8 is by no means in its simplest form. In fact, without considerable experience, the real meaning of the expression may not be apparent. To determine which (if either) of the full-adder outputs is represented by the output of G_8, we must convert the NAND form to another form, perhaps the AND–OR (logical sum of logical products) form.

One of the first things that must be done when converting from the NAND to the AND–OR form is to remove the *vincula* (using DeMorgan's Law) *from all but single variable terms.* Expansion to the AND–OR form is then accomplished. *One* of the many methods of algebraic simplification is shown here.

$$
\begin{aligned}
\overline{[\overline{XY}][\overline{(X)(Z)(\overline{XY})}][\overline{(Y)(Z)(\overline{XY})}]} &= \overline{\overline{[\overline{XY}]}} + \overline{\overline{[(X)(Z)(\overline{XY})]}} + \overline{\overline{[(Y)(Z)(\overline{XY})]}} \\
&= XY + [(X)(Z)(\overline{XY})] + [(Y)(Z)(\overline{XY})] \\
&= XY + [(X)(Z)(\overline{X}+\overline{Y})] + [(Y)(Z)(\overline{X}+\overline{Y})] \\
&= XY + XZ\overline{X} + XZ\overline{Y} + YZ\overline{X} + YZ\overline{Y} \\
&= XY + X\overline{Y}Z + \overline{X}YZ
\end{aligned}
$$

The expression is now in the AND–OR form, but it still doesn't seem to resemble either of the full-adder outputs. Further algebraic simplification could be accomplished, but this is a good place to show a practical application of logic maps. The three terms of the simplified expression are shown entered and grouped in the 3-variable map in Fig. 7–2. Evaluation of the three loops results in the expression $XY + YZ + XZ$, which is the *exact* equivalent of the CARRY OUT output of the full adder. Thus the orderly application of the proposed procedure and use of the common tools available for logic-circuit analysis has resulted in identification of the function of a circuit that showed no resemblance

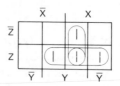

Figure 7–2   Map simplification of $XY + X\overline{Y}Z +$
$\overline{X}YZ = XY + YZ + XZ$

to previously encountered circuits. Documentation should include the *outputs of a all gates encountered* in addition to the simplification effort.

Having gained the knowledge that U is the CARRY OUT output removes some of the mystery about the logic circuit. Output V must be the SUM output, so all that remains is to check whether the logic circuit provides the correct output. If the Boolean algebra expression for the logic circuit matches the Boolean expression for the full adder function, then the complete circuit will have been analyzed.

Three inputs are furnished to the output gate (G_7): G_4 provides one input; G_5 provides the second input; and G_6 furnishes the final input. Examining G_4 first shows that the three inputs are X, $[(X)(Z)(\overline{XY})]$, and $\overline{XY}$. Combining these inputs in the NAND gate results in $[X][(X)(Z)(\overline{XY})][\overline{XY}]$ as the output expression for G_4. G_5 and G_6 are approached similarly, providing a G_5 output of $[Z][(X)(Z)(\overline{XY})][(Y)(Z)(\overline{XY})]$ and a G_6 output of $[Y][(Y)(Z)(\overline{XY})][XY]$. The result of combining the G_4, G_5, and G_6 outputs in G_7 is the somewhat awesome expression

$$\overline{\overline{\{[X][\overline{(X)(Z)(\overline{XY})}][\overline{XY}]\}}\,\overline{\{[Z][\overline{(X)(Z)(\overline{XY})}][\overline{(Y)(Z)(\overline{XY})}]\}}\,\overline{\{[Y][\overline{(Y)(Z)(\overline{XY})}][\overline{XY}]\}}}$$

that once again in no way resembles what is known to be the SUM output of a full adder! Both algebraic and map methods are used to determine the equivalent logic expression that will be compared with the full-adder expression.

Since any expression that is in simplified form has all vincula removed except those over single variables, we may appropriately apply DeMorgan's Law until this criterion is met. The first step splits the full-length vinculum at major divisions in the G_7 output expression, resulting in

$$\overline{\overline{\{[X][\overline{(X)(Z)(\overline{XY})}][\overline{XY}]\}}} + \overline{\overline{\{[Z][\overline{(X)(Z)(\overline{XY})}][\overline{(Y)(Z)(\overline{XY})}]\}}} + \overline{\overline{\{[Y][\overline{(Y)(Z)(\overline{XY})}][\overline{XY}]\}}}.$$

Note that the expression is now composed of three separate subexpressions that are ORed together — each subexpression topped by double vincula. It is known that double negation (defined by the double vincula) is equivalent to the original expression, so all double vincula are removed. The expression simplifies to

$$\{[X][\overline{(X)(Z)(\overline{XY})}][\overline{XY}]\} + \{[Z][\overline{(X)(Z)(\overline{XY})}][\overline{(Y)(Z)(\overline{XY})}]\} + \{[Y][\overline{(Y)(Z)(\overline{XY})}][\overline{XY}]\}.$$

Due to the complexity of the total expression, each of the three terms is simplified separately, and the simplified versions are recombined to form the new expression.

The left subexpression is attacked first. DeMorgan's Law removes the vincula from the center and last factor, resulting in simplification to $[[X][\overline{X} + \overline{Z} + (\overline{XY})][\overline{X} + \overline{Y}]$. Of course, the double vincula over the term XY may be removed to reach our first goal of having no vincula extending over more than a single variable. Perhaps, however, rearrangement of this expression and regrouping to $[X][\overline{X} + \overline{Y}][\overline{X} + \overline{Z} + XY]$ will make it easier to visualize the expansion steps. The first two factors (X and $\overline{X} + \overline{Y}$) are expanded to $X\overline{X} + X\overline{Y}$ and are then further expanded using the $[\overline{X} + \overline{Z} + XY]$ factor. $X\overline{X}$ cancels, and combining $X\overline{Y}$ with $\overline{X} + \overline{Z} + XY$ makes the AND–OR form ($X\overline{Y}\,\overline{X} + X\overline{Y}\,\overline{Z} + X\overline{Y}XY$) available for any further simplification.

Examination of the AND–OR form shows that, in this case, algebraic simplification is relatively simple. The first term is rewritten in the form $\overline{X}X\overline{Y}$. $\overline{X}X = 0$, simplifying the term to $0 \cdot \overline{Y}$. The theorems of Boolean algebra state that conjunctively combining any variable with 0 results in a logic value of 0. Therefore the first term simplifies to 0. The last term in the AND–OR form of the expression is $X\overline{Y}XY$. This term is rearranged to the form $XXY\overline{Y}$, which may be easier to understand. $XX = X$, of course, and $Y\overline{Y} = 0$. $X \cdot 0 = 0$, so the last term in the AND–OR form is 0. Now the AND–OR form of the expression is $0 + X\overline{Y}\,\overline{Z} + 0$ or, simply, $X\overline{Y}\,\overline{Z}$. Quite a reduction in complexity from the original $[X]\,\overline{[(X)(Z)(\overline{XY})]}\,[\overline{XY}]$ isn't it?

Close persual of the right-hand subexpression in the original expression reveals that it is of the same form as the left-hand subexpression. It will also simplify to a single 3-variable term ($\overline{X}Y\overline{Z}$).

The center subexpression does not lend itself quite as well to algebraic manipulation. Two of the three terms are complex, and expansion can become troublesome. Following removal of the vincula by DeMorgan's Law, the center subexpression is further simplified by the use of maps. The map method is not an absolute requirement in this case, but it does serve as an excellent example of map ANDing. The center term ($[Z][\overline{(X)(Z)(\overline{XY})}][\overline{(Y)(Z)(\overline{XY})}]$) becomes $[Z][\overline{X} + \overline{Z} + (XY)][\overline{Y} + \overline{Z} + (XY)]$, after application of DeMorgan's Law and the double negation theorem. Each of the three terms is mapped separately, and the maps are ANDed as described in Ch. 6. Z is shown mapped in Fig. 7–3a; $\overline{X} + \overline{Z} + XY$ in Fig. 7–3b; and $\overline{Y} + \overline{Z} + XY$ in Fig. 7–3c. Superimposing the maps reveals that only two common squares are marked on all three maps. These two squares, which are identified as $\overline{X}\overline{Y}Z$ and XYZ, represent the ANDing of the three terms in the center subexpression (Fig. 7–3d). Therefore

$$[Z]\,\overline{[(X)(Z)(\overline{XY})]}\ \overline{[(Y)(Z)(\overline{XY})]} = \overline{X}\overline{Y}Z + XYZ.$$

All of the subexpressions have now been examined, and simplified forms have been developed. Substitution of the simplified forms for the original sub-

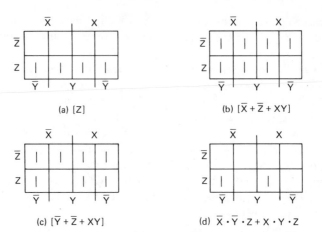

Figure 7—3   Logic maps for $[Z][\overline{X} + \overline{Z} + XY][\overline{Y} + \overline{Z} + XY]$

expressions results in $X\overline{Y}\overline{Z} + \overline{X}\overline{Y}Z + XYZ + \overline{X}Y\overline{Z}$ as the equivalent expression for the output of G_7. Comparison of this expression with the SUM output expression for the full adder in Fig. 7—1 shows that the two expressions are the same *except* for the arrangement of terms. Both the SUM and the CARRY OUT outputs of the NAND gate full adder have been shown to be equivalent to the outputs developed from the standard truth table for the full adder. Thus the circuit being studied is indeed capable of performing the functions of a full adder, and except for documentation of the results, the analysis is complete. Figure 7—4 shows typical documentation of a logic-analysis problem using the full-adder circuit and methods of analysis just completed.

Another possible method of investigating a logic circuit, such as the full-adder, is to try all possible combinations of the inputs and then record the output for each of the conditions tried. This has effectively been done by constructing the truth table, but it may be enlightening to actually try the eight combinations that can exist for the 3-input full-adder and determine if the truth table is really a valid presentation of full-adder operation. The use of a logic-signal source to furnish the input signals is one way to implement such a test. However the logic levels defined by the truth table are assumed to be present, and the logic levels are followed throughout the circuit to determine the effect of each combination. Certain input combinations will produce an almost obvious output. Other combinations will require tracing logic levels through the complete circuit to establish output conditions.

Prior to beginning this investigation, therefore, a review of the operation of the logic element concerned would be helpful. Remember that the NAND gate functions as an AND gate followed by an inverter. Thus when all inputs to the NAND gate are HIGH, the output is LOW. Another viewpoint is that if any input is LOW, the output is HIGH. Both viewpoints will prove valuable while tracing logic levels through the full-adder. It should also be recalled that logic 1

FULL ADDER ANALYSIS

1. Logic diagram (see Figure 7–1a)
2. Gate output equations:

$$G_1 = \overline{X}\,\overline{Y}$$
$$G_2 = \overline{(X)\,(Z)\,(\overline{X\,Y})}$$
$$G_3 = \overline{(Y)\,(Z)\,(\overline{X\,Y})}$$
$$G_4 = \overline{[X]\,[\overline{(X)\,(Z)\,(\overline{X\,Y})}]\,[\overline{X\,Y}]}$$
$$G_5 = \overline{[Z]\,[\overline{(X)\,(Z)\,(\overline{X\,Y})}]\,[\overline{(Y)\,(Z)\,(\overline{X\,Y})}]}$$
$$G_6 = \overline{[Y]\,[\overline{(Y)\,(Z)\,(\overline{X\,Y})}]\,[\overline{X\,Y}]}$$
$$G_7 = \overline{\left\{[X]\,[\overline{(X)\,(Z)\,(\overline{X\,Y})}]\,[\overline{X\,Y}]\right\}\,\left\{[Z]\,[\overline{(X)\,(Z)\,(\overline{X\,Y})}]\,[\overline{(Y)\,(Z)\,(\overline{X\,Y})}]\right\}\,\left\{[Y]\,[\overline{(Y)\,(Z)\,(\overline{X\,Y})}]\,[\overline{X\,Y}]\right\}}$$
$$G_8 = \overline{[\overline{X\,Y}]\,[\overline{(X)\,(Z)\,(\overline{X\,Y})}]\,[\overline{(Y)\,(Z)\,(\overline{X\,Y})}]}$$

3. Algebraic simplification:

$$G_8 = \overline{[\overline{X\,Y}]\,[\overline{(X)\,(Z)\,(\overline{X\,Y})}]\,[\overline{(Y)\,(Z)\,(\overline{X\,Y})}]}$$
$$= [\overline{\overline{X\,Y}}] + [\overline{\overline{(X)\,(Z)\,(\overline{X\,Y})}}] + [\overline{\overline{(Y)\,(Z)\,(\overline{X\,Y})}}]$$
$$= [X\,Y] + [(X)\,(Z)\,(\overline{X\,Y})] + [(Y)\,(Z)\,(\overline{X\,Y})]$$
$$= [X\,Y] + [(XZ)\,(\overline{X}+\overline{Y})] + [(YZ)\,(\overline{X}+\overline{Y})]$$
$$= [X\,Y] + [X\overline{X}Z + X\overline{Y}Z] + [\overline{X}YZ + Y\overline{Y}Z]$$
$$= XY + X\overline{X}Z + X\overline{Y}Z + \overline{X}YZ + Y\overline{Y}Z$$
$$= XY + X\overline{Y}Z + \overline{X}YZ$$

$$G_7 = \overline{\left\{[X]\,[\overline{(X)\,(Z)\,(\overline{X\,Y})}]\,[\overline{X\,Y}]\right\}\,\left\{[Z]\,[\overline{(X)\,(Z)\,(\overline{X\,Y})}]\,[\overline{(Y)\,(Z)\,(\overline{X\,Y})}]\right\}\,\left\{[Y]\,[\overline{(Y)\,(Z)\,(\overline{X\,Y})}]\,[\overline{X\,Y}]\right\}}$$

$$\overline{\left\{[X]\,[\overline{(X)\,(Z)\,(\overline{X\,Y})}]\,[\overline{X\,Y}]\right\}} + \overline{\left\{[Z]\,[\overline{(X)\,(Z)\,(\overline{X\,Y})}]\,[\overline{(Y)\,(Z)\,(\overline{X\,Y})}]\right\}} + \overline{\left\{[Y]\,[\overline{(Y)\,(Z)\,(\overline{X\,Y})}]\,[\overline{X\,Y}]\right\}}$$

$$\left\{[X]\,[\overline{(X)\,(Z)\,(\overline{X\,Y})}]\,[\overline{X\,Y}]\right\} + \left\{[Z]\,[\overline{(X)\,(Z)\,(\overline{X\,Y})}]\,[\overline{(Y)\,(Z)\,(\overline{X\,Y})}]\right\} + \left\{[Y]\,[\overline{(Y)\,(Z)\,(\overline{X\,Y})}]\,[\overline{X\,Y}]\right\}$$

$$\left\{[X]\,[\overline{X}+\overline{Z}+\overline{X\,Y}]\,[\overline{X\,Y}]\right\} + \left\{[Z]\,[\overline{X}+\overline{Z}+XY]\,[\overline{Y}+\overline{Z}+\overline{X\,Y}]\right\} + \left\{[Y]\,[\overline{Y}+\overline{Z}+\overline{X\,Y}]\,[\overline{X\,Y}]\right\}$$
$$\left\{[X]\,[\overline{X}+\overline{Z}+XY]\,[\overline{X\,Y}]\right\} + \left\{[Z]\,[\overline{X}+\overline{Z}+XY]\,[\overline{Y}+\overline{Z}+XY]\right\} + \left\{[Y]\,[\overline{X}+\overline{Y}]\,[\overline{Y}+\overline{Z}+XY]\right\}$$
$$\left\{[X]\,[\overline{X}+\overline{Y}]\,[\overline{X}+\overline{Z}+XY]\right\} + \left\{[\overline{X}Z+Z\overline{Z}+XYZ]\,[\overline{Y}+\overline{Z}+XY]\right\} + \left\{[Y]\,[\overline{X}+\overline{Y}]\,[\overline{Y}+\overline{Z}+XY]\right\}$$
$$\left\{[X\overline{X}+X\overline{Y}]\,[\overline{X}+\overline{Z}+XY]\right\} + \left\{[\overline{X}Z+XYZ]\,[\overline{Y}+\overline{Z}+XY]\right\} + \left\{[\overline{X}Y+Y\overline{Y}]\,[\overline{Y}+\overline{Z}+XY]\right\}$$
$$\left\{[X\overline{Y}]\,[\overline{X}+\overline{Z}+XY]\right\} + \left\{\overline{X}Y\overline{Z} + \overline{X}Z\overline{Z} + \overline{X}ZXY + XY\overline{Y}Z + XYZ\overline{Z} + XYZXY\right\} + \left\{[\overline{X}Y]\,[\overline{Y}+\overline{Z}+XY]\right\}$$
$$\left\{[X\overline{X}Y + X\overline{Y}\,\overline{Z} + X\overline{Y}XY]\right\} + \left\{[\overline{X}Y\overline{Z} + XYZ]\right\} + [\overline{X}Y\overline{Y} + \overline{X}Y\overline{Z} + \overline{X}YXY]$$
$$X\overline{Y}\overline{Z} \qquad + \qquad \overline{X}Y\overline{Z} + XYZ \quad + \quad \overline{X}Y\overline{Z}$$

$$G_7 = X\overline{Y}\,\overline{Z} + \overline{X}\,\overline{Y}Z + XYZ + \overline{X}Y\overline{Z}$$

4. Map Simplification

For the G_7 output there is no need to map the first and last terms.
Simplifications are immediately apparent as soon as the logical sum of
logical products form is obtained.
The G_7 middle term is mapped in Figure 7–3.
Figure 7–2 maps the G_8 expression.

Figure 7–4   Documentation of a logic circuit analysis

is often used in place of the term HIGH, and logic 0 is used in place of the term LOW. In a device such as the full-adder, it is advantageous to use logic 1s and 0s in place of HIGHs and LOWs, due to the correlation with actual binary numbers that are processed in the adder. So 1s and 0s are used here, although the reader may substitute HIGHs and LOWs if it makes the circuit more understandable.

As with any task, it is best to organize and plan what is to be done prior to actually starting. So much is already known about the full adder that little preparation (other than determining how to record the data) is necessary. Since the results of this analysis will be closely allied with the truth table for the full-adder function, we can record data in tabular form. Therefore the usual physical layout of the truth table is used, and the same sequence of investigating inputs is employed. Instead of providing output columns for only the SUM and CARRY OUT outputs, *each* gate is shown. As each combination of inputs is investigated, it is necessary to place only the proper 1 or 0 in the column for each gate. When the analysis is completed, the table contains the output of each gate for all possible combinations of the three input variables. The completed truth table for "1 and 0" analysis of the full adder is shown in Table 7–1.

Just as with the algebraic/graphic investigation, the starting point is the inputs, and each gate is discussed as encountered. By examining the circuit and from our previous analysis, we see that both outputs require signals from G_1, G_2, and G_3. Although it does not result in *immediate* insight into the circuit operation, this slight deviation from the procedure of analyzing gates that are not in a specific path from input to output will speed the analysis process. G_1 is examined first – not only because it is the first gate with inputs applied but also because it is probably the simplest to understand. Only two inputs, X and Y, are provided. By definition, only when all inputs are 1s (HIGH) will the output be 0 (LOW). When either input is 0, the output is 1. Using the X and Y input columns in the table, we see that both X and Y are 1 at the same time *only* in rows 3 and 7 – all other combinations of X and Y have at least one 0 present. The output of G_1 is 0 at rows 3 and 7; all other rows result in 1 outputs.

G_2 is studied next: it is a 3-input NAND gate, and it requires X to be 1, Z to be 1, and the output of G_1 to be 1 *simultaneously* to be activated. Z is 0 in rows 0 through 3, so obviously the output of G_2 will be 1 for at least these four combinations – no matter what the other variables are. G_1 is 0 at row 7, so G_2

Table 7–1   Complete truth table for full adder of Figure 7–1a

Row	Z	Y	X	G_1	G_2	G_3	G_8	G_4	G_5	G_6	G_7
0	0	0	0	1	1	1	0	1	1	1	0
1	0	0	1	1	1	1	0	0	1	1	1
2	0	1	0	1	1	1	0	1	1	0	1
3	0	1	1	0	1	1	1	1	1	1	0
4	1	0	0	1	1	1	0	1	0	1	1
5	1	0	1	1	0	1	1	1	1	1	0
6	1	1	0	1	1	0	1	1	1	1	0
7	1	1	1	0	1	1	1	1	0	1	1

is marked 1 for that row. Rows 4, 5, and 6 remain to be checked. X is 0 in rows 4 and 6, leaving only row 5. Careful perusal of row 5 shows that X, Z, and G_1 outputs are all 1 in this case. The output of G_2 will then be 0, which is desired.

Analysis of G_3 proceeds in a similar manner. It also is a 3-input NAND gate requiring the output of G_1 to be 1, Y to be 1, and Z to be 1 simultaneously to activate the gate. Rows 0 through 3 provide a 0 input from Z with the resultant 1 output from G_3. Y is 0 in rows 4 and 5, while still keeping the output at 1. In row 6, Y is 1, Z is 1, and according to the truth table, the output of G_1 is also 1. G_3 is activated under these circumstances, and its output becomes 0. Row 7 shows that the output of G_1 is 0, so the G_3 output becomes 1 again. All three gates common to both the SUM and the CARRY OUT outputs have now been checked, and their logic values determined for all possible combinations of inputs. G_1 through G_8 may now be evaluated in a relatively simple manner without numerous trips back to develop new inputs.

The CARRY OUT output from G_8 is checked now that G_1, G_2, and G_3 outputs have been determined. Evaluation of the G_1, G_2, and G_3 columns (one row at a time) allows selection of the proper entry in the G_8 column. Perhaps an easier method is to look for 0s in either the G_1, G_2, or G_3 columns for each row. Any 0 input to a NAND gate results in a 1 output, and the 0s tend to "stand out" from the 1s, due to their relative scarcity in the table. Just locate a 0 in any of the three columns representing inputs to G_8, and the row that contains the 0 has a 1 placed in the G_8 column. This effectively recognizes that DeMorgan's Law works, since $\overline{ABC} = \overline{A} + \overline{B} + \overline{C}$. After all eight rows have been examined and the G_8 outputs recorded, the full-adder truth table of Fig. 7—1b should be consulted and the CARRY OUT column compared with the G_8 column. One-to-one correspondence between the two tables should be observed, indicating once again that the NAND version of the full adder performs the same CARRY OUT function as other versions previously used.

The SUM output is developed from G_7. With inputs supplied from G_4, G_5, and G_6, we must now develop the outputs of each of these gates. There is no particular advantage to selecting a specific gate to be examined, so G_4 is arbitrarily evaluated first. G_4's inputs are X, G_1, and G_2. (Remember, any 0 input results in a 1 output.) Rows 0, 2, 4, and 6 contain 0s in the X column, so the corresponding locations in the G_4 column contain 1s. The output of G_1 is 0 in rows 3 and 7, and the output of G_2 is 0 in row 5. Ones are entered in the same locations in the G_4 column. Row 1 is the only location left without an entry, and evaluation of the X, G_1, and G_2 columns indicates that they all have a value of 1. Therefore the row 1 location in column G_4 has a 0 inserted.

G_6 is approached in a similar manner. In this case, Y is combined with G_1 and G_3. The result is but one case of the eight possible input combinations where the output is 0 (row 2). All other combinations of inputs to G_6 produce a 1 output, due to at least one 0 input for each row on the table. G_5 combines the G_2 and G_3 outputs with Z, and a different form of output appears. Investigation of the columns for the three inputs reveals that at least one 0 exists for all rows *except* rows 4 and 7. Ones appear then in all of the rows that provide at least *one* 0 input, and 0s appear in rows 4 and 7 because *all* inputs are 1 in these two rows.

Final evaluation of the full-adder circuit is now in sight. The SUM output is formed by combining the G_4, G_5, and G_6 outputs in G_7. Using the three input columns, once again scan from row 0 to row 7, placing a 1 in the G_7 column for any row that has at least one 0 in the input columns. Rows 1, 2, 4, and 7 have at least one 0 in the G_4, G_5, or G_6 column. The other four rows have 1s as inputs, and 0s are placed in the G_7 column. Comparison of the G_7 column with the SUM output column in the truth table for the full adder (Fig. 7-1b) once again shows a one-to-one correspondence. Hence the NAND gate full adder has been proven functionally identical to other versions of the full adder.

If truth table analysis has been used, Table 7-1 should be included in the documentation of the logic-diagram evaluation.

Still another approach using 1s and 0s can be shown. This method is not as rigorous in its requirements to *record the outputs of all gates*. The basic concept that any 0 input to a NAND gate results in a 1 output is used to trace data flow through the NAND gate full adder. Each of the input combinations is examined, and based on the value of the inputs, a decision about the value of the outputs is made *directly* — without recording the intermediate values.

For example, the first combination to be evaluated is row 0 on the table: $X = 0$, $Y = 0$, and $Z = 0$. (Remember, any 0 input results in a 1 output.) Investigating X first reveals that it is fed directly to G_1, G_2, and G_4. Regardless of any other inputs, G_1, G_2, and G_4 will have 1 outputs. Thus the input from G_2 to G_8 is 1, and the input from G_4 to G_7 is 1. Now we must determine the other inputs to G_8 and G_7 to decide what outputs are present. The Y input is 0 at the same time the X input is 0, so G_3 and G_6 have their outputs forced to 1. Now two inputs to both G_8 and G_7 are 1, so we have to determine only the last input to completely define the SUM and CARRY OUT outputs of the full adder for the first combination of inputs. In fact, the third input to G_8 (the output of G_1) has already been defined as 1, due to either the X or Y input. Therefore G_8 has each of its three inputs defined as 1, and its output is 0. Thus when $X = 0$, $Y = 0$, and $Z = 0$, the CARRY OUT output of the NAND gate full adder is 0.

The third input to G_7 is supplied by G_5. Actually, two of the three inputs to G_5 are 1s because they come from G_2 and G_3, respectively. So the third input to G_5 determines its output. Working back toward the input, it can be seen that Z is the third input to G_5, and this input was originally defined as 0. The output of G_5 is forced to 1, providing the third 1 input to G_7 and causing its output to become 0. Just as with the CARRY OUT outputs, when $X = 0$, $Y = 0$, and $Z = 0$, the SUM output is 0.* These results correspond to the similar entry in the truth table for the full adder.

---

* If difficulty is encountered keeping the 1s and 0s in mind, writing them down may be of assistance. Placing a sheet of clear plastic over the diagram being evaluated and marking with a crayon or grease pencil provides a means of recalling previously determined values. When the problem is finished, the markings may be wiped from the plastic, and the next problem attacked in the same manner.

With $X = 1$, $Y = 0$, and $Z = 0$ (row 1), different results can be expected. G_1, G_3, and G_6 have an output of 1, due to $Y = 0$. Additionally, G_2 and G_5 have an output of 1, due to $Z = 0$. All of the inputs for G_8 have been derived (G_1, G_2, and G_3), and they are all 1s. The G_8 output is forced to 0 to present no CARRY operation when binary 1 is added to binary 0. Only the G_4 output to feed G_7 has not been ascertained at this time. Investigation shows that G_4 requires X, G_2 output, and G_1 output. All of these signals are 1, forcing the output of G_4 to 0. Now G_7 has one 0 input, and its output becomes 1. Thus when binary 1 is added to binary 0, the SUM is binary 1.

A little additional effort allows some very useful *general statements* to be made. They are listed here and may be used to complete the remaining steps of the analysis.

    1. When $X = 0$, G_1, G_2, and G_4 outputs are 1.
    2. When $Y = 0$, G_1, G_3, and G_6 outputs are 1.
    3. When $Z = 0$, G_2, G_3, and G_5 outputs are 1.

Consider row 2, where $X = 0$, $Y = 1$, and $Z = 0$. The general statements just listed allow establishment of the G_1, G_2, G_3, G_4, and G_5 outputs at the 1 level. Since G_1, G_2, and G_3 supply the inputs to G_8, the CARRY OUT output is 0. Two of the three inputs to G_7 are already described (G_4 and G_5), and all that is necessary is to find the G_6 output to determine the G_7 output. The G_1 and G_3 inputs to G_6 are both 1, leaving the final decision to the value of Y. Since $Y = 1$, all of the inputs of G_6 are 1, resulting in a 0 output. G_7 now has two 1s and one 0 as inputs, and the output is forced to the 1 condition. Thus when $X = 0$, $Y = 1$, and $Z = 0$, the SUM output of the full adder is 1 and the CARRY OUT output is 0. The remaining five rows in the table are left to the reader as an exercise.

Waveform analysis of the full adder may also be used to demonstrate that the logic diagram of Fig. 7—1a actually performs the required operations. The eight different input-signal combinations required for investigation are developed from the logic-signal source used in earlier chapters. Recall that the logic-signal source is made up of *interconnected flip-flops* (FFs), arranged so that the outputs of the individual FFs furnish input signals for logic circuits to be tested. The relationship of the FF-output signals is such that with two FFs all four possible combinations required are generated, with three FFs all eight possible combinations required are generated, and so on. The waveform outputs (X, Y, and Z) of a 3-stage logic-signal source are shown at the top of Fig. 7—5. Recognition of the characteristics of the logic gates used in the full adder allows investigation of the output of each gate under varying input conditions. Waveform analysis of the full adder of Fig. 7—1b is shown in Fig. 7—5. Note that the H and L markings are not provided in Fig. 7—5. (The reader should have sufficient experience with waveforms to be able to recognize the HIGHs from the LOWs, if all furnished problems and examples have been investigated.) The reader should examine each input condition and satisfy himself that the outputs are shown correctly.

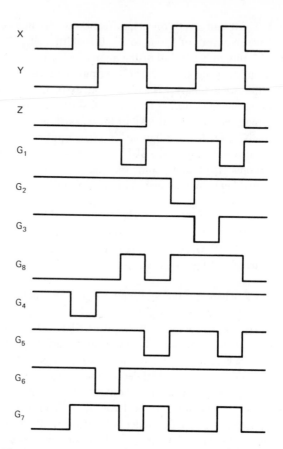

Figure 7—5   Waveform analysis of a full-adder

7—3 TROUBLESHOOTING COMBINATIONAL LOGIC

The Systems Concept

Few digital systems are strictly combinational in nature. In subsequent chapters, the sequential aspects of digital logic will be considered, and combinational and sequential logic will be combined to form complete systems. The *complete system* (see Fig. 7—6) occupies the highest level of the hierarchy. A number of *subsystems* make up the next level. Each one of these subsystems is usually identifiable by its specific functions, and often, it is physically packaged separately. These subsystems are, in turn, made up of *assemblies* that contain a number of *subassemblies*, which are usually removable and capable of being

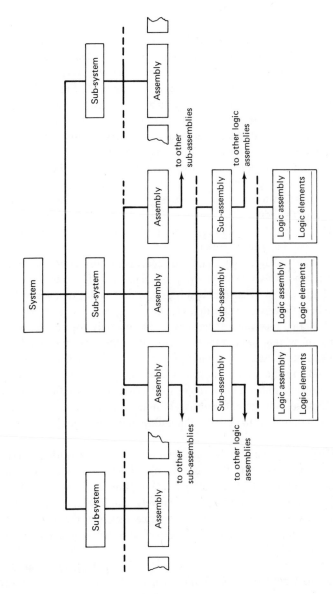

Figure 7–6   The systems concept

tested as a single unit. The subassemblies consist of *logic assemblies* (such as multifunction integrated-circuit packages), *logic elements* (such as gates), and perhaps even *discrete components* (such as transistors, diodes, resistors, and capacitors).

A modern electronic *telemetering system*, although not completely digital in nature, is a good example of the systems concept. *Telemetering systems acquire data at relatively inaccessible locations, then process, transmit, recover, and display the data at the location where it is to be used.* Figure 7—7 applies the systems concept to a typical telemetering system. Six subsystems are combined to form the complete telemetering system: the *transducer subsystem* (1) senses and converts physical and electrical parameters into suitable electrical quantities for application to the *multiplexer-programmer subsystem* (2), which electrically times and encodes the sensed data from the transducer subsystem into a form that may be transmitted; the *transmitter subsystem* (3) conveys the encoded information to the *receiver subsystem* (4), which acquires and processes the data; within the *demultiplexing-decoding subsystem* (5), the originally encoded data is separated and decoded so that the information originally sensed at the source may be visually presented by the *display subsystem* (6).

Each of the subsytems is identified according to *function*, and, in the case of the telemetering system, is *physically* separated. The transducer, multi-plexer-programmer, and transmitter subsystems may be located within a space vehicle, while the receiver, demultiplexing-decoding, and display subsystems are at widely separated locations on earth. Considering only the multiplexer-programmer subsystem shows that it is made of a number of individual assemblies, such as a timer assembly, a multiplexer assembly, an analog-to-digital converter, and a memory. The multiplexer assembly is composed of one digital multiplexer subassembly and one analog multiplexer subassembly. Finally, an analog multiplexer subassembly consists of numerous integrated-circuit gate packages interconnected to perform the multiplexing of a number of analog signals.

The logic technician can become involved in troubleshooting at any level from the complete system to the logic-assembly level. Working at the lowest level of complexity (the logic-assembly level) requires basic knowledge of individual logic functions combined in relatively simple relationships. As the level of complexity increases toward the complete system, the relationships between logic functions become more abstract, and the technician must become more and more skillful.

The methods discussed in this chapter are aimed at the isolation of mal-functions at the *logic-element level of a subassembly.* It is at this level that combinational logic is most likely to be encountered. As sequential logic is covered in subsequent chapters, additional methods will be developed so that the overall concepts of system troubleshooting will evolve.

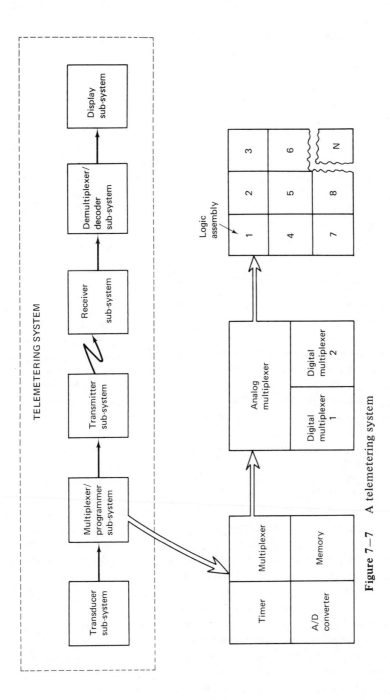

Figure 7-7 A telemetering system

Troubleshooting Philosophy and Concepts

The concepts of troubleshooting combinational logic are little different than the concepts of troubleshooting any type of electronic equipment. Certainly, different troubleshooting equipment is employed, and slightly different viewpoints on interpretation of indications exist, but if the reader is currently skilled in the techniques of troubleshooting, he will find little new in this section other than a means of *organizing the task.*

Organization of the task is not only the first but probably the most important of all the concepts associated with logic-circuit troubleshooting. If a traveler desires to go from point A to point B, he must first know the location of point A! Then he must determine where point B is and how to get there. Effectively, the same situation exists in troubleshooting logic circuits — the troubleshooter must know from where he is starting, and he must also know where he is going and how to get there. He must organize his task so that he may efficiently determine what information he has available to him, what additional information is needed, how to get that additional information, and how to analyze the logic circuit to determine the point(s) of malfunction. The method of organization selected must include all of these points, and it must provide a logical plan to accomplish the troubleshooting.

Preliminary to any troubleshooting activity must be the identification of the fact that a malfunction *really* exists. In many cases, an apparent malfunction turns out to be "pilot error": that is, improper use or lack of knowledge of what the equipment should really do. Pertinent questions should be asked: what did it do? what didn't it do? under what circumstances did the apparent malfunction occur? did it smoke? did it blow fuses? what does the operator think is wrong? These types of questions will not only help pinpoint the malfunction but will also reveal potential cases of improper use, perhaps saving many hours of troubleshooting time.

If an *instruction manual* is available, consult it. Become familiar with the theory and operation of the device under examination. Follow the manual when troubleshooting if such procedures are provided. On much modern digital equipment, a self-test provision is available. Perform the self-test to determine if an actual malfunction exists. If no self-test is provided, use the equipment or device to perform an operation with known results. For example, if a digital voltmeter is being checked, measurement of a known voltage will furnish information that can be used to verify proper or improper operation.

At the subsystem, assembly, circuit-card, or subassembly and logic-function levels of effort other approaches must be used. Of course, the same preliminary questions are asked. However it may not be practical to verify the malfunction in quite such an easy manner. Unless special test sets are available to perform detailed checks, we have to assume that the malfunction does exist and proceed with troubleshooting procedures. In cases where a large volume of subsystems or logic cards are encountered, *special test set-ups* are commonly built so that a specific unit may be tested quickly and efficiently.

As equipment becomes more and more modularized (contained in packages or on cards), *automatic test equipment* is being used. Complete systems and subsystems are connected to the test equipment and digital logic circuits sequence tests, providing the correct stimuli for inputs and measuring/recording the output responses. Limits are provided, and when output responses fall outside of the limits, the defective subassembly is identified. If the subassembly is repairable, it is also subjected to automatic testing in another area with another test set performing detailed checks that isolate the malfunction on the subassembly to a replaceable/repairable item.

But there is no need to despair! The automatic test equipment is not about to replace the troubleshooter! Automatic operations are applicable only where *large-volume testing* must be performed on a relatively small variety of subassemblies or assemblies. This situation occurs in some digital computer and data processing applications. Even if automatic testing expands tremendously, the technician is still needed to troubleshoot and repair the test equipment!

The complexion of troubleshooting is changing with the changing technology. Where the electronic technician previously isolated troubles to the discrete-component level (e.g., resistors, capacitors, and tubes), now the trouble may be isolated to complete subsystems. The trend seems to be *packaging complete functions on single, replaceable units.* Failure of the unit results in replacement with a new one — rather than repair. Equipment reliability must be high with this concept, and this appears to be the goal of digital equipment and module manufacturers. From a troubleshooting viewpoint, this means that the technician must begin to think "system level" rather than discrete-component or simple logic-function level. Analysis at this level requires the ability to understand the operation of subsystems and smaller assemblies and to logically evaluate the inputs and outputs of the larger system in terms of what is happening in the smaller assemblies. Skills learned and abilities acquired while evaluating simple logic functions result in more adequate troubleshooting of the subassemblies and complete systems that are commonplace today.

Once the malfunction is verified and it becomes apparent that actual troubleshooting is required, it is necessary to determine *what information is needed* to perform the troubleshooting tasks. Actually, the information required is somewhat dependent on the troubleshooting methods to be applied. For example: if an oscilloscope is used, waveforms that will be encountered will surely be needed. If some type of logic-level indicator such as a voltmeter or indicator lamp device is used, the logic levels must be available. In both cases, the descriptive Boolean equations may prove quite helfpul, and in many cases, it becomes advisable to develop the equations even if they are not immediately available. And, of course, the logic diagram of the device being tested is needed.

Information available for troubleshooting can vary from practically nothing to considerably more than is ever required in day-to-day operation. Instruction manuals may contain everything from English-language descriptions of equipment operation to complete logic diagrams with waveforms and Boolean equations. Troubleshooting guides may be provided to lead to the precise point of the malfunction. The technician should investigate all available information to determine if sufficient data is available to allow troubleshooting of the malfunctioning equipment.

The Flowchart Concept

One way of organizing a troubleshooting task into a logical plan is to borrow a method from the digital computer field. The digital computer, despite all the publicity to the contrary, is not capable of any appreciable problem-solving effort by itself. Every task that it undertakes must be carefully thought out by someone who understands the task. Each step in the task must be identified in terms of actions that the computer is capable of performing and then presented to the computer in the proper sequence for storage and later use. The person who understands the task and can "talk" to the computer is called a *programmer.* When the programmer is called upon to tell the computer to perform some task, he is faced with a situation similar to that facing the technician who is troubleshooting a logic circuit. Both the programmer and the technician must first organize the task and then develop a logical plan to follow. Of course a number of organizational methods exist, but the programmer usually develops a *flowchart* to aid him in logically thinking out the solution to his problem. Remembering that the computer must be told step by step what to do to solve a problem, the programmer develops a list of instructions for solving the problem. In mathematical terms, this list of instructions is commonly called an *algorithm.* To make his list of instructions relatively understandable by other programmers and to help him organize and check his algorithm, the programmer makes a diagram to represent the algorithm. This diagram is called a "flowchart."

Flowcharts are applicable in areas other than computer programming. In fact, any time a list of instructions has been developed for a task, it can be put into flowchart form. Even a simple recipe from a cookbook (which could be considered as an algorithm) can be put in flowchart form for greater understanding. Organization of troubleshooting procedures for combinational logic is presented by the use of flowcharts in this text. Following the initial presentation of the special symbols and their meanings, a number of typical troubleshooting situations are shown. It will soon become apparent that much effort is saved by the use of flowcharts, and greater assurance that the complete procedure has been accomplished is provided.

In flowcharts, *symbols define the general type of operation that is to take place at each step in the procedure.* Just as with logic diagrams, the *shape* of the symbol is meaningful. Entrance into and exit from a particular flowchart is shown by the symbol in Fig. 7—8a. This symbol is seen only at the beginning or at the end of a procedural flowchart and signifies by the words written inside

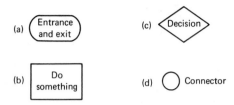

Figure 7—8   Flowchart symbols

which of the meanings is to be assigned to it. Usually, the word START signifies the beginning, and the word STOP or EXIT signifies the end (see Figs. 7—9 and 7—10). The rectangular shape of Fig. 7—8b is used to imply that "something is to be done" in that specific step. Typical legends to be seen inside the "do something" shape might be "determine logic level" or "check system power."

Decisions are signified by the diamond-shaped symbol of Fig. 7—8c. As in troubleshooting of conventional electronic circuits, many decisions must be made. Every time a measurement is performed, the technician must determine whether that measurement is within tolerance. Of course, in binary-based equipment such as is implemented with digital logic, measurement decisions are based on simple criteria, e.g., whether the measured parameter is HIGH or LOW or whether the HIGH signal is present for the required period of time. Typical statements seen inside the decision block could read "power supply voltage in tolerance?" or "output gate level LOW?"

The last symbol to be used in troubleshooting flowcharts in this text is the circle in Fig. 7—8d. No action is performed as a result of this symbol, since it serves merely as a connector from one place on the flowchart to another. Due to the physical constraints of paper size, it is often impractical to draw a flowchart in the easily visualised straight line flow of information. When the end of the paper is reached and additional flowcharting is required, the connector symbol is drawn and an identifying letter is placed inside. Another circle is drawn elsewhere on the paper or on another page to signify the restarting of the flowchart at that particular point. Figure 7—9 demonstrates this technique. If the circle is shown without an identifying symbol, it is merely used to connect or combine two or more branches of the flowchart.

Finally, flowchart symbols are connected by straight lines with arrows at the receiving end of the line to signify direction of information flow. Using combinations of the flowchart symbols and straight lines allows the technician to develop an easily visualized, organized means of troubleshooting combinational logic circuits.

The general concepts behind the construction and use of flowcharts are shown by investigating a flowchart that is relatively common to all troubleshooting situations. Figure 7—9 represents a procedure often followed at the beginning of a troubleshooting problem. Each symbol is numbered for reference

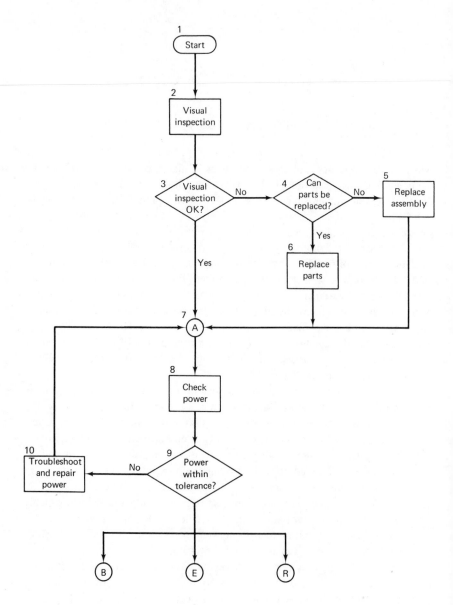

Figure 7—9　"Preliminary" Flowchart

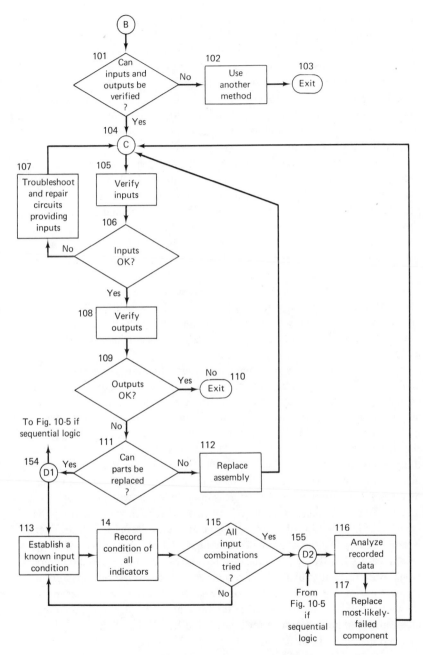

Figure 7—10 "Built-in Indicators" flowchart

213

purposes so that the following explanation may be easily traced. As we noted, most flowcharts begin with the START symbol (Step 1). The shape of Step 2 indicates "doing something," i.e., visually inspecting the logic device that is malfunctioning. Problems commonly detected by visual inspection are burned or broken discrete components, improperly soldered connections, and loose or broken interconnections. Modern circuit boards are precision devices, and very slight discrepancies in the position of components or interconnections can result in malfunctions. The use of a magnifier helps locate problems that would otherwise go undetected visually.

Decisions concerning the physical condition of the logic device are made in Step 3. Note the shape of the symbol, denoting decision-making. If physical damage was detected during the visual inspection, the NO path is followed to Step 4. Another decision is made at this point. If the maintenance concept for the system of logic being used allows, repair or replacement of damaged components is accomplished according to Step 6. The logic device is replaced (Step 5) if the maintenance concept does not allow repair and/or replacement of damaged or defective components. Many manufacturers feel that replacement of the complete logic device enhances system reliability, so repair and/or replacement of damaged or defective components is accomplished at a facility where special precision repair equipment is maintained. The digital logic technician encounters both maintenance concepts (plus many variations) during his troubleshooting assignments.

Step 7 merely serves as a connector, where the results of Steps 3, 5, and 6 come together. If the visual inspection of Step 2 was satisfactory, as indicated by the YES path from Step 3, then the entrance to the connector would be from decision Step 3. If it was not satisfactory, then the entrance to Step 7 would be from either Steps 5 or 6, depending on the maintenance concept. Regardless of the entrance path into the connector, the exit is to Step 8, another "do something" step. All power provided to the logic device under investigation is measured at this time. Some of the discrete-component logic devices use as many as three separate voltages, whereas modern integrated-circuit logic devices may use only one voltage supply. In any case, all applied voltages should be measured. The decision about whether the applied voltages are within tolerance is made in Step 9. One or more power sources out of tolerance can send the technician to Step 10, which indicates that the power supply should undergo troubleshooting and repair. Upon completion of repair, the flowchart returns to connector 7 so that the normal path through the flowchart can be resumed. If the power had been out of tolerance the first time through Step 9, now the decision would be "not out of tolerance," and the YES exit from Step 9 would be to connector B, E, or R.

The combination of Steps 7, 8, 9, and 10 form what is known in flowchart terminology as a "loop." *Loops allow modification of a sequence based on the results of a decision.* In this case, the decision that a defective power supply existed caused the action of "troubleshoot and repair" instead of proceeding

on with a known defect. The original action and decision that detected the malfunction are repeated to verify the adequacy of the repair action.

This flowchart demonstrates the use of all the basic symbols and provides
an introduction to flowchart techniques. Actually, the flowchart has been
generalized so that it may serve as a guide and preliminary chart to all troubleshooting situations. The only prerequisite is that the logic-circuit description be
in a form that is compatible with the logic-troubleshooting charts to follow.
The flowchart just discussed is not specifically a logic-troubleshooting chart. It
is applicable to any electronic troubleshooting situation when it is desired to
establish that the malfunction is within the equipment being checked. This chart
will account for repair of physically damaged or defective components and
improperly operating power supplies.

Common Troubleshooting Situations

In all of the situations we shall investigate, certain general assumptions are
made. The logic diagram is assumed to be available. In addition, the logic circuit
is assumed to be inoperative, i.e., not performing its required operations as
designed. Furthermore, it is also assumed that the logic circuit *had been* opera-
tive and *had been* performing its required operations properly.

Test points are usually provided on logic assemblies at circuit locations that
allow isolation and replacement of defective components without actual soldering operations. For example, if the logic assembly is so constructed that diodes
and transistors are removable and replaceable without soldering, sufficient test
points are provided to isolate difficulties to that degree. When a plug-in logic
board contains a number of different functions on the same board but none of
the components on the board can be replaced, at least the output of each logic
circuit (e.g., an AND gate or an OR gate) is made available for test purposes.
Isolation of the defective logic circuit by use of the test points identifies the
defective board, which is removed and replaced. The defective board is then
repaired at a facility with the necessary equipment. Similar approaches are used
with integrated circuit boards. Only sufficient test points are made available to
identify the defective integrated circuit, which is replaced at a separate repair
facility. The complete integrated-circuit board is usually replaced at the equipment installation and operation location to reduce the requirement for specialized repair equipment at each location.

The choice of which flowchart to use next depends on what type of
troubleshooting aids are to be used: one chart assumes the use of built-in indicators; another assumes the use of external logic-level indicators; and the third
considers the use of an oscilloscope to troubleshoot malfunctions.

BUILT-IN INDICATORS

Depending on the type of digital equipment and its degree of complexity,
built-in indicators may be available to assist troubleshooting operations. At the

system level, many types of digital equipment provide panel indicators to allow monitoring of system status. Such indicators are usually gross indications of system malfunction and isolate problems only to major subassemblies. From that point, it becomes necessary to resort either to automatic test equipment to isolate to more detail or to individual logic analysis to determine the amount of repair necessary.

Some manufacturers of digital equipment provide built-in indicators on each subassembly. When this troubleshooting aid is available, it usually allows isolation of malfunctions to a replaceable component — whether it is a discrete component such as a transistor, resistor, or capacitor or an individual integrated-circuit assembly. The general trend, however, is to provide only sufficient indicators to isolate malfunctions to replaceable major subassemblies. Instruction manuals commonly contain charts relating combinations of status indicators to malfunctioning subassemblies, so troubleshooting tasks are relatively simple. Skillful analysis of logic circuitry will usually reveal numerous additional combinations of indicators that can extend the malfunction isolation capabilities of the built-in indicators. Figure 7–10 is the flowchart that could be used to troubleshoot digital equipment using only built-in indicators.

Entrance to the flowchart in Fig. 7–10 is from connector B, which previously appeared on the "Preliminary" Flowchart in Fig. 7–9. The first step encountered is a decision (101): can the inputs and outputs be verified? Logically, this is a reasonable question. If the built-in indicators are not properly located to allow verification of the inputs and outputs, there is no need to continue the procedure. The path to follow is from the NO side of the decision block to Step 102 (use another method). Either the "External Logic-Level Indicators" or the "Waveforms" procedure must be accomplished to troubleshoot the malfunctioning equipment if indicators are not available to verify the inputs and outputs. Exit from this procedure is via Step 103.

If the logic circuit inputs and outputs can be verified with internal indicators, the YES path is followed. Connector C, which is the next symbol in the YES path, allows interconnection of a number of additional paths within the flowchart. The logic circuit inputs are checked in Step 105, and the decision as to the correctness of the inputs is made in Step 106. If the correct inputs are not being provided, it is necessary to troubleshoot and repair the circuits providing the inputs (Step 107) before proceeding. The NO path from Step 106 through Step 107 terminates at connector C so that the circuit inputs may be checked again after accomplishing Step 107. When the correct inputs are obtained, the YES path from Step 106 is followed to Step 108 (check outputs). If the outputs are correct after performing Step 109, the YES path is followed to an exit from the procedure at Step 110. Verification of correct outputs with correct inputs is an indication of proper circuit operation.

A number of actions have been performed since the start of the trouble-shooting procedures, and any of these actions could have cleared the malfunction. Repair/replacement of defective components, assurance that power inputs are within tolerance, and/or verification of correct inputs could have rectified the original difficulty. Or — and this is not uncommon — handling of the logic assembly during troubleshooting and maintenance has temporarily "fixed" a problem. Extreme care must be exercised to prevent such an occurrence. When replacing components, only those requiring replacement should be disturbed. All other components should be carefully avoided to prevent physically displacing them and perhaps causing an intermittent condition to temporarily disappear. Power should be applied to the circuit under test in a manner that closely approximates power application during actual use. Transients that develop due to sudden removal and application of circuit power often cause an intermittent condition to vanish. In addition, shorting of power inputs to each other or to common power should be avoided when making measurements. Additional damage to the circuit under test can result along with a possible "masking" of the real malfunction, due to the power transients. If no actual maintenance actions have been performed on the circuit under test and the exit from the procedure has been taken at Step 110, it is likely that the circuit under test will appear for troubleshooting and repair at some later time. The actual problem may have just been concealed by *any* of the situations described earlier.

Selection of the NO route from Step 109 (outputs *not* OK) results in another decision at Step 111. It is necessary to decide whether components, logic elements, or logic assemblies may be replaced on the assembly being tested. If the answer is NO, there is no need to troubleshoot further. The assembly should be replaced (Step 112) and the flowchart should be reentered at connector C for recheck. This time through, the flowchart should result in YES answers at all decisions (through Step 109) and exit at Step 110 with a properly operating assembly.

If components, logic elements, or logic assemblies can be replaced, the YES path from Step 111 is taken to connector D1, which allows a branch to Step 113 if combinational logic is being repaired or to Fig. 10—5 if sequential logic is being repaired. Reentry to Fig. 7—10 (if necessary) is via connector D2 (155). At Step 113, a known input condition is established. The condition of all built-in indicators is recorded at Step 114, and a question (have all input combinations been tried?) is asked at Step 115. If all input combinations for the number of input variables have been tried, the flowchart continues on to Step 116. If they have not, the flowchart returns to Step 113, and a new combination of inputs is tried. The indicator conditions are recorded, and the decision is made again as to whether all input combinations have been tried. The loop continues until all combinations have been tried and all indicators recorded. At Step 116, the recorded data is analyzed by using *at least* the logic diagram and all other available information. The most likely failed component, logic element, or logic assembly is replaced (Step 117), and the flowchart is reentered at connector C for a complete recheck. If the only difficulty in the logic circuit under test has been caused by the replaced component, logic element, or logic assembly, the YES path is followed through Steps 106 and 109, and the flowchart is completed by exit with a repaired assembly at Step 110.

By using two flowcharts, a complete troubleshooting procedure has been developed — one that allows identification and repair of a malfunctioning logic assembly. The Preliminary Flowchart aided in identification and repair of physically damaged parts and established circuit power within tolerance. Application of the procedures of the Built-in Indicators Flowchart guided the technician through input and output checks and established the conditions for analysis and identification of the most likely failed component, logic element, or logic assembly.

While there is no guarantee that flowcharts will cover every possible case of troubleshooting logic assemblies with built-in indicators, most of the malfunctioning assemblies will submit to these procedures. Modifications may be necessary to cover special cases, but the concepts shown in these and subsequent flowcharts should prepare the reader to construct most flowcharts with a minimum of effort.

EXTERNAL LOGIC-LEVEL INDICATORS

The scope of logic-circuit troubleshooting may be extended to greater detail by using *external logic-level indicators.* Such devices consist of one or more indicator(s) in a hand-held test instrument that is applied to interconnections in logic circuits to determine the logic level at the point of application. Simple logic-level indicators provide the logic level at only one point at a time. Figure 7–11a is a typical device for this application. The conventional multipurpose voltmeter and the high-impedance voltmeter should not be overlooked as simple logic-level indicators. More complex logic-level indicators are manufactured to fit over complete integrated-circuit packages and indicate the logic level at each pin simultaneously (see Fig. 7–11b).

The flowchart for troubleshooting logic circuits using external logic-level indicators has many points in common with the flowchart for built-in indicators. Since the external logic-level indicator may be applied directly to the input and output connections of the logic circuit assemblies, we will assume that there is no difficulty in verifying the input and output signals. Therefore the decision block that appeared at the beginning of the Built-In Indicators Flowchart will not be required. In other words, the External Logic-Level Indicators Flowchart (Fig. 7–12) may be directly entered at the "verify inputs" block and developed from that point.

Steps 201 through 208, as a group, perform the same functions as Steps 105 through 112. Inputs are checked and restored if necessary; outputs are verified; and a decision is made concerning replacement or repair of the assembly being tested. The entry to the External Logic-Level Indicators Flowchart is via

Figure 7—11 External logic level indicators; (a) logic probe,
(b) logic clip (courtesy Hewlett-Packard Company, Palo Alto, CA.)

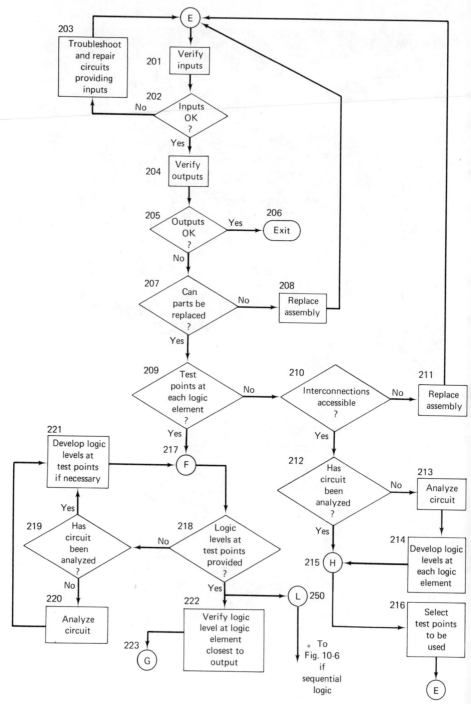

Figure 7–12 "External Logic Level Indicators" flowchart

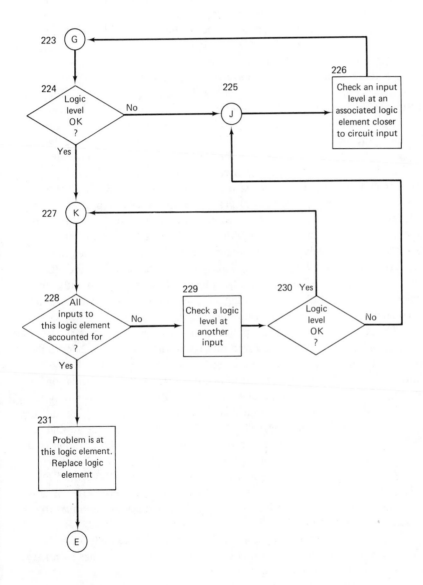

223 G

224 Logic level OK ? — No → **225** J → **226** Check an input level at an associated logic element closer to circuit input

Yes

227 K

228 All inputs to this logic element accounted for ? — No → **229** Check a logic level at another input → **230** Yes / No Logic level OK ?

Yes

231 Problem is at this logic element. Replace logic element

E

Figure 7−12 (cont'd)

connector E on the Preliminary Flowchart. If components, logic elements, or logic assemblies can be replaced, the YES path from Step 207 is directly to Step 209. Here the question "are test points available at each logic element?" is asked. If test points are not available at each logic element, it is then necessary to determine if internal connections are accessible (Step 210). Lacking built-in test points, it is necessary to have access to the *wires* or *printed circuit conductors* connecting the various components, logic elements, or logic assemblies. If neither test points nor circuit connection accessibility is provided, then there is no further need to troubleshoot. Malfunctions cannot be isolated to replaceable units. The complete assembly must be replaced (Step 211) and a retest initiated at connector E to verify that the new assembly is operating satisfactorily.

Accessibility of circuit connections leads to Step 212 from the YES path of Step 210. This path is used to select test points to be used in the troubleshooting portion of the flowchart. To select test points, it is first necessary to determine the logic levels (HIGHs and LOWs) at each logic element. If the logic circuit has been analyzed (Step 212), then the logic levels have already been determined, and the YES path from Step 212 leads through connector H to Step 216 (select test points to be used) through connector 215. The NO path from Step 212 leads through Step 213 (analyze circuit) and Step 214 (develop logic levels at each circuit element) to Step 216 via connector H. Either entry to Step 216 via connector H has required logic-circuit analysis. Any of the circuit-analysis methods previously discussed are adequate for application at Step 216. The *truth table form* is probably the most informative, since the information is presented in a way that can be easily correlated with indicators depicting HIGHs and LOWs. Following selection of test points, the flow returns via connector E to the major path through the chart.

The second time through Step 209 test points will be available at each logic element, and the YES path is followed through connector F (Step 217) to the question at Step 218 (are logic levels at each test point provided?). Perhaps test points were provided by the equipment manufacturer, but the information concerning logic levels at the test point either was not furnished or was not available.

The NO path from Step 218 is then followed to Step 219, where the question "has the circuit been analyzed?" is asked. A procedure similar to that discussed in the previous paragraph is followed in Steps 220 and 221, and control of the flowchart is returned to the major path through connector F.

Actual troubleshooting begins with an exit from Step 218 via the YES path. If the logic being evaluated is sequential, an exit via Step 250 (connector L) to Fig. 10–6 is made. Combinational logic troubleshooting continues to Step 222.

A short review of what we know to this point will materially aid us in following the remainder of the flowchart. To get to Step 222, it has been necessary to verify the following conditions:

1. No visible damage.
2. Circuit power is within tolerance.
3. Inputs have been verified.
4. Outputs are not correct.
5. Repairs can be performed when malfunctioning components are located.
6. Test points are available at each logic element.
7. Logic levels are available at each logic element.

Thus the malfunction has been pinpointed to the logic assembly under test, and sufficient information is available to further isolate the problem to a replaceable component, logic element, or logic assembly.

Checking the logic level at an *input to the logic element closest to the output* (Step 222) provides the first bit of information that will lead to detection of the faulty portion of the logic assembly. The decision as to the validity of the logic level just checked is made in Step 224, which has been entered from Step 222 via connector G (Step 223). Assuming a correct logic level, Step 224 is exited via the YES path through connector K (Step 227) to Step 228, where it is determined whether all inputs to the logic element being examined have been accounted for. A single-input logic element such as an inverter amplifier has only one input, and under these circumstances, all inputs would have been accounted for. If the output of a device is incorrect and the input is correct, the device must be faulty. Therefore Step 228 can be exited via the YES route to Step 231, which contains direction to replace the defective logic element. After replacement of the malfunctioning component, the flowchart returns via connector E to completely recheck assembly operation. If the only defective component was the one replaced, the output check is now passed through the flowchart, and an exit from the procedure occurs at Step 206.

If the logic element closest to the output is a multiple-input device such as a gate and if the first input checked is correct, then the NO path from Step 228 is taken. At Step 229, the logic level at another input to the logic element under test is checked. A proper logic level at the Step 229 test causes exit from Step 230 on the YES path to connector K. The output of connector K feeds the decision block 228, and the resultant loop is continued until all inputs to the logic element have been verified as correct. Once again, if all inputs are correct and the output is incorrect, the difficulty must be with the logic element being tested. Replacement is made at Step 231, and retest is accomplished by returning to the main flow via connector E.

Any time that a logic-level check is not satisfactory, a loop is entered that causes *measurements to be made at an associated logic element closer to the circuit input.* Step 226 provides the directions for this measurement and is entered as a result of either Step 224 or 230 through connector J (Step 225).

The loop entered through connector J to Step 226 allows the troubleshooting to advance toward the actual basic inputs to the complete unit under test. Multiple entry is provided to this loop, and the flow is quite complex. A complete explanation is "wordy" — it is better left to a troubleshooting example that will appear later in this chapter. However following the procedure provided will result in detection and isolation of components that have caused the overall unit under test to develop improper outputs. In all cases, the detection and replacement of the faulty components at Step 231 cause recheck via connector E, and an exit at Step 206 with a properly functioning logic assembly.

OSCILLOSCOPE

In strictly combinational logic applications, the *oscilloscope* provides little to gain when compared with other external logic-level indicators. Instead of an illuminated indicator or the position of a pointer on a meter, the position of a trace on the oscilloscope provides the indication of logic level at a particular point in the circuit. The oscilloscope is also useful when it is necessary to determine the action of a logic circuit at the time that circuit inputs are changing. *Most external logic-level indicators are relatively slow acting, and transient conditions are difficult to detect without the speed of the oscilloscope.*

The flowchart for troubleshooting with the oscilloscope (Fig. 7—13) is almost identical to the flowchart for troubleshooting with the other external logic-level indicators. Generally, the word *waveforms* replaces the term "logic level," and the overall flow remains the same. The use of the oscilloscope will be readily apparent when analysis and troubleshooting of sequential circuits is covered in subsequent chapters.

Exit to Fig. 10—7 is provided if sequential logic is being repaired.

7—4 A TYPICAL TROUBLESHOOTING PROBLEM

With the completion of the flowcharts for troubleshooting combinational logic, the guidance is now available for working with actual logic-circuit problems. The sample troubleshooting exercise provided in this section allows the reader to use flowcharts and develop an appreciation for this method of organization. One example alone, however, is not enough to firmly implant the complete concepts. The reader should apply these and subsequently developed flowcharts at every opportunity to fully realize their tremendous potential.

Preliminary Information

When extreme reliability is required in logic applications, as many as three redundant systems may be used. Of course, this requires three times as much hardware, but it does assure accurate and reliable operation. Usually the system is arranged so that when *any* two out of three paths are operating properly, it is considered sufficient indication that accurate information is being provided. A logic arrangement called a *majority circuit* is used to control acceptance of the incoming information, based on the two-out-of-three rule. Thus if three paths called A, B, and C are furnished, either A and B, B and C, or A and C (or all three) should provide an output that indicates the incoming information is acceptable. For maintenance purposes, however, it is necessary to know if *any* of the three paths is *not* functioning. This information is not available from the majority circuit, so a *dissent circuit* is required to furnish the information that one of the three paths has failed. The output of the dissent circuit is an indication that one or more of the inputs is *not* operating correctly.

A hypothetical application of these circuits could be found in modern jet aircraft, which use autopilots for much of their routine flying operations. Three different navigational devices can be used to guide the aircraft along a predetermined path. Experience has shown that any two of the three provide an adequate safety margin as far as maintaining accurate navigation is concerned. Therefore when any two of the three navigational devices are operating satisfactorily, the output of the majority circuit enables the autopilot to accept information. However, if one of the devices has failed, the crew must be aware of the fact so that alternate procedures may be instituted should a second device fail. The dissent circuit provides the warning that one of the three has failed, and maintenance personnel can use predetermined procedures to isolate and replace the defective navigational devices.

The three inputs to the majority/dissent circuit are logic levels. If, for example, a device represented by A is operating correctly, A is HIGH. If the device is not operating correctly, A is LOW. Determination of correct operation is made within each navigational device, and a fault warning light is illuminated on the device when it is operating incorrectly. Logic levels A, B, and C are merely indications of correct or incorrect operation. The majority output (M) is HIGH when any two out of three (or all three) inputs are HIGH, and an indicator lamp called the *majority lamp* is illuminated. If one or more (but not all) of the inputs are LOW, the output of the dissent circuit (D) is LOW, and the *D Warning Lamp* is illuminated. Thus either a LOW M or a LOW D output is indicative of difficulty, and maintenance action is required. Normal indications are the *D Warning Lamp* off, and the *M Lamp* on. This indicates that all three inputs are present (*D Warning Lamp* off), and that the autopilot is connected to the navigational units.

For the purposes of this example, it is assumed that the complete majority/ dissent circuit is contained on a removable printed-circuit card. Each of the logic elements is individually removable without soldering, and repair to the plug-in component level is authorized by the maintenance concept. No built-in indications are available on the circuit card, and no intermediate test points are provided. However it is possible to gain access to the interconnection wiring on the circuit board. It is also assumed that a test fixture is available. As we mentioned previously, such an arrangement is common when a large number of circuit

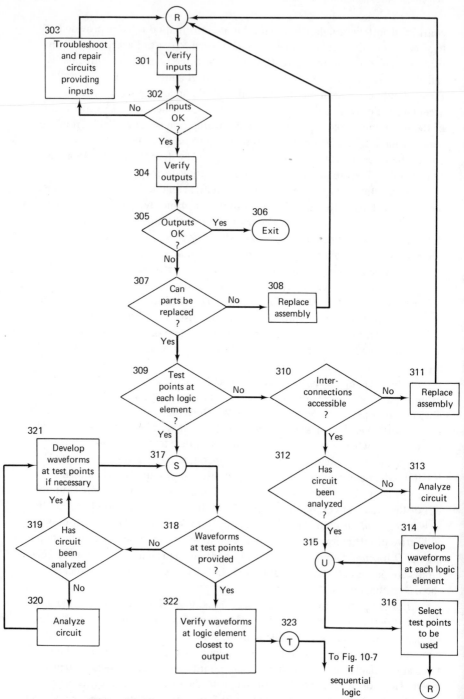

Figure 7–13 "Oscilloscope" flowchart

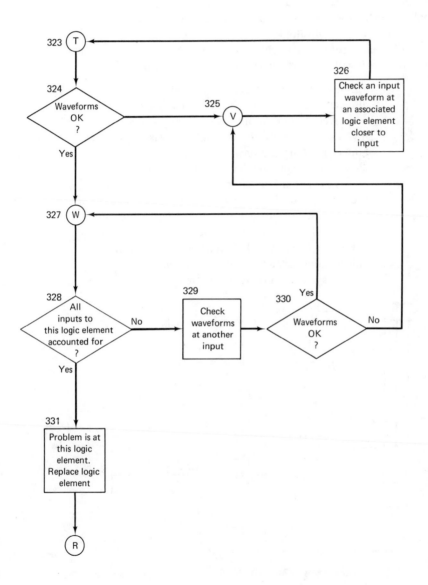

Figure 7-13 (cont'd)

227

cards need to be tested. The test fixture is usually composed of a socket which accepts the circuit card, power supplies to simulate system power, a means to synthesize input signals, and indicators to monitor inputs and outputs. The only other troubleshooting aids available are a logic diagram (Fig. 7—14) and an external logic-level indicator.

The majority/dissent circuit card has been returned to the troubleshooting area for repair with the following information furnished.

> The *M Lamp* went off, and the autopilot automatically disconnected from the navigational units, requiring manual control of the aircraft. After regaining control, it was noted that the *D Warning Lamp* was off and that both *A* and *B* navigational units had fault warning lamps illuminated. Replacing the majority/dissent circuit card caused the *D Warning Lamp* to illuminate with both *A* and *B* navigational units malfunctioning. Replacement of unit *A* caused reillumination of the *M Lamp*, indicating that two of three navigational units were operable. The *D Warning Lamp* stayed illuminated. Replacement of unit *B* extinguished the *D Warning Lamp*, and the *M Lamp* stayed illuminated as expected.

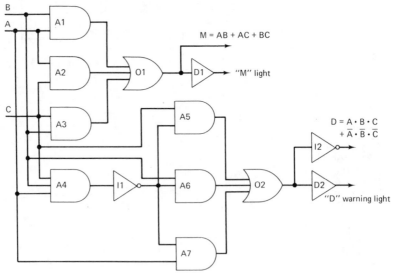

Figure 7—14   Majority/Dissent circuit logic diagram

The Troubleshooting Task

A decision must now be made concerning the actual troubleshooting procedure that is to be used. In all cases, of course, the preliminary flowchart should be followed to repair physical damage and to assure that the power in the text fixture is properly adjusted and connected. The next flowchart to follow is dependent on the troubleshooting aids available. In the case of the majority/dissent circuit card, only the logic diagram and an external logic-level indicator have been furnished. This leads to the 200 series flowchart (Fig. 7–12), which allows the technician to locate and identify the malfunctioning logic element. The complete procedure is now followed from beginning to end to demonstrate the use of flowcharts.

Starting with the Preliminary Flowchart (Fig. 7–9), a visual inspection is performed. With no obvious physical damage apparent, the circuit card is placed in the test fixture, and the applied voltage(s) are checked. When all voltages are properly adjusted, the Preliminary Flowchart is exited, and the External Logic-Level Detector Flowchart is entered. The inputs are checked and placed in the configuration of the original malfunction, i.e., both A and B LOW. Upon checking the outputs, it is determined that they are incorrect for the input conditions, and Step 205 is exited via the NO path. Parts can be replaced according to the maintenance concept, so Step 207 is exited via the YES route to Step 209.

The earlier description of the circuit card indicated that no built-in test points were provided but that interconnections were accessible. A path through Blocks 209 and 210 to Block 212 is established, and the first real trouble-shooting/analysis decision is made. Of course, if the logic diagram has already been analyzed, we need only to select the test points to be used (block 216) and proceed back to Step 209 via connector E and intermediate steps. If the circuit has not been analyzed, the route through Steps 213 and 214 must be followed. Note that the logic diagram does not show the minimum logic implementation of the majority/dissent functions, but the circuit arrangement adapts well to the demonstration of troubleshooting techniques. Any of the common analytical methods discussed in this chapter may be used in *circuit analysis.* The most worthwhile analysis when troubleshooting with an external logic level detector is one in which levels at each logic element result. A truth table provides this information, and the expected logic level can be correlated directly with that indicated on the test device. The truth table for both outputs of the majority/ dissent circuit is shown in Table 7–2.

This time through Block 209 test points are available, and the YES route to Block 218 is taken. If the logic levels at the test points are not known, it is necessary either to analyze the circuit and develop the logic levels or to develop the logic levels from the already analyzed circuit. Blocks 219 through 221 provide the required procedural steps. When test-point logic levels are known, Block 218 exits to Block 222, and actual troubleshooting begins. Although it is not shown on the flowchart, in a logic circuit with more than one output, it is necessary to decide which output to start with. Random selection could be used, but in many cases, sufficient information is available to make one choice better than others. The clue in this case is the malfunction description that came with

Table 7–2 Majority/Dissent circuit truth table

Row	A	B	C	A_1	A_2	A_3	O1	LD_1	A_4	I1	A_5	A_6	A_7	O2	I2	LD_2
0	0	0	0	0	0	0	0	0	0	1	0	0	0	0	1	0
1	0	0	1	0	0	0	0	0	0	1	1	1	0	1	0	1
2	0	1	0	0	0	0	0	0	0	1	0	1	0	1	0	1
3	0	1	1	0	0	1	1	1	0	1	1	1	0	1	0	1
4	1	0	0	0	0	0	0	0	0	1	0	0	1	1	0	1
5	1	0	1	1	1	0	1	1	0	1	1	1	1	1	0	1
6	1	1	0	1	0	0	1	1	0	1	0	0	1	1	0	1
7	1	1	1	1	1	1	1	1	1	0	0	0	0	0	1	0

the circuit card. Apparently the portion of the logic circuit feeding the *D Warning Lamp* is the culprit, since the *D Warning Lamp* did not illuminate even though both *A* and *B* navigational units were malfunctioning. The description of the dissent circuit states that *D* is HIGH, and the *D Warning Lamp* is off when all three inputs agree. With both *A* and *B* LOW and *C* HIGH, obviously the three inputs do not agree. Therefore the *D Warning Lamp* should be on, and the *D* output should be LOW. It is with this information that Block 222 is entered.

.The input to the logic element closest to the output (INVERTER 2) is obtained from OR 2. If the *D* output should be LOW, then the input to INVERTER 2 should be HIGH. With the input combination that causes the malfunction, the information in Table 7–2 (specifically row 1) is used to determine the required logic level at each test point. The logic level indicator shows the level to be LOW, which is incorrect. This is confirmed by the nonillumination of the *D Warning Lamp*. The NO path is taken from Block 224 to Block 226, and an input to an associated logic element closer to the input (OR 2) is checked. Arbitrarily selecting the output of AND 5, it is found to be LOW. According to row 1 on the truth table, the output of AND 5 should be HIGH with the given input combination. This causes the selection of the NO path again from Block 224 and the checking of an input level at an associated logic element closer to the input (AND 5). AND 5 has two inputs, one of which comes from input *C*. It is known that *C* is HIGH under the given input conditions, so the only way that the AND 5 output can be LOW is for the other input (from INVERTER 1) to be LOW or for AND 5 to have failed with its output LOW.

The output of INVERTER 1 is checked and is found to be LOW. Table 7–2 indicates that the given input combination should result in a HIGH output, so the input to INVERTER 1 is investigated. The input to INVERTER 1 is from AND 4, and the logic-level indicator reveals a HIGH level. Note that the loop consisting of Blocks 223, 224, 225, and 226 is being constantly repeated, striving to locate a logic level that is valid. The HIGH level at the output of AND 4 is incorrect according to row 1 of Table 7–2, and the loop is again repeated. The complete dissent circuit from output to input has been traversed, and when an input to AND 4 is checked, it will be found to be correct, as indicated in the truth table. Block 224 is exited through connector K via the YES path to Block 228.

But all inputs have not been accounted for, and the NO exit from Block 228 requires checking of a logic level at another input. Block 230 makes the decision on the validity of the logic level (which is correct according to the truth table), and an exit along the YES path leads to connector K and Block 228 for a repeat of the loop through Blocks 228, 229, and 230. As soon as all inputs to the logic element (AND 5) have been checked, Block 228 exits via the YES route to Block 231. Since all of the inputs to AND 4 have been verified, and the output is still incorrect, AND 4 is defective, having failed in the HIGH output condition. AND 4 is replaced by direction of Block 231, and the flowchart returns via connector E to recheck the complete majority/dissent circuit card.

PROBLEM APPLICATIONS

7−1. What information should be obtained prior to starting an investigation of a combinational logic circuit?

7−2. Where does the analysis of a combinational logic circuit begin?

7−3. What must be done when a level of logic is reached that requires other inputs?

7−4. What is the most effective method of logic analysis?

7−5. What is meant by the term "systems concept"?

7−6. Why is it desirable to understand the systems concept?

7−7. Discuss the philosophy and concepts of combinational logic trouble-shooting.

7−8. What is a flowchart?

7−9. Why are flowcharts used?

7−10. What flowchart symbols are used in each of the following cases?
(a) Begin at this point.
(b) Begin at this point as defined on the previous page.
(c) Is the logic level HIGH?
(d) Check the power supply voltage.
(e) Set the input switches to all 0s.
(f) Go back to the input-level check.

7−11. What information must be obtained before beginning troubleshooting?

7−12. Develop a flowchart for troubleshooting a complete logic system, including a status panel consisting of indicator lamps that are capable of displaying the inputs and outputs of each assembly.

7−13. What is the most effective method of troubleshooting?

7−14. Assume that a full adder is constructed from two half-adders and an OR gate as shown below. The half-adders are implemented in the manner of Fig. 5−17. Analyze the operation of the full adder using
(a) truth tables   (b) waveforms   (c) logic levels (H and L)

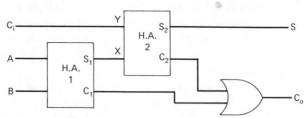

Document the analysis thoroughly.

7−15. Assuming that the inputs are correct, the outputs of the first half-adder are correct, and the C_0 output remains HIGH at all times, troubleshoot the logic circuit in Problem 14 using
(a) the external logic-level indicator method.
(b) the oscilloscope method.
Document the analysis thoroughly.

7−16. What symptoms would appear if in Fig. 7−1
(a) G_1 fails in the HIGH state?
(b) G_1 fails in the LOW state?
(c) G_2 fails in the HIGH state?
(d) G_3 fails in the LOW state?
(e) G_5 fails in the LOW state?

8

Sequential
Logic

The preceding chapters have been primarily devoted to developing an understanding of combinational logic circuits. Combinational logic circuits have outputs that are solely dependent on the input conditions at the time that outputs are being observed. A particular set of input conditions always establishes the same output, regardless of the past history of the circuit. But changing or removing the inputs also changes the outputs, and no record is provided to show the state of the previous output before the input change. In other words, *the combinational logic circuit has a very poor memory*!

In Chapter 3 we pointed out that two general types of logic circuits exist in digital equipment: combinational and sequential. And, *in contrast to combinational circuits, sequential circuits do possess memory.* Furthermore, *the outputs of sequential circuits* depend not only on *the present input condition* but also on *past input history.* It is this memory capability that allows the sequential circuit to perform its function.

8–1 AN EXAMPLE OF SEQUENTIAL LOGIC USE

A simple example of memory capability requirements may be found in the digital control of machinery in an industrial application such as the drilling and shaping of precision metal parts for high-speed aircraft. The metal-working machines must be told precisely where to position their cutting tools in order to perform each operation. Positioning commands are often as simple as "move left" or "move right." In fact, to take advantage of digital operation, only *simple commands*, capable of being represented by binary notation, may be used. Information is commonly fed to the digitally controlled machine by means of a punched paper tape with hole location and number of holes arranged so that numerous unique combinations exist. (Chapter 14 discusses the use of such digital codes.)

But, since only *one* command may be issued at a time, some sort of *memory capability* is definitely required. If the machine is told to "move the tool right," this information must be stored until it is told "how *far* to move." A combinational logic device cannot remember which direction the tool is to move when it is told how far to move. Issuance of the "how far to move" command removes the "move the tool right" command, and the machine could move the tool in the wrong direction. In just such a simple situation at this, it is apparent that a requirement exists for retention of information in digital systems.

Although similar, the memory requirement discussed in this chapter should not be confused with the meaning of the term "memory" as applied to digital computers. In describing the memory function in sequential logic, a more general description is applied. *Memory* in sequential logic serves the purpose of retaining information about the *immediate past* so that the basic sequential-logic element may perform based on that information and the current incoming information. Having furnished information on the immediate past, the sequential-logic element then stores the results of the interaction of that information and the current incoming information as "immediate past" information and awaits the next sequence of incoming information. This suggests a *time dependency* for sequential logic. It is the memory and time dependency properties of sequential-logic circuits that differentiate them from combinational logic circuits. Thus *in a sequential logic circuit, the output(s) depend not only on the input(s) immediately present but also on the past history of both.*

8–2   THE GENERAL SEQUENTIAL ELEMENT

The "Latch" Function

Combinational-logic circuits may be *interconnected* to provide a sequential-logic function. The term *latch* is used to describe a logic function that is similar to a mechanical operation. A mechanical latch, such as that used on a fence gate, requires two separate actions, usually applied at two separate physical locations. One action occurs upon closing the gate, when the latch is physically raised to allow the latch pin to enter and then falls into position to lock the latch pin in position. Further attempts to close the gate have no effect, since the latch pin is locked in position and cannot move. The only way to unlatch the gate is to raise the lock and allow the latch pin to be released. The unlatch operation occurs at a different physical location than the latch operation, i.e., raising the latch lock to unlatch was a different operation than raising the latch lock to allow entry of the latch pin.

A logic latch operates in a similar manner – *an input action must be applied to a specific input to cause the device to assume one of two logic conditions.* Further application of the latch input signal has no effect. The latch remains in the latched condition. Release of the latch must be accomplished by application of an input signal to another input location.

A very primitive latch (Fig. 8−1) may be constructed using only two inverting amplifiers. Single-input NOR gates are used to demonstrate the principle of operation. (Actually, the circuit shown is not a latch at all but is merely a circuit that exhibits some of the properties of a latch.) The single-input NOR gate acts as an inverter. When the input of the top inverter is HIGH, the output is LOW. The LOW output of the top inverter fed to the input of the bottom inverter causes its output to be HIGH. Thus the circuit is locked into a specific configuration. Which inverter initially possesses the HIGH output on power application depends on individual circuit difference, but one *or* the other is HIGH and initiates the locked-up condition. As we can see, no provision is made to control this circuit. It is merely representative of the general concept of the latching principle.

A more practical latching circuit is shown in Fig. 8−2. Cross-coupled NOR gates are still used, but another input is shown on each gate. The initial feedback configuration is retained for both NOR gates, and the added inputs are used for control purposes.

For the purposes of discussion, it is assumed that the circuit of Fig. 8−2 has the LATCHED output LOW and the UNLATCHED output HIGH, following initial power application. Furthermore, it is assumed that both the UNLATCH and LATCH inputs are at a LOW level. With the given inputs and outputs, the

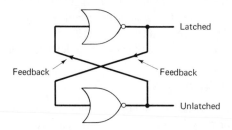

Figure 8−1   A primitive latch

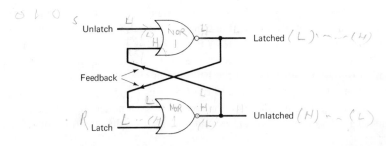

Figure 8−2   A practical latch

feedback input to NOR 1 is HIGH, due to its connection directly to the UNLATCHED output. And, since any HIGH input to a NOR gate results in a LOW output, LATCHED is LOW. If the UNLATCH input transitions from LOW to HIGH (either permanently or momentarily), no change occurs at the outputs. (Any or all HIGH inputs to a NOR gate result in a LOW output.) Examination of NOR 2, however, shows another situation. The feedback input to NOR 2 from the LATCHED output is a LOW level. As in the primitive latch, the LOW input forces the output HIGH if no other inputs are present. In Fig. 8-2, however, the state of the LATCH input must be considered. As initially defined, the LATCH input is LOW, and when combined with the LOW input from the LATCHED output in the NOR gate, furnishes a HIGH level at the UNLATCHED output. If the LATCH input now transitions from LOW to HIGH (either permanently or momentarily), NOR 2 has one LOW and one HIGH input. Any HIGH input to a NOR gate results in a LOW output, and the UNLATCHED output transitions to a LOW level. The change from HIGH to LOW at the UNLATCHED output is fed back to the input of NOR 1, and removal of the HIGH input causes the output of NOR 1 to transition to the LOW level. The outputs are now reversed from the original state – the LATCHED output is HIGH and the UNLATCHED output is LOW.

Return of the LATCH input to a LOW level does not affect the state of the circuit, since the feedback from the LATCHED output effectively locks the circuit into the LATCHED state. The only means of changing the state of the circuit is to apply a HIGH-going input to the UNLATCH input. NOR 1 operates in exactly the same manner as NOR 2, and the circuit returns to the UNLATCHED state. Thus, just as in the mechanical example, actions must occur at *different positions* to affect change from a latched to an unlatched condition and vice-versa.

Nomenclature and Symbology

A generalized approach to the nomenclature and symbology* of *flip-flops* (FFs) is almost mandatory in discussing the different types and many means of operational implementation. Perhaps the most basic requirement is to be able to discuss FFs without becoming confused about the relationship between inputs, outputs, and the "state of the circuit."

The basic shape of the FF symbol is retained, but the input and output designations are modified to fit the specific type of FF being discussed. One of the simplest configurations, the $R-S$ FF, is used as an example. The symbols shown in Fig. 8-3 can be recognized as $R-S$ FFs by the input labels. When this symbol is used, it indicates that the FF represented behaves in accordance with a specific set of rules peculiar only to the $R-S$ FF.

Defining the rules of operation for the FF in a nonambiguous manner is somewhat of a problem. Just classifying the state of the circuit is not easy, due

---

* Slight modifications must be made to MIL-STD-806B symbology to accommodate the many different circuit configurations.

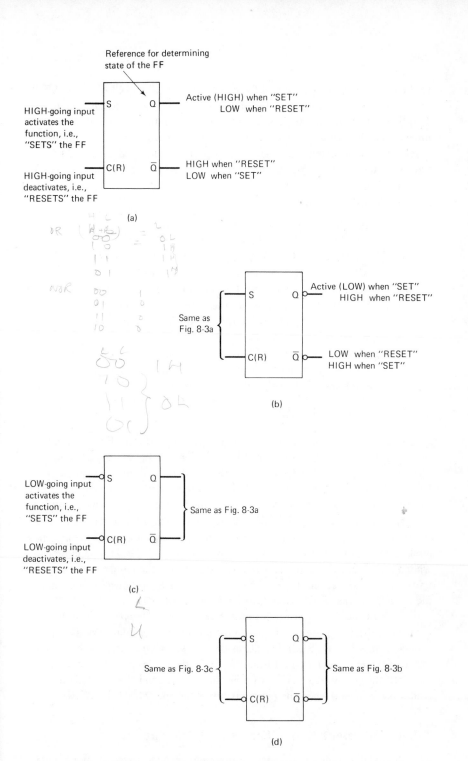

Reference for determining state of the FF

(a)

S   Q — Active (HIGH) when "SET"
         LOW when "RESET"

HIGH-going input activates the function, i.e., "SETS" the FF

C(R)   Q̄ — HIGH when "RESET"
            LOW when "SET"

HIGH-going input deactivates, i.e., "RESETS" the FF

(b)

Same as Fig. 8-3a {
S   Q — Active (LOW) when "SET"
         HIGH when "RESET"

C(R)   Q̄ — LOW when "RESET"
            HIGH when "SET"
}

(c)

LOW-going input activates the function, i.e., "SETS" the FF

S   Q
C(R)   Q̄ } Same as Fig. 8-3a

LOW-going input deactivates, i.e., "RESETS" the FF

(d)

Same as Fig. 8-3c { S   Q } Same as Fig. 8-3b
                 { C(R)   Q̄ }

Figure 8–3   Flip-flop symbols

237

to the many different terms used for this purpose. SET and CLEAR (RESET) are excellent descriptive terms, but they tend to imply a particular type of FF. 1 and 0 work well, but what about the inevitable confusion generated with binary 1 and binary 0? TRUE and FALSE are good but are more commonly applied to Formal Symbolic Logic. Q and $\overline{Q}$ are usually used to define FF outputs, not states of FFs. What, therefore, is the least ambiguous means to define the rules of operation for a FF?

Quoting from MIL-STD-806B:

> The flip-flop is a device which stores a single bit of information. It has three possible inputs, set (S), clear (reset) (C), and toggle (trigger) (T), and two possible outputs, 1 and 0. When not used, the trigger input may be omitted. The two outputs are normally of opposite polarity. A "1" is stored in the flip-flop when the "1" output level is active, and the "0" output level is inactive. A "0" is stored in the flip-flop when the above condition is reversed. The flip-flop assumes the "1" state when an active signal appears at the "S" input regardless of the original state. It assumes the "0" state when an active signal appears at the "C" input regardless of the original state. . . .

One possible approach is to use the SET and CLEAR (RESET) notation to describe FF states, despite the possible confusion that could be generated when discussing other types of FFs. Although the initial rules described in Ch. 2 indicated that HIGH-going signals activated the FF and caused the Q output to be HIGH when the FF was SET and the $\overline{Q}$ output to be HIGH when the FF was CLEAR (RESET), MIL-STD-806B offers some alternatives. The use of the terms *active* and *activating*, along with the basic symbology of the attached circle indicating the LOW level as the activating or active level (depending on whether it is attached to the input or the output of the symbol) allows almost universal representation of the FF function. Figure 8–3 shows the most common combinations of FF symbols normally encountered.

One further important point must be discussed. *Despite* the resulting confusion, the SET state of a FF is almost universally considered as the "1" state and the CLEAR (RESET) state as the "0" state. Of course, such convention may generate further confusion. As far as the FF is concerned, it is storing a 1 when the Q output is active, whether that active condition is HIGH or LOW! And it is storing a 0 when the $\overline{Q}$ level is active and the Q level is inactive.

Tabular Representation of Sequential-Logic Circuits

Sequential-logic circuits may be represented in tabular form in a manner similar to combinational circuits. However conventional truth tables cannot be directly used, due to the memory aspects of sequential circuits, the dependence on previous circuit conditions, and the feedback requirements. A variation of the conventional truth table may be employed to describe sequential logic circuits. Before showing the most commonly used form of the sequential-logic truth table, however, we should collect all of the known information about the circuit being examined. This has been done in Table 8-1.

Rows 1 through 4 examine the response of the latch to various combinations of input signals, when the assumed condition following power application is the latched state. Latch operations described earlier are reproduced in rows 1, 2, and 3. Row 4 poses a problem. Note that when both LATCH and UNLATCH are HIGH, the transition from the latched state is indeterminate. The actual resulting condition may not be the same each time. When both inputs are HIGH, both outputs will be LOW. The final state is determined by which of the inputs is removed (or returned to the LOW level) first and by circuit constants. Thus additional logic circuits must be provided to guard against both inputs going HIGH at the same time. Without such safeguards, a latch with both inputs going HIGH at the same time might be looked upon as a defective light switch — every time it was turned on, the light *might* or *might not* come on! In digital logic, *maybe* is not allowable! Rows 5 through 8 duplicate the operations of rows 1 through 4, except the initial condition is assumed to be unlatched.

Table 8-1 seems to duplicate functions and information. Combining functions alone reduces the size of the table, but another point must be taken into consideration. The latch's state following an input-signal change is dependent upon its state prior to application of the input signal. A conventional truth table does not provide this type of information, so modifications must be accomplished. The information from Table 8-1 is rearranged into the form of

<h4 align="center">Table 8-1   Latch Operation</h4>

	A Assumed conditions (LATCHED)		B Signal Inputs		C Resulting conditions	
	LATCHED	UNLATCHED	LATCH	UNLATCH	LATCHED	UNLATCHED
1	H	L	L	L	H	L
2	H	L	L	H	L	H
3	H	L	H	L	H	L
4	H	L	H	H	Indeterminate	
	(UNLATCHED)					
5	L	H	L	L	L	H
6	L	H	L	H	L	H
7	L	H	H	L	H	L
8	L	H	H	H	Indeterminate	

Table 8–2 and is renamed a *detailed truth table*. HIGH is replaced by 1, and LOW is replaced by 0. Column A is renamed $latched_n$, and column C is renamed $latched_{n+1}$.* Only the LATCHED output is used, since it is known that the UNLATCHED output is always opposite. Evaluation of Table 8–2 reveals that the rules previously defined for operation of the latch are valid.

Duplication still exists in the detailed truth table, and further simplifications may be made. In the *limited truth table*, Table 8–3, the redundant entries are not shown. Also, since it is known that $latched_n$ represents the LATCHED output prior to the change in inputs defined in the table, no need exists for the $latched_n$ column. Table 8–3 is the most commonly encountered form of table used to describe sequential-circuit operation. The reader should be able to expand this table to detailed truth-table form, if clarification of certain circuits is required.

As you may have observed, the tables used to describe the latch are not really "truth tables" in the strict sense of the word. The memory and time dependency of sequential-logic elements place other requirements on tabular representations. Other names commonly encountered include *characteristics table*, *state table*, *modified truth table*, and *flow table* (usually reserved for specific types of tabular displays). The term *state table* is used in this text to describe the tabular representation of sequential-logic element characteristics. Justification for this choice appears during initial discussion of the $R-S$ FF.

Table 8–2 Detailed truth table, Latch

Unlatch	Latch	$Latched_n$	$Latched_{n+1}$
0	0	0	0
0	0	1	1
0	1	0	1
0	1	1	1
1	0	0	0
1	0	1	0
1	1	0	*
1	1	1	*

* Indeterminate

Table 8–3 Limited truth table, Latch

Unlatch	Latch	$Latched_{n+1}$
0	0	$Latched_n$
0	1	1
1	0	0
1	1	*

* Indeterminate

---

* The subscripts n and $n+1$ are used to indicate a time differential. If $latched_n$ is considered to be the logic condition at some time n, then $latched_{n+1}$ is the logic condition at some later time, usually following a change in input conditions.

Algebraic Representation of Sequential Logic Circuits

Boolean equations may also be used to represent sequential-logic elements. Determination of the *Boolean expression* for sequential-logic elements (from state tables) follows closely the extraction of combinational-logic expressions from truth tables. The state table for the latch circuit (Table 8–4) is used as a starting point, and necessary modifications are made to allow easy development of the Boolean expression for operation of the latch.

Replacement of the input names on the left side of the state table with letter designations is the first step. The unlatch column is labeled U, and the latch column labeled L. Since the input columns define the states of the inputs at a reference time n, the labels are called U_n and L_n. At this point, the inputs are in a form that can be easily used in Boolean expressions. The output column is now relabeled in a similar manner. It is common practice to provide sequential-logic elements with a "name" and to use that name as an identifier in Boolean expressions. When such a "name" technique is used, the true or HIGH output is assumed. For example, if the latch circuit is called "A", then when the circuit is in the latched position, it is so identified by calling the output A. When in the unlatched position, it is identified as $\overline{A}$. The output column is then labeled A_{n+1} to denote that it is the state of the output *following* the change in input conditions defined on the same row. Table 8–4 is the relabeled state table.

Examination of the A_{n+1} column discloses that a 1 exists in the second row; therefore A_{n+1} is true (HIGH). The input requirements necessary to obtain this condition are read from the input side of the state table as $U_n = 0$ and $L_n = 1$, or $\overline{U}_n L_n$. Thus one possible Boolean equation describing a means of obtaining the true state for the latch called A is $A_{n+1} = \overline{U}_n L_n$. Another possibility exists. Note that $A_{n+1} = A_n$ if $U_n = 0$ AND $L_n = 0$. If $A_n = 1$ prior to U_n becoming equal to 0 and L_n becoming equal to 0, then A_{n+1} would equal 1 at the time of transition. Thus A_{n+1} could equal 1 if $U_n = 0$, $L_n = 0$, and $A_n = 1$, or if $A_{n+1} = \overline{U}_n \overline{L}_n A_n$.

Now two means of obtaining the true state for the latch called A are available, that is $A_{n+1} = \overline{U}_n L_n + \overline{U}_n \overline{L}_n A_n$. But one additional fact must be taken into consideration. According to the truth table, at no time can both U_n and L_n be 1 at the same time. As we noted, the results are indeterminate, and such a situation is not allowed in digital logic. Another way of showing this condition is $U_n L_n = 0$. It may be verified from the state table that all of the allowable combinations of U_n and L_n do meet this criterion.

Table 8–4   State table, Latch circuit

U_n	L_n	A_{n+1}
0	0	A_n
0	1	1
1	0	0
1	1	*

* Indeterminate

The two equations that fully describe the operation of the latch circuit are

$$A_{n+1} = \overline{U}_n L_n + \overline{U}_n \overline{L}_n A_n \qquad\qquad (8\text{--}1)$$

$$U_n L_n = 0 \qquad\qquad (8\text{--}2)$$

Algebraic simplification of Eq. (8–1) would be

$$
\begin{aligned}
A_{n+1} &= \overline{U}_n L_n + \overline{U}_n \overline{L}_n A_n && \text{Original expression}\\
&= \overline{U}_n L_n + \overline{U}_n \overline{L}_n A_n + 0 && \text{Theorem 7}\\
&= \overline{U}_n L_n + \overline{U}_n \overline{L}_n A_n + U_n L_n && \text{Substitution Eq.(8--2)}\\
&= U_n L_n + \overline{U}_n L_n + \overline{U}_n \overline{L}_n A_n && \text{Commutative Law}\\
&= L_n(U_n + \overline{U}_n) + \overline{U}_n \overline{L}_n A_n && \text{Factoring}\\
&= L_n \quad\quad + \overline{U}_n \overline{L}_n A_n && \text{Theorems 12 and 13}
\end{aligned}
$$

Waveform Representation of Sequential Logic Circuits

Waveform diagrams are also an accepted method of displaying sequential-logic element operation. In fact, the waveform *diagram* was the method encountered in Ch. 2, where a sequential-logic element, called the FF, was used to demonstrate the binary counting process. The waveform diagrams of Fig. 8–4 should be studied carefully and compared with other methods of sequential-logic element presentation to assure that correlation between methods is established.

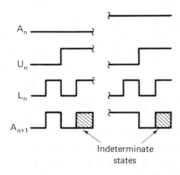

Figure 8–4   Sequential circuit waveform diagram

Physical Configuration of Sequential Elements

FFs may be obtained in any of the physical configurations previously mentioned during discussion of combinational-logic element packaging. Vacuum tube FFs are almost a rarity today, and discrete component semiconductor FFs are rapidly disappearing from the digital scene. The round metal can package (TO-5 type) and the Dual In-line Package (both plastic and ceramic) are in common use; they generally appear permanently mounted on epoxy-glass circuit boards. Interconnections are of the "printed or etched-conductor" type and lead to a multicontact plug, which is an integral part of the circuit board.

Circuit boards may be from 1/16 to 1/4 in. thick and vary from 2 in. x 3 in. to larger than 4 in. x 5 in. More than 70 external connections are found on some circuit boards with upwards of 35 separate integrated-circuit (IC) assemblies mounted. Determination of the actual physical configuration of circuit boards is a task to be undertaken by the design engineer. However many manufacturers make available standard circuit boards of uniform size and plug configuration. Some boards contain sequential-logic elements, while others contain combinational circuits. The design engineer then determines the interconnections necessary to perform the required logic functions. In fact, many logic hardware manufacturers provide the mechanical design, layout, and interconnection of a logic device directly from a user's logic diagram.

In subsequent chapters we shall see sequential logic elements (and combinational elements also) combined in IC form in such a manner that complete functional-logic devices are contained in a single physical package. Numerous FFs, already interconnected to form a device such as a counter, are readily available. The future promises even greater advances in packaging of logic functions. Logic for a complete desk calculator on a single silicon IC-chip may seem fantastic today but is merely a harbinger of things to come.

8–3 THE *R—S* FLIP-FLOP

Just as the word *gate* is universally used to describe combinational-logic elements, so is the word *flip-flop* (FF) universally used to describe sequential-logic elements. Different kinds of gates are identified by additional words describing the function or kind of operation performed (e.g., AND gate, OR gate). FFs also employ this type of designation. Terms like *R—S* and *J—K* are used to *define how a sequential-logic element responds to input signals,* just as AND gate defines how a specific combinational-logic element responds to input signals. New words and definitions must be learned, but in so doing, sequential-logic operations become clearer and easier to use.

Characteristics

The *R—S* FF (Fig. 8–5a) usually consists of electronic circuitry with two input and two output connections. One input connection is labeled R (RESET) and, when activated, causes the *R—S* FF to assume the RESET (CLEAR) condition. Activation of the second input, labeled S (SET), causes the FF to assume the SET condition. As we indicated previously, the outputs are labeled Q and $\overline{Q}$ and are complements of each other, i.e., when one output is HIGH the other is LOW and vice versa. By generally-agreed-upon definitions, the *R—S* FF behaves according to Table 8–5.

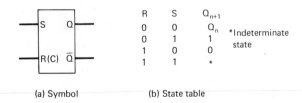

	R	S	Q_{n+1}	
	0	0	Q_n	*Indeterminate
	0	1	1	state
	1	0	0	
	1	1	*	

(a) Symbol   (b) State table

Figure 8—5 $R-S$ flip-flop, symbol and state table

Table 8–5   General Operating Characteristics, $R-S$ Flip-Flop

Input		If the Q output			
R	S	was	it becomes	was	it becomes
Inactive	Inactive	Inactive → Inactive		Active → Active	
Inactive	Active	Inactive → Active		Active → Active	
Active	Inactive	Inactive → Inactive		Active → Inactive	
Active	Active	*		*	

*   Indeterminate

Implementing the $R-S$ Flip-flop with Combinational-Logic Elements

NOR gate implementation of the latch function has already been discussed and the logic diagram shown in Fig. 8–2. By merely relabeling the inputs and outputs, this same logic diagram may be used to describe the $R-S$ FF. Of course, many other combinational-logic element arrangements perform the same function. The only requirement is that the circuit perform in accordance with the general rules and the table of operating characteristics.

Consider now the $R-S$ FF operation with NAND gates. It seems reasonable that such an implementation is practical, since NOR gates have already been used with good success. Figure 8–6 shows a cross-coupled NAND gate version of the $R-S$ FF and the MIL-STD-806B symbol. Investigation of the symbol reveals the following information.

1. Activation of the S input with a LOW-going signal causes the Q output to become HIGH.
2. Activation of the R input with a LOW-going signal causes the $\bar{Q}$ output to become HIGH.

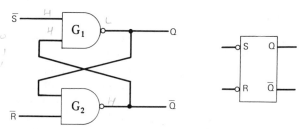

Figure 8−6   NAND gate implementation of the *R−S* flip-flop

The first step in analysis of the NAND gate version of the *R−S* FF is the establishment of an arbitrary starting condition. Choosing the situation where $\overline{Q}$ is HIGH (RESET state) is a conventional approach. If $\overline{Q}$ is HIGH, then Q must be LOW. For Q to be LOW, both inputs to G_1 must be HIGH, since a NAND gate requires all inputs to be HIGH for the output to be LOW. Since one input is derived from the $\overline{Q}$ output, which is HIGH, the second input ($\overline{S}$) must be HIGH at the same time. G_2 then receives a LOW input from the Q output of G_1, and since any LOW input to a NAND gate results in a HIGH output, the circuit is latched into the RESET state. As long as we remain in the RESET state, the $\overline{R}$ input level is not important. It may be either HIGH or LOW, although with the NAND gate version of the *R−S* FF, it is usually maintained in the HIGH level condition when the FF has completed its transition from one state to the other. One of the differences between the NOR gate and the NAND gate version of the *R−S* FF should now be apparent. The static conditions for the NAND gate version require the inputs to be at HIGH levels, while the NOR gate version requires them to be at LOW levels.

With the FF in the RESET state, any change in the $\overline{R}$ input has no effect. The FF is latched, due to the LOW input to G_2 furnished by G_1. The only way to unlatch the circuit is to cause the Q output of G_1 to go HIGH. Changing the $\overline{S}$ input from HIGH to LOW causes this change to occur, since both inputs to G_1 are no longer HIGH. The HIGH Q output combined with the quiescent HIGH $\overline{R}$ input at G_2 causes the $\overline{Q}$ output to go LOW. G_1 now has a permanent LOW input from the G_2 output, and the circuit remains in this (the SET) state. Note that it is the *LOW-going change* that causes the FF to change state, not the HIGH-going change that is usually considered to be the activating signal. The MIL-STD-806B symbol of Fig. 8−6 makes this fact immediately apparent, due to the "inverting" circles at both the S and R inputs.

The NAND gate *R−S* FF thus performs the same functions as the NOR gate *R−S* FF − only the activating input signals are different. (This is easily seen by perusal of the symbol used to represent the FF.) Numerous other arrangements of combinational-logic elements may be used to implement the *R−S* FF, but they all perform the same basic operation. The symbol used describes the input and output conditions necessary to perform the required functions.

$R-S$ **Flip-flop Equations**

Sequential-logic elements may also be represented by Boolean expressions, as demonstrated earlier. The $R-S$ FF is no exception.

A good place to locate the $R-S$ FF expressions is in the state table furnished by the manufacturer of the FF. Generally, this information is in the form of a limited truth table (Fig. 8–7a) and is expanded to the detailed truth table form (state table) in Fig. 8–7b. All possible combinations of the R and S inputs are shown with all possible combinations of the state of the FF at time n. The result of the interaction of the input combinations and the present state of the FF appears in the Q_{n+1} column.

The detailed truth table of Fig. 8–7b shows that the Q_{n+1} output is active (1) if any of the following combinations of inputs exists: $\overline{R}\overline{S}Q_n$ or $\overline{R}S\overline{Q}_n$ or $\overline{R}SQ_n$. Therefore $Q_{n+1} = \overline{R}\overline{S}Q_n + \overline{R}S\overline{Q}_n + \overline{R}SQ_n$, and the FF is "storing" a 1 if any of these input combinations occurs. The input requirements

R	S	Q_{n+1}
0	0	Q_n
0	1	1
1	0	0
1	1	*

*Indeterminate

(a)

R	S	Q_n	Q_{n+1}	"Store" a 1	"Store" a 0
0	0	0	0		$\overline{R}\overline{S}\overline{Q}_n$
0	0	1	1	$\overline{R}\overline{S}Q_n$	
0	1	0	1	$\overline{R}S\overline{Q}_n$	
0	1	1	1	$\overline{R}SQ_n$	
1	0	0	0		$R\overline{S}\overline{Q}_n$
1	0	1	0		$R\overline{S}Q_n$
1	1	0	*		
1	1	1	*		

(b)

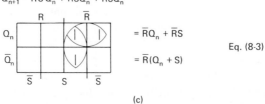

$Q_{n+1} = \overline{R}\,\overline{S}\,Q_n + \overline{R}S\overline{Q}_n + \overline{R}SQ_n$

$= \overline{R}Q_n + \overline{R}S$

$= \overline{R}(Q_n + S)$ Eq. (8-3)

(c)

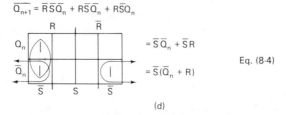

$\overline{Q_{n+1}} = \overline{R}\,\overline{S}\,\overline{Q}_n + R S\overline{Q}_n + R\overline{S}Q_n$

$= \overline{S}\,\overline{Q}_n + \overline{S}R$

$= \overline{S}(\overline{Q}_n + R)$ Eq. (8-4)

(d)

Figure 8–7 $R-S$ flip-flop equations derivations

for "storing" a 0 may also be read from the detailed truth table and are noted as $\overline{Q_{n+1}} = \overline{R}\overline{S}Q_n + R\overline{S}\overline{Q_n} + R\overline{S}Q_n$.

Both of these expressions may be simplified by either algebraic or graphic methods. Figure 8–7c shows map simplification of the equation for storing a 1, while Fig. 8–7d shows similar simplification of the equation for storing a 0. Equations not only provide insight into the operational characteristics of FFs, but they are also useful in understanding operation of devices using sequential-logic elements. Subsequent chapters discuss this matter in greater detail.

R–S **Flip-flop Applications**

Although it is the basic FF, the pure *R–S* FF sees little application in modern digital equipment. It is generally used in situations where only a memory function is required or where sequential operations can be implemented that do not necessitate synchronization with other parts of the sequential circuit. Most other applications of the *R–S* FF require external gating to modify its input/output (I/O) characteristics.

The *R–S* FF could be used in the machine-control example mentioned in Sec. 8–1. One state of the FF is assigned the "move-left" meaning and the opposite state to the "move-right" meaning. The choice of a state for a specific meaning ("move-left" or "move-right") is a matter for the circuit designer to decide.

Another common application for the simple latch, or *R–S* FF, is in the area of mechanical switch replacement. Actually, the mechanical switch is not replaced, but the function is improved by use of the FF. Due to physical construction, most mechanical switches do not provide a "clean" make or break. Spring tensions and mechanical linkages tend to cause a number of intermittent makes and breaks at every activation. When digital equipment is switched, it responds to each of the makes and breaks and often provides unwanted and erroneous operations.

Figure 8–8 shows a typical logic diagram of a device that provides a *bounceless* output for a single-pole, double-throw switch. In the position shown (*A*), the *Q* output is HIGH and the $\overline{Q}$ output is LOW. Any erratic contact with *A* as the switch moves toward *B* can have no effect, since the NAND gates cannot change state until at least one input to G_2 becomes LOW. Upon first contact of the switch arm with *B*, one input to G_2 is LOW, and $\overline{Q}$ becomes HIGH. G_1 now has two HIGH inputs, and the *Q* output is LOW. Any bounce that occurs at contact *B* following initial contact has no effect, since the circuit must be in the opposite state for any change to occur. The operation of the *R–S* FF as a contact-bounce eliminator is exactly as indicated by the state table for the NAND gate implemented function. Once in the SET state, further application of *S* inputs has no effect. Similarly, once in the RESET state, further application of *R* inputs has no effect.

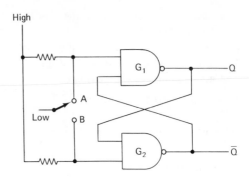

Figure 8—8   Contact bounce eliminator

8—4   THE CLOCKED $R-S$ FLIP-FLOP

Although the $R-S$ FF, or latch circuit, is the basis for almost all other FFs, it does not appear in great numbers in modern digital devices. The basic $R-S$ function is commonly required, but implementing it directly with the $R-S$ FF presents a fundamental problem. Any spurious input signals that appear on either the R or the S inputs cause the FF to change state at a time when such a change is not desired.

To alleviate this problem, the FF is often conditioned (enabled) first with a level and then allowed to make its required transition *after* it receives another input from a different source. The addition of two more NAND gates to the $R-S$ FF of Fig. 8—6 allows the $R-S$ FF to be controlled by *both* the SET/RESET inputs *and* the added input (commonly called the *clock input*). The clock-input signal (called a *clock pulse*) is normally of short duration compared with the time between FF state changes, so the SET and RESET inputs may be changed almost any time between clock pulses (the FF does not change state until the arrival of the clock pulse). Operation of the FF is now isolated from spurious changes of the SET and RESET inputs, except for the extremely short time that the clock pulse is present.

Clock pulses are provided by a clock generator. In *clocked systems*, clock pulses are distributed throughout the system and are used to synchronize equipment operation.

The clocked $R-S$ FF responds exactly as the standard $R-S$ FF, except for the function of the clock inputs. The clock pulse is assumed to be present for the purposes of operation, and the transition from one state to the other occurs upon arrival of the clock pulse. Figure 8—9 shows the clocked $R-S$ FF symbol, the state table, operating equations, waveforms, and a typical implementation using combinational-logic elements. It should be noted that numerous imple-

R	S	CP	Q_{n+1}
0	0	0	Q_n
0	0	1	Q_n
0	1	0	Q_n
0	1	1	1
1	0	0	Q_n
1	0	1	0
1	1	0	Q_n
1	1	1	*

*Indeterminate

$=$

R	S	Q_{n+1}
0	0	Q_n
0	1	1
1	0	0
1	1	*

If the clock pulse is
assumed to be present

$$Q_{n+1} = CP \cdot \overline{R} \, (Q_n + S) \qquad\qquad \text{Eq. (8-5)}$$
$$\overline{Q}_{n+1} = CP \cdot \overline{S}(\overline{Q}_n + R) \qquad\qquad \text{Eq. (8-6)}$$

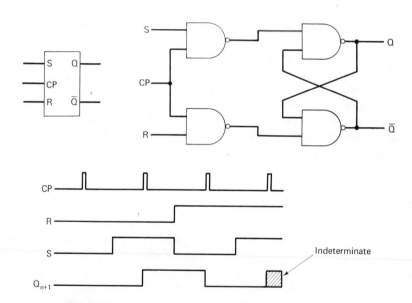

Figure 8–9   The clocked $R-S$ flip-flop

mentations of the clocked $R-S$ FF are available and that the method shown in Fig. 8–9 is merely representative.

The clocked $R-S$ FF finds application most often in *simple* memory circuits, just as the standard $R-S$ FF. Reduction in response to spurious input signals causes the clocked $R-S$ FF to be more acceptable where reliability of stored information is a requirement. In addition, the clocked $R-S$ FF also serves as the basic circuit from which the T, D, and $J-K$ (some versions) FFs are developed.

8-5   THE MASTER-SLAVE PRINCIPLE

Adoption of the clocked $R-S$ concept alleviates the problem of spurious input signals triggering the FF into operation at unwanted times. However, if the $R-S$ FF (either basic or clocked) is used in other than simple storage applications, other difficulties arise. In many counting (Ch. 11) and shift-register (Ch. 12) applications, the Q and $\overline{Q}$ outputs of one FF are connected to the S and R inputs of another FF. The change in state of the first FF tends to cause an immediate and sometimes undesired change in state of the next FF, depending on clock-pulse width and internal delays. External gating may be used to prevent such undesired changes, but common practice is to employ the *master-slave principle.*

The basic concepts of the master-slave principle are seen in Fig 8-10. Information on the S and R inputs is stored in the master FF at the HIGH-going transition of the clock input and transferred to the slave FF at the LOW-going transition of the clock. The S and R inputs are isolated from the Q and $\overline{Q}$ outputs, since the slave control logic is disabled when the master control logic is enabled and vice versa.

Symbology and state tables for the master-slave (sometimes called *dual-rank*) $R-S$ FF are identical to the conventional clocked $R-S$ FF. It is necessary, however, to realize that the actual change in the outputs does not occur until the "trailing edge" of the clock pulse. The waveform diagrams in Fig. 8-10 should be carefully studied to verify operation of the master-slave FF.

8-6   THE T FLIP-FLOP

Although the $R-S$ FF is perhaps the basis for all sequential-logic memory circuits, many other forms of FFs exist. Most of these circuits use the $R-S$ FF as the basic component. For example, a T (*toggle*) FF, which changes state *every* time an activating signal is applied, is often used in counting circuits. T FF variations are numerous, although the basic logic circuit is quite simple. It is constructed from the simple $R-S$ binary element with the addition of gates (or additional inputs to existing gates) to route the incoming signal (trigger) to the appropriate side of the FF. Figure 8-11 shows the logical symbol used for the T FF, its state table, construction using combinational-logic elements, equations, and typical waveforms.

Referring to the logic diagram discloses that NOR 1 and NOR 2 comprise the basic binary element, while AND 1 and AND 2 "steer" the input to the proper NOR gate to operate the T FF. Assuming an initially RESET condition, we see from the waveforms drawing that AND 2 has one input HIGH from NOR 2 and the other input LOW from the normally LOW trigger source. If the

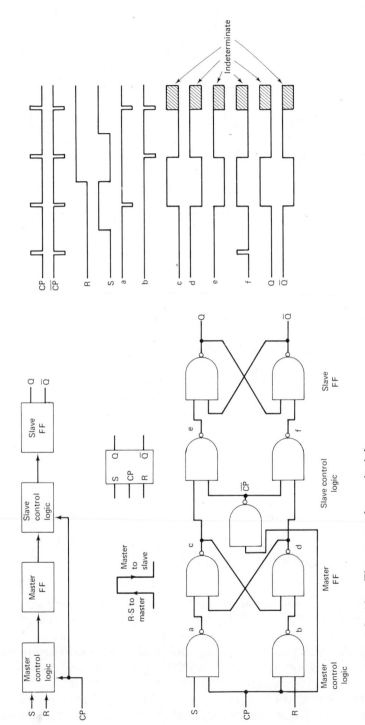

Figure 8–10 The master-slave principle

251

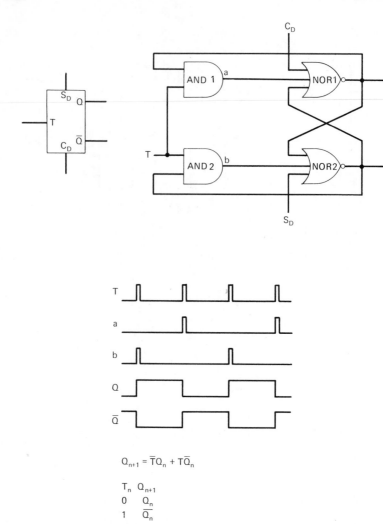

$$Q_{n+1} = \overline{T}Q_n + T\overline{Q}_n$$

T_n	Q_{n+1}
0	Q_n
1	$\overline{Q}_n$

Figure 8–11   The T flip-flop

T input momentarily transitions from LOW to HIGH under these conditions, both inputs to AND 2 are HIGH, and the resulting output supplies a HIGH to NOR 2 input. Any HIGH input to a NOR gate results in a LOW output, and $\overline{Q}$ goes LOW. NOR 1 no longer has any HIGH inputs, and all LOW inputs to a NOR gate provide a HIGH output. Q goes HIGH, locking the binary into the new state (SET) and furnishing a HIGH input to AND 1. AND 2 is now locked out (disabled) with a LOW input from $\overline{Q}$, while AND 1 is enabled with its HIGH input from Q. The next momentary LOW to HIGH transition of the T input is routed through AND 1 to NOR 1, and the binary returns to the RESET state.

Thus the *T* FF changes state for every HIGH-going transition that is received on the input line.

A retractable-point ballpoint pen acts in a manner similar to a *T* FF. If the operating button of the pen is pushed while the point is retracted, the point is lowered to the operating position. If already in the operating position when the operating button is pushed, the point retracts. The "output" changes each time the button is operated.

The *T* FF finds application in simple asynchronous (non-clocked) counting and dividing circuits, such as the logic signal source used to generate truth table combinations in earlier chapters. Many applications require the FF in the logic circuit to be in a specific state prior to the start of the circuit function. Direct SET (S_D) and/or direct CLEAR (C_D) are made available in many FFs. These inputs are used to establish the initial state of the circuit. When S_D and C_D are provided, the state table describing that portion of the FF's operation is identical to the *R—S* FF (Fig. 8—5). The symbol indicates the availability of the direct SET and direct CLEAR as shown in Fig. 8—11.

8—7 THE *D* FLIP-FLOP

In all of the FFs discussed (except the *T* FF), a major problem exists. A specific combination of inputs has resulted in an indeterminate output condition, which must be guarded against with external circuitry. One way of ensuring that no indeterminate state occurs in the circuit operation is to provide only *one* conditioning input, which can be either HIGH or LOW. This input is called the *D* (*DATA*) input. Whatever logic level is present at the *D* input prior to and during the clock pulse appears at the *Q* output when the clock pulse occurs. Since the *Q* output does not assume the *D* input level until *after* the arrival of a clock pulse, this configuration is often called a *DELAY* FF.

Figure 8—12 shows the symbol, state table, equations, logic diagram, and waveforms of the *D* FF.

The *D* FF is used most often to delay a change in operation so that the change can occur in synchronization with the clock pulse. For example, the *D* FF may be used to store output information from a shift register or counter until readout has occurred. This allows the register or counter to begin its next cycle of operation while readout from the previous cycle is taking place.

8—8 THE *J—K* FLIP-FLOP

The *J—K* FF is perhaps the closest approach available to a truly *universal* FF. It was developed primarily to overcome the ambiguity of the *R—S* operation and is available in both the *asynchronous* (nonclocked) and *synchronous* (clocked) versions. Since the synchronous version is commonly used, it is the subject of

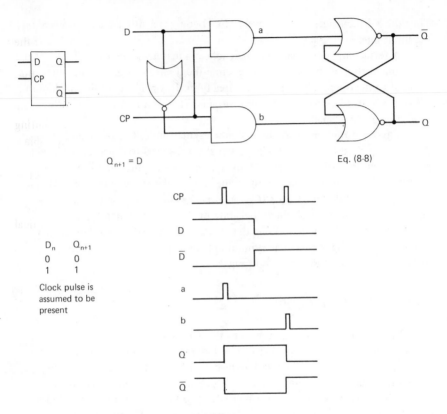

Figure 8–12   The *D* flip-flop

this discussion. The same state table applies for both versions (the clock pulse is assumed to be present in the synchronous mode).

The characteristics of the *R–S* and *T* FFs are combined in the *J–K* FF. New designations for inputs are chosen so that the *J–K* characteristics will not be confused with the *R–S* characteristics, but the *J* input may be considered equivalent to the *S* input and the *K* input the same as the *R* input. However the *J–K* FF may have both inputs activated at the same time — in contrast to the *R–S* circuit. Activation of the *J* and *K* inputs simultaneously causes the FF to change state upon activation by the clock pulse, just as though the *T* input of a *T* binary had been activated. The indeterminate state of *R–S* operation is not present in the *J–K* FF.

J–K symbology, state table, equations, logic diagrams, and waveforms appear in Fig. 8–13. Note the availability of the direct SET and direct CLEAR operations on the symbol.

Due to the flexibility of the *J–K* FF, it may be found in almost all applications of sequential logic. In fact, many designers use only *J–K* FFs to take advantage of quantity purchase prices and to reduce spares requirements.

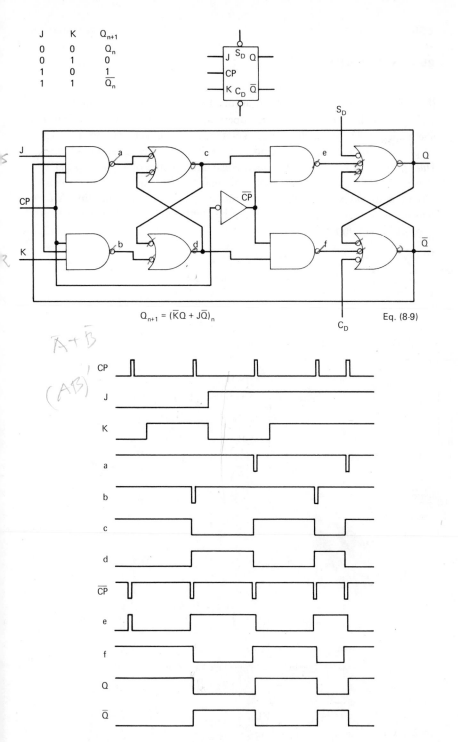

J	K	Q_{n+1}
0	0	Q_n
0	1	0
1	0	1
1	1	$\overline{Q_n}$

$$Q_{n+1} = (\overline{K}Q + J\overline{Q})_n \qquad \text{Eq. (8-9)}$$

$\overline{A} + \overline{B}$

$(AB)'$

Figure 8–13 The $J–K$ flip-flop

255

8–9 SUMMARY

Each of the major classifications of FFs have been discussed in this chapter. Different methods of presenting the operating characteristics were used so that the reader could choose the most understandable approach. However one method of presentation is common to almost all specification sheets, textbooks, etc. The *state table* precisely defines the operating characteristics of any sequential logic device. Table 8–6 is a master state table that combines all of the information from the individual state tables. It summarizes the information in this chapter and should be carefully studied so that the identifying characteristics of each type of FF are understood.

Table 8–6   Master State Table

Inputs							R–S	Clocked R–S	T	D	J–K	Clocked J–K
S	R	T	D	J	K	CP	Q_{n+1}	Q_{n+1}	Q_{n+1}	Q_{n+1}	Q_{n+1}	Q_{n+1}
0	0						Q_n					
0	1						0					
1	0						1					
1	1						*					
0	0					0		Q_n				
0	0					1		Q_n				
0	1					0		Q_n				
0	1					1		0				
1	0					0		Q_n				
1	0					1		1				
1	1					0		Q_n				
1	1					1		*				
		0							Q_n			
		1							$\overline{Q}_n$			
			0							0		
			1							1		
				0	0						Q_n	
				0	1						0	
				1	0						1	
				1	1						$\overline{Q}_n$	
				0	0	0						Q_n
				0	0	1						Q_n
				0	1	0						Q_n
				0	1	1						0
				1	0	0						Q_n
				1	0	1						1
				1	1	0						Q_n
				1	1	1						$\overline{Q}_n$

* Indeterminate

One additional fact is worth mentioning: *most FFs have provision for direct* SET *and* RESET *without the need for clock pulses or internal gating.* This allows preconditioning of FFs when used in counter and shift-register applications so that all stages start in the required logic state.

PROBLEM APPLICATIONS

8—1. Define sequential logic as opposed to combinational logic.

8—2. Why is sequential logic required?

8—3. What is a logic latch?

8—4. Explain the response of a latch circuit to LATCH and UNLATCH inputs.

8—5. Is it possible to determine the operating characteristics of a FF from its MIL-STD-806B symbol? How?

8—6. What is a state table? How is it used?

8—7. List two additional methods of representing sequential functions and explain how they are used.

8—8. Show the characteristics of the $R-S$ FF and explain how these characteristics may be applied to a practical sequential circuit.

8—9. Explain why the $R-S$ FF locks into one or the other state when both the R and S inputs are activated.

8—10. Define asynchronous sequential circuit. List the advantages and disadvantages.

8—11. Define synchronous sequential circuit. List the advantages and disadvantages.

8—12. Is the clocked $R-S$ FF synchronous or asynchronous? Why?

8—13. Discuss the operation of the clocked $R-S$ FF, explaining the operation of each of the gates in Fig. 8—9.

8—14. In the circuit of Fig. 8—9, what is the effect of changing the R and/or S input levels when the clock pulse is not present?

8—15. Why is the master-slave $R-S$ FF preferable to the standard $R-S$ FF?

8—16. Discuss the T FF and its applications.

8—17. What are the minimum number of inputs required for operation of the synchronous D FF? The asynchronous D FF? Explain.

8—18. Does changing the input level to the D FF during the activating time of the clock pulse affect the output after the transition? Explain.

8—19. What is a $J-K$ FF? What are its advantages and disadvantages?

8—20. Why are there no indeterminate states in the operation of the $J-K$ FF?

9

Sequential
Logic
Analysis

Due to the *time dependency* of sequential circuits, their analysis is more complex than combinational circuits. Thus we must develop and apply more advanced procedures. Fortunately, some of the techniques already masterd in combinational logic may be modified to suit sequential-circuit investigations.

9–1 SEQUENTIAL LOGIC ANALYSIS METHODS

The Concepts of Feedback

As we showed in Ch. 8, combinational logic circuits are *interconnected* to provide sequential-logic operations. The sequential-logic circuit includes not only combinational-logic elements but also some type of *memory circuit*. The memory circuit provides feedback to the input of the combinational circuit so that it can "remember" the previous state. This is shown in block diagram form in Fig. 9–1. The external inputs are labeled $X_1, \ldots, X_i$, and the outputs are labeled $Z_1, \ldots, Z_j$.

Memory characteristics are realized by *feedback circuits*, which possess a time delay between the time they receive a particular input and the time that input is applied as feedback to the combinational logic. The time delay may be an actual *delay element*, or it may be *inherent* in the design of the circuit. Regardless of its source, *the inputs to the delay elements become the outputs of the delay elements*, following the delay time. Feedback inputs are labeled $Y_1, \ldots, Y_n$, and feedback outputs are labeled $y_1, \ldots, y_n$. Thus $y = Y$, after a specific time delay. Note that the outputs and the feedback signals are functions of *both* the inputs and the fed-back signals so that the fed-back signals enter into their own control.

Figure 9–1 may be correlated with the basic NOR-implemented R–S flip-flop (FF) of Fig. 9–2. The inputs (Xs) are R and S; the outputs (Zs) are Q and $\overline{Q}$, and the feedback paths (Ys) are the connections from Q to NOR 2 and $\overline{Q}$ to NOR 1.

In Ch. 8, waveform diagrams made it quite apparent that outputs were dependent not only on inputs but also on feedback paths. However, in the initial discussion, a very important point was overlooked: *logic gates are not perfect – they tend to delay the reaction of the output signal to the input signals.* Such delays *may* cause stability problems.

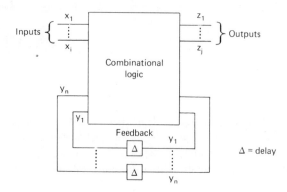

Figure 9–1   The general sequential circuit

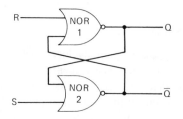

Figure 9–2   A NOR-implemented flip-flop

Stability

If the output of any gate is what we would expect from an examination of the inputs and the gate's truth table, the gate is *stable*. Delays tend to introduce inconsistencies, and short periods of time *may* exist when the truth table is violated. Such deviations from the truth table occur when the input changes and it has not had time to affect the output. *The period of time when the truth* *table is violated is called a period of instability.* Since the inherent delay in a gate is finite (on the order of 10 nanoseconds* in modern high-speed logic), the unstable condition soon disappears, and the truth table is again satisfied.

When logic gates are connected into a network such as Fig. 9–2, the network is stable if *all gates* in the network are stable. The entire network is unstable if *any gate* within the network is unstable. Due to interconnections, logic networks may move from one unstable state to another and may *never* become stable. (An oscillator is a good example of a constantly unstable network.)

The concept of stability is best demonstrated by the use of a *relay*. Figure 9–3 shows the coil and a normally open contact of a typical relay. The following variable assignments are made for the purpose of developing a circuit explanation.

$y = 0$,   Contacts not operated $y = 1$,   Contacts operated
$Y = 0$,   Coil de-energized $Y = 1$,   Coil energized

Initially, it is assumed that the switch is open, the coil is de-energized, and the contacts are operated. At this time, $y = 0$ and $Y = 0$. The circuit is stable; that is, with the coil de-energized and the contacts not operated, these are the conditions that would be expected from the circuit. By definition then, the circuit is stable when the inputs (y and Y) provide the expected results.

The switch is now moved to the closed position. For a brief instant the coil is energized, but the magnetic field has not yet built up, so the contacts remain unoperated. This is shown as $y = 0$ and $Y = 1$. *The circuit is unstable because the contacts should be operated when the coil is energized.*

When the contacts operate, the unstable condition ends. Following the very short unstable condition, $y = 1$ and $Y = 1$, and the circuit is stable. Returning the switch to the open position establishes another unstable situation. For a brief period following switch opening, the contacts are still operated, although the coil is de-energized. (The magnetic field does not decay to zero instantaneously.) During this transitory time, $y = 1$ and $Y = 0$, which is an unstable state. The unstable condition ends when the contacts return to their normally open state, and $y = 0$ with $Y = 0$ (a stable state).

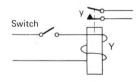

Figure 9–3   Relay coil and contacts

---

* A nanosecond is 10^{-9} seconds or $1/1,000,000,000$ of a second.

Flowtables and Their Construction

In Ch. 8 waveform diagrams were used to explain the operation of sequential circuits (e.g., Fig. 9–2). Such diagrams are useful but somewhat restrictive when it comes to a detailed discussion of FF operation. They are intended to explain the operation of the sequential network under *one pattern of input signals*, but they become impractical when *all* input-signal sequences must be examined. Specifically, analysis of the effect of nonstable states becomes difficult using waveform diagrams.

The ambiguity of English-language descriptions makes their use of questionable value. However tabular methods of sequential-network representation have been developed that seem to offer reasonable display of all possible states (both stable *and* nonstable).

A flowtable shows all possible inputs to all gates contained in a sequential network. It is arranged similar to a logic map, providing one square for each possible combination of inputs to each gate. The physical configuration of a typical flowtable is shown in Fig. 9–4b, where it is compared with the conventional 4-variable logic map (Fig. 9–4a). Note the method of labeling the flowtable, since it differs from the logic map. This labeling allows easier identification of the effects of input-signal changes. Each column differs from the adjacent column by only *one* variable, just like the logic map.

The sequential network of Fig. 9–2 may be analyzed by using the flowtable in Fig. 9–4b. Since the NOR implemented FF has two external inputs (*R* and *S*) and two other inputs that are internally generated (one input to NOR 1 and another input to NOR 2), *sixteen* possible combinations exist. Again, like the logic maps, 2^n combinations must be accounted for, where n = the total number of inputs (both external *and* internal). *The flowtable is so arranged that all external combinations are shown across the top of the table and all internal combinations are displayed along the side of the table*

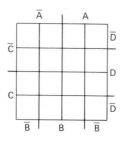

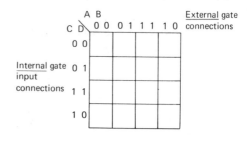

(a) 4-variable logic map   (b) Flow table for a 2-input, 2-gate sequential network.
                                  Each gate has only 2 inputs.

Figure 9–4   Flowtable development

Classification of Sequential Circuits

Sequential logic circuits fall into two general classifications. The *asynchronous sequential circuit* (previously called a *nonclocked circuit*) *responds to changes in input levels through changes in output levels.* The time interval between input signal changes is not constant — an activating signal may be applied to the asynchronous circuit within microseconds of the previous activating signal, or seconds may elapse before the next activating signal arrives. *Since output-level changes rather than rapidly changing output signals* (pulses) *are characteristic of the asynchronous circuit*, it is sometimes referred to as a *dc circuit.*

In contrast, *synchronous sequential circuits* are often called *ac circuits*, since signals received at the input of a synchronous circuit may be level changes but at least one of the inputs is a *pulse.* The pulses occur at a standard *clock rate* and cause the synchronous circuit to react, thus supplying either level changes or pulse outputs. *The synchronous circuit provides an output synchronized with the clock pulses*, while *the asynchronous circuit* (which has no clock pulses) *provides an output almost instantaneously with any input change.* We should note that in synchronous circuits, input levels may change *anytime* between clock pulses, but the output changes do not occur until the arrival of the clock pulses.

9–2 *R–S* FLIP-FLOP ANALYSIS

The first detailed sequential-logic circuit analysis investigates the asynchronous *R–S* FF of Fig. 9–2 and thereby demonstrates a *method of analysis.* General procedures are developed from the demonstration.

Boolean Expressions

For the purpose of discussion, the outputs of each gate are relabeled from Q and $\bar{Q}$ to G_1 and G_2, respectively. At this point such an operation will prevent confusion in the use of flowtables. The Boolean expressions for G_1 and G_2 are

$$G_1 = \overline{R + G_2} = \bar{R}\bar{G_2} \qquad (9–1)$$

$$G_2 = \overline{S + G_1} = \bar{S}\bar{G_1} \qquad (9–2)$$

Entering Information in the Flowtable

Armed with Eqs. (9–1), (9–2), and the blank flowtable of Fig. 9–4, our actual investigation begins. Remember, the flowtable requires 2^n squares, where n is the total number of inputs (both external *and* internal). Two *external inputs* are provided, S and R, and one column must be provided for each combination of S and R. Each combination of *internal inputs* (G_1 to NOR 2 and G_2 to NOR 1) is assigned a *row*. Thus four columns for external inputs and four rows for the internal inputs combine to form the 16-square flowtable for the circuit of Fig. 9–2.

For explanatory purposes, each of the squares in the flowtable is divided into a left-hand and a right-hand section (Fig. 9–5a). The logic value for the Boolean expression representing G_1 is entered in the left-hand section, since the rows are labeled with G_1 preceding G_2. In other words, Eq. (9–1) is evaluated for its logic value by substituting the logic value of R and the logic value of G_2, which appear in each column and row, and by then placing the result in the appropriate square. For example, the square in the upper left-hand corner has an R value of 0 and a G_2 value of 0. Substituting in the equation $G_1 = \bar{R}\bar{G_2}$,

$$\bar{0}\bar{0} = 1 \cdot 1 = 1$$

Thus a 1 is placed in the left-hand side of the upper left-hand square. Systematically evaluating the remaining fifteen combinations results in the flowtable in Fig. 9–5b.

Equation (9–2) is now evaluated in the same manner. Since G_2 follows G_1 in the row labeling scheme, the G_2 values are placed in the right-hand side

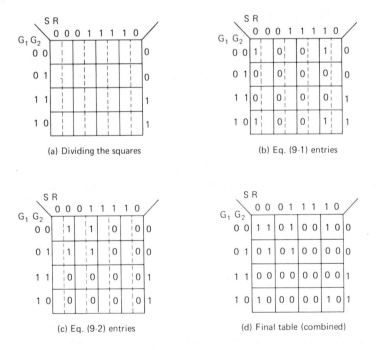

(a) Dividing the squares

(b) Eq. (9-1) entries

(c) Eq. (9-2) entries

(d) Final table (combined)

Figure 9–5   Flowtable construction

of the flowchart (Fig. 9–5c). The two flowtables are now combined to form the chart in Fig. 9–5d. The combined table represents all outputs that result from all possible combinations of the two internal and the two external inputs. If G_1 is arbitrarily assumed to be the output of the network, then the G_1 column is reproduced on the right-hand side of the flowtable.

Identification and Labeling of States

Stable states and unstable states may now be identified. *For the network to be stable, the binary number within a square must be equal to the binary number labeling that row.* This is equivalent to saying that the inputs to all gates (G_1 and G_2, the row labels) must match the value of the truth table for the output of all gates (the numbers in the squares). Remember the numbers in the squares represent the outputs G_1 and G_2, which are a result of evaluation of all inputs to NOR 1 and NOR 2, respectively. As stable states are discovered, they are circled for reference (Fig. 9–6a) and given an *arbitrary identification number*, as shown in Fig. 9–6b.

All squares that are *not* labeled with an identification number (circled) are in unstable states. The network cannot stay in an unstable state for any extended period of time; so assuming that there are no *external-input changes*, the binary

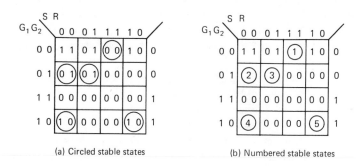

(a) Circled stable states

(b) Numbered stable states

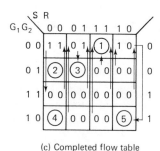

(c) Completed flow table

Figure 9–6   Identification of stable states

number within the square indicates what the network will do while attempting to become stable. Since external inputs are assumed *not* to be changing at this time, the state of the network at any given instant of time is defined by one of the rows within the column defining the external-input states. Movement within the column is indicated by arrows (Fig. 9—6c).

The Operating Point

The term *operating point* describes *the state of the network at any given instant.* It moves from column to column, as the input conditions change, and up and down within the column, according to arrows drawn on the chart. Movement within the column continues until a stable state is reached. The operating point remains at the stable state until an input condition moves it to another column. Once again, it follows the arrows in the new column until a stable state is reached. Tracing the operating point in the flowtable for the NOR-implemented FF demonstrates this concept (see Fig. 9—6c).

Reading the Completed Flowtable

In the first column, both inputs (S and R) are LOW, and the network has two stable states (labeled 2 and 4). Stable state 2 has an output of 0, while stable state 4 has an output of 1. The starting point for discussion is the first column ($S = 0$ and $R = 0$), operating at stable state 2. Changing S to 1 momentarily and then returning it to 0 causes the following actions to take place.

1. The operating point moves horizontally from stable state 2 in the first column to column 10.
2. The operating point follows the arrows in column 10 to the top row and then to stable state 5.
3. It remains at stable state 5 until the S input is returned to 0, at which time it moves horizontally to stable state 4 in column 00.
4. No further movement occurs, since the network is stable and the inputs are not changing.

Note the change in output signal as the operating point changes. Starting at 0, it changes to 1 when stable state 5 is reached and remains at 1, even after the S input is returned to 0.

Another input change has similar results. Changing the R input to 1 while retaining the S input at 0 causes the output to return to 0 by moving the operating point to column 01 and stable state 3. Returning the R line to 0 merely causes the operating point to move horizontally to stable state 2, where the output is still 0.

The indeterminate outcome of the truth table entry of $R = 1$ and $S = 1$ is easily demonstrated by flowtables. Actually, the $R = 1$ and $S = 1$ condition is quite predictable. No matter what the previous input condition, horizontal movement to column 11 will eventually result in vertical movement within the column to stable state 1 and an output of 0. Now, however, returning the input conditions to $R = 0$ and $S = 0$ presents a problem. The operating point moves horizontally to the left and reaches the upper left-hand square (column 00). The arrows show that a transition to row 11 then back to row 00 occurs, and that an oscillation sets in with the state transferring back and forth between rows 00 and 11. A continuous transition between rows 00 and 11 requires G_1 and G_2 to change *simultaneously*. Since two gates never change (predictably) at the same speed, either G_1 or G_2 arrives at the 1 level slightly ahead of the other. Under these circumstances, the network moves to either row 01 or 10. Once either of these rows is reached, a stable condition (either 2 or 4) is established, and the oscillation never occurs. If stable state 2 is reached first, the output is 0; if stable state 4 is reached first, the output is 1. Thus the action of raising both R and S to 1 and returning them to 0 simultaneously results in an indeterminate output. This type of operation is called a *critical race condition*, and it is apparent that the output is unpredictable. When such circumstances are encountered during an investigation of sequential networks, they point to a potential source of problems.

When the outcome of a race condition is predictable, the race is called *noncritical*. All races are characterized by arrows that skip rows on flowtables, since more than one gate must change at the same time to allow the operating point to follow the arrow. Arrows that point to adjacent rows (e.g., in combinational logic maps, top and bottom rows are adjacent) present no problems, since only one output changes at a time. All race conditions may be classified as either critical or noncritical. In terms of the flowtable, *a noncritical race is one in which the outcome is always the same no matter what path the operating point takes.* The critical race in column 00 has already been discussed. A noncritical race appears in column 11, where the operating point must move from row 11 to row 00.

The Compressed Flowtable

In most cases, the flowtables discussed in this chapter are more detailed than what is actually necessary for an understanding of properly designed and checked sequential-logic networks. By selecting certain critical internal-input lines, we can reduce the scope of the flowchart and still show the important information. The previous example started by writing the Boolean expressions for the outputs of both NOR 1 and NOR 2. The same network may be investigated by writing *only* the Boolean expression for the output of G_1; that is,

$$G_1 = \overline{R + (\overline{S + G_1})} \qquad (9\text{–}3)$$
$$= \overline{R + \overline{S}\overline{G_1}}$$
$$= \overline{R}(S + G_1)$$
$$G_1 = \overline{R}S + \overline{R}G_1 \qquad (9\text{–}4)$$

Note that the expression for G_2 is directly substituted.

Although Eq. (9–4) is more complex than either of the two expressions it replaces, it does encompass the operation of both gates. The new expression is entered in the flowtable in the usual manner. Since only one expression is required, the table needs only two rows. Figure 9–7 shows this *compressed flowtable* and should be compared with Fig. 9–6 to ensure that all important factors have been displayed.

The compressed flowtable displays the same number of stable states with the same related outputs – but in a much simpler manner. However the critical race in column 00 is completely masked and could be troublesome in an untried sequential network. In this specific case, no difficulty will occur if we remember that the only path into the race is from column 11 to column 00. This requires a change in *two* inputs, and as we mentioned earlier, it is highly improbable that two signals will change simultaneously. As long as this restriction on input changes is recognized, the compressed flowtable adequately displays the operation of the NOR-implemented $R–S$ FF.

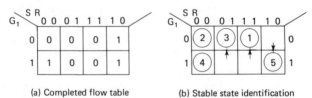

| | (a) Completed flow table | | | (b) Stable state identification | | | |

Figure 9–7   Compressed flowtable for NOR-implemented R-S flip-flop

A Summary of Analysis Procedure

The investigation of a sequential-logic network implemented with combinational logic elements can be summarized in the following steps.

1. Label all gates.
2. Write the Boolean expressions for each gate output. If desired, write the Boolean expression for only the major output (to simplify flowtable construction and analysis).
3. Construct the basic flowtable, using the form shown in Fig. 9–5 or 9–7. The number of squares required is 2^n, where n is the total number of inputs (both external *and* internal).
4. Starting with the gate represented in the left-hand column of the internal-inputs labels, enter the logic value of the Boolean expression for the output of this gate in the left-hand section of each square.
5. Repeat Step 4, working toward the right, until the output of each gate is shown in each square.

6. Circle all binary numbers within squares that are equal to the binary numbers labeling those rows. (The stable states are now identified.)
7. Assign arbitrary numbers to each stable state and redraw the flowtable using only the stable state identification numbers. Draw operating-point movement arrows to show the logic transitions between states.
8. Assure that no critical races exist.
9. Check the FF for operation in accordance with the flowtable to determine proper operation.

9—3 ANALYSIS USING ONLY A LOGIC DIAGRAM

Establishing Initial Conditions

Now for a typical sequential circuit. We know that the logic diagram of Fig. 9—8a represents a sequential network. Troubleshooting has isolated a malfunction to the module that contains this circuit. However we do not know what the module is supposed to do or how it functions. So we must first explore the circuit and establish its normal operation, before we can identify its abnormal operation.

If the Boolean expression is not easily ascertained through the logic diagram, alternate methods may be used. In the case of the sequential network of Fig. 9—8a, a number of feedback connections exist, thus complicating the derivation of the Boolean expression. The method shown herein is helpful under these conditions.

One approach is to make some assumptions about the inputs and outputs. Hence we may assume that the input is initially at a LOW level and that the output of G_2 is at a HIGH level. The inputs and outputs of all gates in the circuit are developed using these assumptions.

With the input initially LOW, the output of G_3 is HIGH, and when combined with the HIGH output of G_2 in G_1, it results in a LOW output from G_1. G_5 receives inputs from G_1 (LOW) and G_3 (HIGH), and a HIGH G_5 output is noted. One input to G_6, then, is a HIGH from G_5. The other input to G_6 is from G_4, where the LOW input level has generated a HIGH output. Therefore G_6 has both HIGH inputs and also furnishes a LOW input to G_4, thus assuring that its output remains HIGH. G_2 has a LOW input from G_1 and a HIGH input from G_4, so the initially assumed HIGH G_2 output is verified. Although many circuits provide outputs that are complementary, this is also *one* of the characteristics of the Q and $\bar{Q}$ outputs of a FF. Thus at least a small piece of evidence exists to justify identification of the circuit of Fig. 9—8a as a sequential network.*

---

* Note the direct SET and CLEAR inputs. Examination reveals that the use of these inputs results in a typical $R-S$ operation, discussed in Section 9—2.

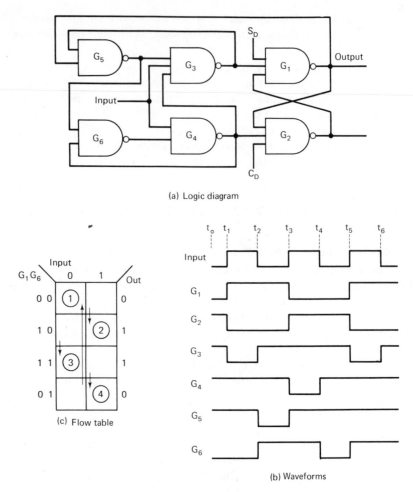

(a) Logic diagram

(c) Flow table

(b) Waveforms

Figure 9–8   A sequential circuit

Using Waveforms

The preceding information may be displayed on a waveform diagram; it is seen in the time period t_0 to t_1 in Fig. 9–8b. Reaction of both outputs to changes in inputs is shown in the waveform diagram. Only one external input is provided, and an investigation of the internal inputs reveals that the output of G_1 is indirectly applied to both G_3 and G_4. One additional internal input should be considered so that the complete flowtable may be developed. The choice lies between G_4, G_5, and G_6. G_6 is selected because it derives its output from *both* G_4 and G_5. The reader should investigate each of the input transitions and verify that the waveform diagram adequately describes the circuit operation.

From the information furnished, it is apparent that the outputs are complementary and that they change state with every activating signal input. The circuit thus performs according to the definition of a T FF.

Boolean Expressions

Although the waveform method does show the general operation of a sequential circuit, the critical action of the circuit during transition from one state to another does not appear easily. The additional details necessary to define the action of the circuit during transition may be obtained using the flowtable method. Derivation of the Boolean expressions is not exactly "intuitively obvious," but the reader should now be able to perform such derivations. Each gate output is listed, and the selected internal inputs (G_1 and G_6) are derived as follows.

$$G_1 = \overline{G_2 G_3} = \overline{G}_2 + \overline{G}_3$$
$$G_3 = \overline{T G_4 G_5} = \overline{T} + \overline{G}_4 + \overline{G}_5$$
$$G_5 = \overline{G_1 G_3} = \overline{G}_1 + \overline{G}_3$$
$$G_1 = \overline{G}_2 + \overline{G}_3$$
$$= \overline{\overline{G_1 G_4}} + \overline{\overline{T G_4 G_5}}$$
$$= G_1 G_4 + T G_4 G_5$$
$$= G_1 G_4 + T(G_4 G_5)$$
$$= G_1(\overline{T} + \overline{G}_6) + T\overline{G}_6$$
$$\therefore\quad G_1 = \overline{T} G_1 + T\overline{G}_6 + G_1 \overline{G}_6$$
$$\text{(Eq. 9–5)}$$

$$G_2 = \overline{G_1 G_4} = \overline{G}_1 + \overline{G}_4$$
$$G_4 = \overline{T G_6} = \overline{T} + \overline{G}_6$$
$$G_6 = \overline{G_4 G_5} = \overline{G}_4 + \overline{G}_5$$
$$G_6 = \overline{G}_4 + \overline{G}_5$$
$$= T G_6 + G_1 G_3$$
$$= T G_6 + G_1(\overline{T G_4 G_5})$$
$$= T G_6 + G_1[\overline{T} + (\overline{G}_4 + \overline{G}_5)]$$
$$= T G_6 + G_1(\overline{T} + G_6)$$
$$\therefore\quad G_6 = T G_6 + \overline{T} G_1 + G_1 G_6$$
$$\text{(Eq. 9–6)}$$

Flowtable

Figure 9–8c is the flowtable for the logic diagram of Fig. 9–8a. Once again, the reader should verify each entry to assure that the table adequately represents the sequential circuit.

With the information developed, troubleshooting may now continue. An input level may be supplied and the output of each gate compared with the waveforms and flowtable to identify the defective element. Following replacement of the defective element, the inputs and outputs should be verified to ensure correct operation of the module.

9–4 THE CLOCKED *R–S* FLIP-FLOP

Boolean Expressions

The clocked $R–S$ FF of Fig. 8–9 is now explored by using flowtables. Synchronous operation (e.g., demonstrated by the clocked $R–S$ FF) may be considered similar to asynchronous operation *except* for the clock input (which acts as another input). From the technician's viewpoint, the circuit has three inputs $(CP, S,$ and $R)$ and two outputs $(Q$ and $\bar{Q})$. The S and R inputs are not allowed to change during CP. They must be present and stable prior to the occurrence of CP. As in previous examples, an output is stated in terms of the inputs *and* outputs. Derivation of the Q and $\bar{Q}$ outputs, based on the initial equations as determined by an inspection of the logic diagram, is presented now.

$$
\begin{aligned}
Q &= \overline{(\overline{S \cdot CP})\,(\bar{Q})} \\
&= \overline{(\overline{S \cdot CP})\,[\overline{(\overline{R \cdot CP})\,(Q)}]} \\
&= \overline{(\overline{S \cdot CP})} + [\overline{(\overline{R \cdot CP})\,(Q)}] \\
&= S \cdot CP + [\overline{(\overline{R \cdot CP})\,(Q)}] \\
&= S \cdot CP + [(\bar{R} + \overline{CP})\,(Q)] \\
&= S \cdot CP + (\bar{R} \cdot Q + \overline{CP} \cdot Q) \\
\therefore\ Q &= S \cdot CP + \bar{R} \cdot Q + \overline{CP} \cdot Q
\end{aligned}
$$
(Eq. 9–7)

$$
\begin{aligned}
\bar{Q} &= \overline{(\overline{R \cdot CP})\,(Q)} \\
&= \overline{(\overline{R \cdot CP})\,[\overline{(\overline{S \cdot CP})\,(\bar{Q})}]} \\
&= \overline{(\overline{R \cdot CP})} + [\overline{(\overline{S \cdot CP})\,(\bar{Q})}] \\
&= R \cdot CP + [\overline{(\overline{S \cdot CP})\,(\bar{Q})}] \\
&= R \cdot CP + [(\bar{S} + \overline{CP})\,(\bar{Q})] \\
&= R \cdot CP + [\bar{S} \cdot \bar{Q} + \overline{CP} \cdot \bar{Q}] \\
\therefore\ \bar{Q} &= R \cdot CP + \bar{S} \cdot \bar{Q} + \overline{CP} \cdot \bar{Q}
\end{aligned}
$$
(Eq. 9–8)

The Flowtable

The complete flowtable is shown in Fig. 9–9a and gives no evidence of critical race problems. (Column labeling is such that only one variable changes from column to column. This simplifies the interpretation of the flowtable after it is completely filled in.) On the basis of no critical races, the compressed flowtable (Fig. 9–9b) is developed, using only the Q output. An interesting observation may be made from these flowtables. The left half of the table represents the conditions under which the clock pulse is logic 0. Note that *no matter what the other inputs do when the clock pulse is logic 0, the output remains the same.* Changing S or Q (or both) merely moves the operating point to another column, which has an identical output. In effect, both before *and* after the occurrence of the extremely short-time duration clock pulse, the output of the circuit is stable. Or, stated another way, *the FF is stable at rest.*

Ambiguous Operation

Although it is not apparent from the flowtable, a potential problem exists when both S and $R = 1$. Just like the asynchronous R–S FF, this combination is not allowed. When the clock pulse changes from 0 to 1 with both R and $S = 1$, the outputs of both input NAND gates go LOW. This causes both Q and $\bar{Q}$ to go HIGH, and there they remain until the clock pulse returns to 0. At this point, the ambiguity appears. Both input gates cannot be expected to change simultaneously, and whichever gate reaches a critical level first releases one of the NAND gates, G_1 and G_2. The regenerative action of this connection causes one output to go HIGH, while the other goes LOW. Which goes HIGH first is indeterminate, and the $R = S = 1$ input combination is to be discouraged. The other

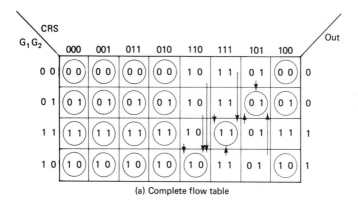

(a) Complete flow table

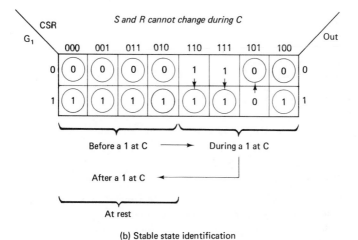

(b) Stable state identification

Figure 9–9   Clocked *R–S* flip-flop flowtable

input combinations are conventional, and the clocked *R–S* FF follows the rules and state table of Fig. 8–9b.

Master-Slave Flip-Flops

A master-slave FF, like that shown in Fig. 8–10, may be viewed as two *separate* sequential networks. The master behaves as a simple clocked *R–S* FF at the leading edge of the clock pulse. After transition, the outputs of the master are used as inputs to the slave FF. At the trailing edge of the clock pulse, the information at the inputs to the slave are transferred to the slave, and its outputs reproduce the information stored in the master FF. Thus *the master FF is a conventional clocked R–S binary*, and *the slave FF is a clocked R–S binary with a* LOW-*going clock input.*

9–5 A TYPICAL SYNCHRONOUS CIRCUIT

The detailed operation of a synchronous sequential-logic circuit is demonstrated by the logic diagram in Fig. 9–10. Once again, we will assume that only the logic diagram is provided, so we must determine the operating characteristics of the circuit without benefit of any circuit description. In this case, the development of the circuit description is undertaken by using Boolean algebra.

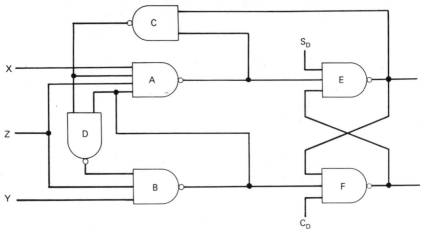

Figure 9–10   A synchronous sequential logic circuit

Boolean Expressions

The cross-coupling of gates E and F is sufficient reason to suspect that this circuit is a FF. But we must decide how to categorize the circuit, i.e., what kind of FF is it? Once again, conventional R–S operation is provided through the S_D and C_D inputs.

The first step is to describe the output of each gate in terms of the input. Investigation of gate A reveals that it has four inputs and is a NAND gate. As we noted previously, all of the inputs of the NAND gate are ANDed together, and the resulting term is then negated. The inputs to gate A are X, Z, B (the output of gate B), and C (the output of gate C). ANDed together the term becomes $XZBC$; when negated, it appears as $\overline{XZBC}$. By applying DeMorgan's Law, $\overline{XZBC}$ may also be expressed as $\overline{X} + \overline{Z} + \overline{B} + \overline{C}$. Each of the remaining gates can be approached in a similar manner, and the Boolean equations describing the outputs of all gates follow.

$$A = \overline{XZBC} = \overline{X} + \overline{Z} + \overline{B} + \overline{C}$$
$$B = \overline{YDZ} = \overline{Y} + \overline{D} + \overline{Z}$$
$$C = \overline{AE} = \overline{A} + \overline{E}$$
$$D = \overline{BC} = \overline{B} + \overline{C} = YDZ + AE$$
$$E = \overline{AF} = \overline{A} + \overline{F}$$
$$F = \overline{BE} = \overline{B} + \overline{E}$$

If the flowtable is used to determine circuit operation, the internal and external inputs must be discovered. The external inputs are relatively apparent: X, Y, and Z. The E output is considered one of the internal inputs, since it feeds back through gate C to gate A and also through gate D to gate B. One additional internal input should be considered so that a complete flowtable may be developed. The choice lies between the B, C, and D outputs. A good selection might be the D output, since it is derived from both B and C. Thus all gates enter into the feedback path.

Derivation of the Boolean equations for E and D is straightforward.

$$
\begin{aligned}
E &= \bar{A} + \bar{F} & D &= \overline{BC} \\
  &= \overline{\overline{XBCZ}} + \overline{\overline{BE}} & &= \bar{B} + \bar{C} \\
  &= XBCZ + BE & &= YDZ + AE \\
  &= (X)(Z)(BC) + BE & &= YDZ + (\overline{XBCZ})(E) \\
  &= XZ\bar{D} + BE & &= YDZ + (\bar{X} + \bar{B} + \bar{C} + \bar{Z})(E) \\
  &= XZ\bar{D} + (\bar{Y} + \bar{D} + \bar{Z})(E) & &= YDZ + (\bar{X} + \bar{Z} + D)(E) \\
\therefore\ E &= XZ\bar{D} + \bar{Y}E + \bar{D}E + \bar{Z}E & \therefore\ D &= YDZ + \bar{X}E + DE + \bar{Z}E \\
 & \text{(Eq. 9--9)} & & \text{(Eq. 9--10)}
\end{aligned}
$$

Flowtable Construction

The E and D logic values are entered in the flowtable just as they were in previous examples. Since three external and two internal inputs are used, a total of 32 squares must be provided ($2^5 = 32$), which are arranged as shown in Fig. 9–11a. The headings for columns are the same as in the clocked R–S FF flowtable. Stable states are identified according to previously defined rules, and arbitrary state identification numbers are provided. Note that identification numbers 10 and 11 are not used − to preclude unnecessary confusion with the binary designations used in each square.

Flowtable Evaluation

Evaluation of the completed flowtable (Fig. 9–11b) may be simplified by investigating *groups of columns* rather than individual columns. The first four columns are investigated as a group, since one of the external inputs (Z) remains at the same logic level for each of these columns. Note that in these columns, the circuit is stable with *either* a 0 or a 1 output. This means that either X or Y (or both) may change, but no change will be seen in the output.

From the flowtable viewpoint, a change in X or Y (or both) results in movement to a different column from the reference column ($Z = 0$, $X = 0$, and $Y = 0$) with no change in row. A change in row *could* occur, so long as no change in output resulted. A change from row 00 to row 01 would not change the output value and would be acceptable, although no such changes are called for in this flowtable. In all cases, in the first four columns, any change in X or Y merely moves the operating point to another stable state in the same row. If the

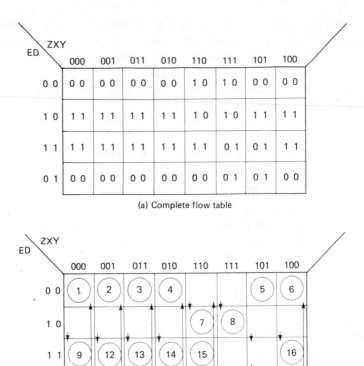

(a) Complete flow table

(b) Stable state identification

Figure 9—11   Flowtable for **Figure 9—10**

output is 0 prior to a change in X or Y input, the output remains 0 at the new combination of inputs. For example, if only the X input changes, the operating point moves from stable state 1 to stable state 4, both of which provide a 0 output. Stable state 9 might have been the initial state (a 1 output), and changing the X input would have moved the operating point to stable state 14 (still with a 1 output). In either case, returning the X input to 0 returned the operating point to the initial stable state with no change in output. The reader should verify the movement of the operating point for all possible cases in the first four columns to ensure that no output change occurs so long as the Z input is 0.

Although any of the eight possible input combinations may be used as a reference point, column 000 is a convenient point of departure. A typical input-signal change is where the Z input is changed to 1 and then, a short time later, returned to 0 with no change in X and Y. As soon as Z becomes 1, the operating point moves to column 100 and stable state 6. The output is still 0 and remains 0 both when Z is 1 and also when Z returns to 0, since the operating point returns to stable state 1 (a 0 output). A similar situation occurs when the point of departure is stable state 9 (1 output) in column 000. The output remains 1, even though Z makes a transition to 1 and then back to 0.

Another possibility is where X is 1 when Z goes through the $0-1-0$ transition. At the 0-to-1 transition of Z, the operating point moves to unstable state 10 in column 110, then to stable state 7. When Z makes a transition back to 0, the operating point moves to unstable state 11 in column 000 and then to stable state 9, where the output is still 1. The output has changed from 0 to 1 by application of the Z-input transition with X input at 1.

Complete Characteristics

Further research, which the reader should verify, shows the following characteristics.

1. A Z-input transition with both X and Y inputs at 0 shows no change in output.
2. A Z-input transition with $X = 1$ and $Y = 0$ causes a 1 output.
3. A Z-input transition with $X = 0$ and $Y = 1$ causes a 0 output.
4. A Z-input transition with $X = 1$ and $Y = 1$ causes a change in state. If the output is 0, it becomes 1; if it is 1, it becomes 0.
5. When $Z = 0$, a change in either X or Y (or both) has no effect on the output.

Comparison of these characteristics with the various FFs discussed in Ch. 8 reveals that the circuit under investigation* behaves as a $J-K$ FF operating in the synchronous (clocked) mode with Z as the clock input. The logic circuit schematic of Fig. 9–10 can therefore be replaced by the $J-K$ FF symbol, and an analysis of the remaining circuitry is much simplified.

9–6 SUMMARY

The basic concepts of sequential-logic analysis have been presented in this chapter. In particular, the use of flowtables to show detailed operation of sequential circuits has been stressed. The examples used are sufficient to demonstrate the general concepts, but more detailed study is required to become proficient in this type of investigation. The reader is urged to apply flowtable techniques not only to the problems provided here but also to as many other sequential circuits as possible.

---

* Note that this investigation has been performed completely without the aid of waveform diagrams. The reader may wish to examine the circuit waveforms with various inputs to verify the Boolean algebra results.

PROBLEM APPLICATIONS

9—1. What other factor must be present with combinational-logic circuits to form sequential-logic circuits?

9—2. Define instability as it is used when discussing sequential-logic circuits.

9—3. How long can a gate remain in an unstable condition?

9—4. List three methods of explaining sequential-logic circuit operation.

9—5. What is a flowtable?

9—6. Where are external inputs shown on a flowtable?

9—7. Where are internal inputs shown on a flowtable?

9—8. How many squares are required on a flowtable that is to be used to explain the operation of a sequential logic circuit with three external and two internal inputs?

9—9. What is a synchronous sequential-logic circuit?

9—10. What is an asynchronous sequential-logic circuit?

9—11. Explain how information is entered in a flowtable.

9—12. How are stable and unstable states identified in a flowtable?

9—13. What is the operating point in a flowtable?

9—14. How is a critical race detected? Explain the meaning of the term critical race.

9—15. How is a noncritical race detected? Explain the meaning of the term noncritical race.

9—16. What is a compressed flowtable? Where does it find application?

9—17. Summarize the steps required in analyzing a sequential logic circuit.

9—18. Develop the operating waveforms for the clocked $R-S$ FF analyzed in Sec. 9—4.

9—19. Develop the Boolean equations for the $R-S$ master-slave FF shown in Fig. 8—10.

9—20. Develop the flowtable for the $R-S$ master-slave FF shown in Fig. 8—10. Discuss any circuit difficulties that appear as a result of applying the flowtable.

9—21. Develop the Boolean equations for the D FF shown in Fig. 8—12.

9—22. Develop the flowtable for the D FF shown in Fig. 8—12. Discuss any circuit difficulties that appear as a result of applying the flowtable.

10

Troubleshooting Combinational / Sequential Logic

Having developed combinational logic in Chs. 3 through 7 and sequential logic in Chs. 8 and 9, our next step is an examination of circuits containing *both* combinational and sequential logic elements. In practical applications, most functions performed by digital logic combine both types of circuits. The analysis and troubleshooting of *combined circuits* closely parallels the techniques discussed in Ch. 7. Due to the increased complexity, however, some modifications are necessary.

10–1 A PROPOSED PROCEDURE

Chapter 7 (Sec. 7–1) describes the preparatory steps required to investigate a combinational logic circuit. These preparatory steps apply to practically all situations, whether they are combinational, sequential, or a combination of both. So before we begin our actual analysis, we must obtain *the overall logic-function description, define the inputs and outputs, establish timing* (if any), and *define the rules of logic* (if nonstandard symbols are used).

At this point, a deviation from the procedure in Ch. 7 is required. The sequential and combinational functions should be separated *first*. This permits each type of logic circuit to be discussed separately and reduces the confusion of working with circuits that operate in different manners. It may be difficult, however, to decide what portions of the circuit are sequential and what portions are combinational. No hard and fixed rules are used, but some clues appear in most logic diagrams.

If the symbol for the flip-flop (FF) is noted, that portion of the circuit is definitely sequential. Logic circuits that have outputs connected *back* to inputs within the same general circuit *could* easily be sequential and should be perused closely. In many cases, the sequential functions will be grouped separately from

279

the combinational functions — an obvious advantage. Most of all, *experience* with a generous serving of common sense must be liberally applied to adequately separate the two types of functions. The reader should investigate as many logic circuits as possible to gain this experience.

Actual analysis of combinational logic functions presents little difficulty if the material in Chs. 3 through 7 has been carefully studied and well understood. Sequential logic circuits may be a bit more challenging. Within Chs. 8 and 9, sequential logic was approached as interconnections of combinational logic circuits. But little was mentioned concerning the interconnection of sequential logic circuits in performing time-dependent operations. Such discussions properly belong in the chapters concerned with counters and shift registers. Simple time-dependent sequential circuits are discussed in this chapter, however, to point out their application in everyday logic circuitry.

After the combinational and sequential circuits have been investigated separately, the functions may be recombined to develop an overall description of the circuit. All of the common descriptive methods may be applied to combined circuits.

10–2   A TYPICAL MIXED FUNCTION

The Logic Diagram

Practically all logic-circuit analysis requires a *logic diagram*. Usually, the analysis is performed because of a circuit malfunction, and from a troubleshooting standpoint, the problem must be isolated to a specific portion of the overall system. The logic diagram for this part of the system is then reviewed, and if other descriptions are not available, investigation begins at this point. Figure 10–1 is a logic diagram that must be analyzed. It represents a control unit for positioning a worktable carrying metal parts to be machined into shape for use in aircraft. A description of the function of the control unit follows.

The control unit responds to two limit switches (A and B) which are used to detect the position of the worktable. In addition, the unit reacts to an internally generated function (C_n), which is a function not only of the internal function's immediate past history (C_{n-1}) but also of the present state of the limit switches. Operationally, the control unit provides an output I_n (an *indexing pulse*) whenever any one (and *only* one) of the three inputs (the two limit switches and the internally generated input) are logic 1; also an output is produced when all three inputs are logic 1. The state of the internally generated function C_n is determined by the requirement that it be logic 1 when any two of the inputs provide a logic 1 and also when all three inputs are logic 1. The control unit also provides an indexing direction capability; that is, the table must be moved four steps to the left, then reversed and moved six steps to the right. Each index pulse must therefore be counted, and a direction-change signal generated after every four and every six index pulses.

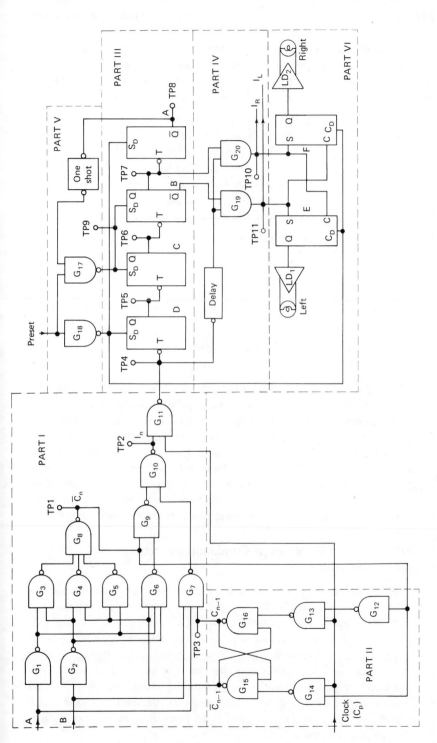

Figure 10–1 Control unit logic diagram

The logic description of the circuit was just provided in English-language form. But, as shown previously, such descriptions tend to be ambiguous and open to individual interpretation. So the assignment of letter symbols to each of the variables, common practice in previous chapters, is continued here.

$$A \quad = \text{One limit switch}$$
$$B \quad = \text{The other limit switch}$$
$$C_n \quad = \text{The internally generated function}$$
$$I_n \quad = \text{The index pulse (output)}$$
$$I_L \quad = \text{Index-left direction}$$
$$I_R \quad = \text{Index-right direction}$$
$$C_{n-1} = \text{The immediate past state of } C_n$$

The equations for I_n and C_n become

$$I_n = \bar{A}\bar{B}C_{n-1} + \bar{A}B\bar{C}_{n-1} + A\bar{B}\bar{C}_{n-1} + ABC_{n-1} \qquad (10-1)$$

and $\qquad C_n = \bar{A}BC_{n-1} + A\bar{B}C_{n-1} + AB\bar{C}_{n-1} + ABC_{n-1} \qquad (10-2)$

To minimize these equations, either algebraic manipulation or logic maps may be used, yielding

$$I_n = ABC_{n-1} + \bar{C}_n(\overline{\overline{A}\,\overline{B}\,\overline{C}_{n-1}}) \qquad (10-3)$$

and $\qquad C_n = AB + BC_{n-1} + AC_{n-1} \qquad (10-4)$

The inputs and outputs have tentatively been described, but further clarification may be in order. A and B represent switches that may be either on *or* off. For the purpose of this discussion, when a switch is *on* (actuated), it furnishes a true, 1, or HIGH input; when it is *off* (not actuated), it furnishes a false, 0, or LOW input. Further, since *pulsed outputs* are desired, the portion of the circuit associated with the output is actuated by a continuously operating string of clock pulses. Outputs I_L and I_R are pulses occurring at the appropriate time.

Special rules of logic are not necessary, since standard logic symbology is used. Also, the timing functions become apparent as the discussion progresses.

Identification and Breakout of Combinational and Sequential Functions

Determination of a starting point for separation of the combinational and sequential logic circuits may be the most difficult part of mixed-functions investigation. In many cases, the development of a *block diagram* from the English-language description proves helpful. At this point, a complete logic-diagram approach is not necessary and may even complicate matters.

From the initial description, note that the three inputs (A, B, and C_{n-1}) are combined to produce I_n and C_n. The terms used to describe the manner of combination are *combinational terms*, implying that combinational logic is used to perform these functions. Figure 10–2 is the block diagram developed from the description of the control unit.

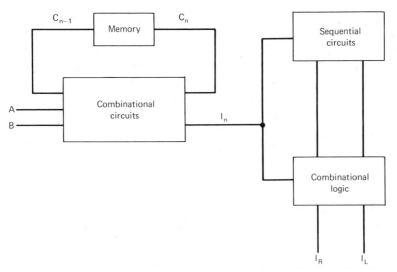

Figure 10—2   Control unit block diagram

To determine the immediate past state of C_n, a memory function is required. C_n is the input, and C_{n-1} is the output. By earlier definition, the basic memory function is sequential in nature, and a separate block is shown for this function. Counting of index pulses I_n is performed with FFs (sequential circuits) in a manner similar to the logic-signal source counter discussed in earlier chapters. Detection of the number of counts is furnished by combinational decoding circuits in much the same manner as the decoder in Ch. 3. Reference to the block diagram should ease the difficulty of identifying the combinational portions of the logic diagram.

The *symbology* used in the logic diagram aids identification of the combinational and sequential portions of the circuit. The FF symbols shown in Part III of Fig. 10—1 are indicative of sequential circuits, and by combining the description and block-diagram knowledge, the major portion of the sequential circuits may be easily isolated. The function of the ONE-SHOT blocks is clarified during the discussion of Part III of Fig. 10—1. Although Part V is actually combinational logic, its close functional tie-in to Part III requires investigation at the same time as Part III. However, due to the implementation of the memory function it *may* not be easily identified.

In any logic circuit, a good indicator of possible sequential circuitry is the *cross-coupled gate arrangement* shown in Chs. 8 and 9. Such a connection may be seen with G_{15} and G_{16}. In addition, a review of the block diagram reveals that an output from the combinational circuits is fed through the memory circuits and then back to the input of the combinational circuits. Close perusal of the logic diagram, tracing back from the signal fed to the sequential circuits, reveals that an output from G_8 feeds gates G_{12}, G_{13}, and G_{14}, which are connected to the cross-coupled gates G_{15} and G_{16}. The cross-coupled gates then feed back into gates G_4, G_5, and G_6, which are used to combine inputs A and B. Thus the circuitry between G_{12} and the output of the cross-coupled gates G_{15} and G_{16} may temporarily be considered to be the memory circuit (see Part II, Fig. 10–1). If needed, this decision may be modified as more information is gathered during the exploration of the combinational circuits.

Parts I, IV, and VI remain to be identified. No apparent cross-coupling exists in Part I, and the gates seem to be combining inputs. Part I will thus be tentatively classified as combinational. Part IV contains only two gates, and the logic diagram shows that they are used to *combine* the Q and $\overline{Q}$ outputs of FF B with the delayed output of G_{11}. The function of the DELAY block is discussed during the analysis of Parts III, IV, and V. Part IV is considered combinational.

Part VI contains FFs and is therefore classified as sequential. Its function is to store the direction of worktable travel and to light indicator lamps that show that direction.

Investigation of Combinational Circuits (Part I)

The combinational circuits that combine A, B, and C_{n-1} are shown in Part I of Fig. 10–1. As might be expected, Eqs. (10–1) and (10–2) are not implemented directly from the truth table. More efficient methods may be found, and the minimized forms of Eqs. (10–3) and (10–4) are used. The reader is urged to use Boolean algebra and/or logic maps to verify the validity of these equations. Each gate's output is shown below so that it may be verified verified that the C_n and I_n equations have been properly implemented.

Gate	Output
1.	$\overline{A}$
2.	$\overline{B}$
3.	$\overline{\overline{A}\,\overline{B}} = A + B$
4.	$\overline{\overline{B}\,\overline{C}_{n-1}} = B + C_{n-1}$
5.	$\overline{\overline{A}\,\overline{C}_{n-1}} = A + C_{n-1}$
6.	$\overline{\overline{A}\,\overline{B}\,\overline{C}_{n-1}} = A + B + C_{n-1}$
7.	$\overline{ABC_{n-1}}$
8.	$\overline{(A+B)(B+C_{n-1})(A+C_{n-1})}* = \overline{C}_n$
9.	$\overline{\overline{C}_n(A+B+C_{n-1})}$
10.	$\overline{\overline{\overline{C}_n(A+B+C_{n-1})}\,(\overline{ABC_{n-1}})} = \overline{C}_n(A+B+C_{n-1}) + (ABC_{n-1}) = I_n$
11.	$\overline{I_n C_p}$

_*_ Note that this is the *negation* of the AND–OR form of Eq. (10–3). Thus it represents $\overline{C}_n$. If C_n is required, it may be obtained by the use of an inverter. However it is used in memory in its negated form.

The clocked I_n output of Part I is also used in its negated form in the remainder of the logic circuit; that is, the FFs in Part III and DELAY in Part IV are activated by a LOW-going input (as indicated by the symbols) rather than a HIGH-going input.

I_n is "clocked" so that the operation of the sequential portions of the logic circuit may be synchronized. Also, unwanted changes in the combinational portion of the circuit will not cause changes in the sequential circuits, except during the clock time. Immunity to noise and external influences is thus provided.

In summary, the combinational logic circuit shown in Part I of Fig. 10−1 combines inputs A and B with an internally derived input C_{n-1}. It provides an output C_n for Part II according to Eq. (10−4). In addition, an output I_n is derived according to Eq. (10−3) and combined with an external clock signal to furnish LOW-going pulses to be used by the sequential circuits of Part III.

Investigation of Sequential Circuits (Part II)

Evaluation of Part II, which has tentatively been identified as a sequential circuit, utilizes the methods in Chs. 8 and 9. Two inputs $(C_n$ and $C_p)$ and two outputs $(C_{n-1}$ and $\overline{C}_{n-1})$ are evident. Boolean equations, waveforms, the state table, and the flowtable for the logic circuitry of Part II are shown in Fig. 10−3. Note that the *compressed flowtable* is used. The cross-coupled gate configuration usually results in the outputs being complements of each other, so it is not necessary to derive the complete flowtable.

The information in Fig. 10−3 reveals that the logic circuitry of Part II functions according to the standard definition of the D FF; that is, the logic state at the input (C_n) is reflected at the output (C_{n-1}) following the arrival of the next clock pulse. The output is thus delayed by *one* clock period, and the output reflects what the input was during the *previous* clock period.

Analysis of Sequential Circuits (Parts III, IV, V, and VI)

With the information given, perhaps the most appropriate method for examining the logic circuitry of Part III is the use of *waveforms*. In fact, since Parts IV and V are so closely aligned operationally with Part III, a composite analysis can be performed. In sequential circuits, it is common to establish a "starting point" from which to start analysis. G_{17} and G_{18} perform this function. A *preset input* is provided, which occurs upon initial power turn-on or when it is necessary to restart the control unit. Application of this LOW-going pulse to G_{17} and G_{18} results in HIGH-going pulses at the direct SET inputs of all four FFs, and the FFs all begin operation in the SET condition. The function of the ONE SHOT is discussed shortly. Figure 10−4 shows the preset waveforms and all other waveforms pertinent to the operation of Parts III, IV, and V of the logic circuit of Fig. 10−1.

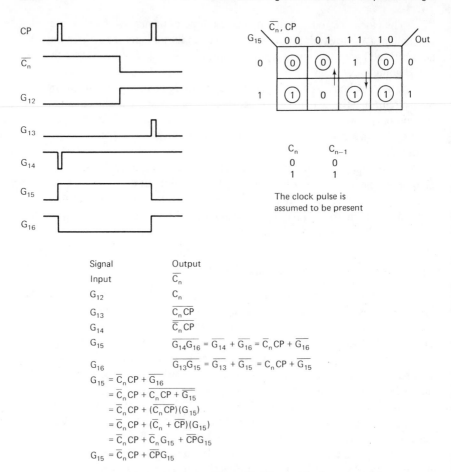

Signal	Output
Input	$\overline{C}_n$
G_{12}	C_n
G_{13}	$\overline{C_n\,CP}$
G_{14}	$\overline{\overline{C}_n\,CP}$
G_{15}	$\overline{G_{14}G_{16}} = \overline{G}_{14} + \overline{G}_{16} = \overline{C}_n\,CP + \overline{G}_{16}$
G_{16}	$\overline{G_{13}G_{15}} = \overline{G}_{13} + \overline{G}_{15} = C_n\,CP + \overline{G}_{15}$

$$G_{15} = \overline{C}_n\,CP + \overline{G}_{16}$$
$$= \overline{C}_n\,CP + \overline{C_n\,CP + \overline{G}_{15}}$$
$$= \overline{C}_n\,CP + (\overline{C_n\,CP})(G_{15})$$
$$= \overline{C}_n\,CP + (\overline{C}_n + \overline{CP})(G_{15})$$
$$= \overline{C}_n\,CP + \overline{C}_n G_{15} + \overline{CP}G_{15}$$
$$G_{15} = \overline{C}_n\,CP + \overline{CP}G_{15}$$

Figure 10–3   Analysis of Part II, Figure 10–1

Investigation of the waveforms and interconnections of the FFs shows that the circuit closely resembles the configuration and operation of the logic-signal source circuit discussed in Ch. 2. The resemblance stops, however, with the arrival of the ninth input pulse. At this time, a rather startling thing happens. FFs B and C are not allowed to assume their natural states for the input sequence; they are forced to the SET state. Since the requirements of the control unit dictate four discrete states for the control of worktable movement to the left and six discrete states for control of movement to the right, only ten states are required. Without the forcing operation, the FF connections shown in Part III would result in sixteen different states, and improper table movement would result.

The FFs of Part III function like a *simple binary counter*, except for the feedback connections. Binary counting was adequately discussed in Ch. 2, so only the forcing function is discussed at this time. After normal operation with the first eight input pulses to be counted, the ninth pulse establishes a sequence of events that ultimately results in the outputs of FFs B and C being forced to the SET state, despite the fact that with the arrival of the 9th pulse, their normal states would be the CLEAR state. The inset shown in Fig. 10-4 shows this sequence on a vastly expanded time scale. FF D cannot change state simultaneously with arrival of the ninth input pulse, due to delays inherent in the gates making up the FF. After the short delay, however, the Q_D output goes LOW and causes FF C to start changing state. Another delay occurs in this FF, but soon its Q output also goes LOW. FF B follows suit shortly thereafter, triggering A into action. $\overline{Q}_A$ goes LOW for the first time in the sequence, causing the ONE SHOT device to operate. *The ONE SHOT is merely a special kind of FF that develops a single output pulse every time it is actuated.*

The output of the ONE SHOT is applied to the input of G_{17}, and the inverted pulse from G_{17} SETs FFs B and C. Thus the result of the ninth input pulse is to place FFs B and C in the SET condition rather than the CLEAR condition. Counting continues in the usual manner, until the equivalent of 15_{10} is reached. The next pulse to be counted CLEARs all FFs by propagating a LOW-going signal to each binary, and the count sequence starts over again at 0000.

Actually, the count sequence in Part III is a bit unconventional. Although it is not absolutely necessary, many binary counters begin the counting sequence with the equivalent of 0 and progress with binary equivalents of decimal numbers in the conventional sequence. Such a method is easy to visualize, and conversion to decimal numbers is simplified. Other sequences result in certain advantages, which will be discussed in Ch. 11. The counting sequence used in Part III behaves in a conventional manner part of the time, and the remainder of the time it is forced to a *different sequence* to fit the requirements of the circuit. Table 10-1 lists each state of the counter with comments (when required).

G_{19} and G_{20} (Part IV) are AND gates with inputs from $\overline{Q}$ and Q outputs of FF B and a common input from the DELAY block. Due to the inherent

Table 10-1   Count progression of counter in Part III

Input pulse	Counter	Remarks
Initial condition	1111	Established by PRESET or counter recycle
1	0000	
2	0001	
3	0010	
4	0011	
5	0100	
6	0101	
7	0110	
8	0111	
9	1110	Forced count to eliminate unnecessary states
10	1111	Initial condition—counter starts over

delays encountered in the FFs of Part III, the gates of Part IV become enabled from FF B at some finite time following arrival of the pulses to be counted. Since we desire to develop output pulses to move the worktable, application of the pulses to be counted to the gates could result in no or limited output, due to the delays. Therefore an element called a DELAY is inserted, and the pulses to be counted are delayed an amount of time sufficient to allow the FFs to stabilize. It may be seen from the symbol (nonstandard functions are symbolized by a rectangle with an aspect ratio of 2:1 or greater according to MIL-STD-806B) that the LOW-going input to the DELAY element results in a HIGH-going output pulse of the same length at some time later than the occurrence of the original pulse. Under these circumstances, it is assured that input pulses to G_{19} and G_{20} arrive *after* the gates become enabled from FF B.

Figure 10–4 shows the development of the four pulses required for left movement of the table and the six pulses for right movement of the table. The complete operation of Parts III, IV, and V should now be reviewed with the complete waveform diagram, and the reader should verify that the sequential requirements of the system have been met.

Part VI provides visual display of the operation of the control unit. FF E stores the fact that the first "move left" command has occurred, while FF F serves a similar function for the "move right" command. The I_L pulse output not only SETs FF E but also CLEARs FF F each time it occurs. I_R pulse output SETs F and CLEARs E. When the LEFT indicator is illuminated, it means that *at least one* "move left" command has occurred and that no "move right" commands have yet been generated. The first "move right" command CLEARs E, turns off the LEFT light, SETs F, and turns on the RIGHT light. Thus the indicator lights show the direction that the worktable should be moving.

Recombination of Circuits

Now that the operation of the combinational and sequential portions of the logic circuit are developed, the complete system operation should be investigated to ensure that the analysis matches the system requirements.

In the case of the *control unit*, it is advantageous to start from the output and work back toward the inputs to integrate the combinational and sequential logic circuits. The overall output requirement was to provide control signals that would move the worktable four steps to the left and then reverse it to move six steps to the right. Both the number of steps and directional movement have been developed by Parts III, IV, and V. Four outputs pulses are seen at the output of G_{19}, while six output pulses are made available at the output of G_{20}. It is now merely necessary to connect these outputs to the electromechanical mechanism that moves the worktable.

Next we must verify *the criteria for development of the pulses* that are counted by Parts III, IV, and V. Comparison of the equation for I_n, developed from the English-language statement, with the equation for I_n, which results

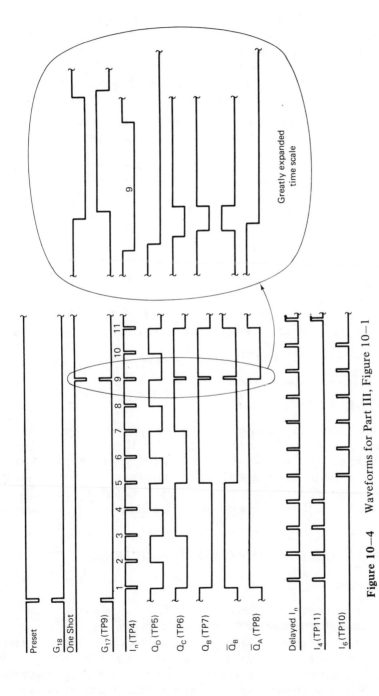

Figure 10—4 Waveforms for Part III, Figure 10—1

from the logic implementation, should be sufficient verification. C_n must be developed in a similar manner prior to development of I_n, and C_{n-1} is necessary for both C_n and I_n. Previously provided detail is sufficient to verify that C_{n-1} actually represents immediately past values of C_n. C_n provided by the logic circuitry is compared with C_n described by the English-language description in the truth table of Table 10−2. A similar proof for I_n is left to the reader.

Table 10−2   Comparison of C_n Requirements with Logic Implementation of C_n

| | | | | From logic diagram | |
| | | | | | |
A	B	C_{n-1}	C_n	$\overline{C_n}$	C_n
0	0	0	0	1	0
0	0	1	0	1	0
0	1	0	0	1	0
0	1	1	1	0	1
1	0	0	0	1	0
1	0	1	1	0	1
1	1	0	1	0	1
1	1	1	1	0	1

Identical

$$C_n = \overline{A}BC_{n-1} + A\overline{B}C_{n-1} + AB\overline{C}_{n-1} + ABC_{n-1} \text{(From English language statement)}$$

$$\overline{C_n} = \overline{(A+B)(B+C_{n-1})(A+C_{n-1})} \quad \text{(From Logic diagram)}$$

$$= \overline{(A+B)} + \overline{(B+C_{n-1})} + \overline{(A+C_{n-1})}$$

$$\overline{C_n} = \overline{A} \cdot \overline{B} + \overline{B} \cdot \overline{C_{n-1}} + \overline{A} \cdot \overline{C_{n-1}}$$

After proving that the logic implementation of I_n matches the initial system requirements, the clocked I_n pulse fed to Part III is used by the sequential portion of the logic circuit and produces the required system outputs. The complete control unit has now been explored, and sufficient information is available to troubleshoot any malfunction that may appear in the logic circuitry. Subsequent sections in this chapter will use this data as the basis for troubleshooting.

10−3   TROUBLESHOOTING MIXED FUNCTIONS

Introduction

Troubleshooting techniques for mixed functions closely approximate the combinational logic techniques. Some minor variations are necessary to compensate for the *time dependency* of the sequential logic circuits. The technique of separating combinational and sequential circuits during analysis allows the use of *flowcharts* (as modified in this chapter) in troubleshooting the combinational circuits. Within this chapter, sequential logic troubleshooting techniques are developed and then integrated with combinational techniques, resulting in an overall digital-logic troubleshooting guide.

The preparatory steps for troubleshooting mixed functions are also similar to those steps used in combinational logic. Chapter 7 supplies the basic information concerning these preparatory steps, which must be implemented prior to any attempt at troubleshooting mixed functions; that is, complete analysis is assumed along with collection of all available technical data concerning the circuit.

Using Built-in Indicators

When troubleshooting logic circuits that contain both combinational and sequential logic, it is quite important to have a complete grasp of the meaning and function of each of the *indicators*. If such information is not furnished with the equipment, analysis should be performed to define each indicator's function.

The flowchart in Fig. 7–10 is followed when troubleshooting logic circuits containing both combinational and sequential logic, but one important fact must be pointed out. When sequential elements are used, *their outputs are a function of not only the present input conditions but also past input conditions.* Complete analysis *must* be performed so that the required combination of inputs and the number of times these combinations must be provided are obtained for the purpose of verifying complete circuit operation.

One of the problems encountered in troubleshooting logic circuits that contain sequential elements is the *speed* at which such elements and circuits operate. Circuit conditions change rapidly — so rapidly that indicator lights may not even have time to illuminate. If the circuit may be operated at a slow speed for test purposes (which is common), then the effect of a single input change may be viewed before going on to the next input change. Certain critical functions, however, may require constant monitoring. In these cases, the indicators are supplied by latching circuits, which maintain the indicator in the ON condition until reset. An example may be seen in the LEFT and RIGHT indicators of Fig. 10–1.

The basic difference in the use of Fig. 7–10 during troubleshooting of the sequential part of a circuit is in the *type of information* that is to be recorded and used to determine the malfunctioning logic element. With combinational elements, the inputs were varied until all possible combinations of the input variables had been tried. With sequential circuits, the input that causes the sequential circuit to change state must be applied a sufficient number of times to cause operation of the circuit throughout its design regime. In other words, if the device is a counter, enough input applications must be furnished to allow the counter to operate completely through its design range and to perform all of the assigned operations.

Due to the fact that the inputs to sequential circuits usually are supplied by combinational circuits (as in the example used in this chapter), it is preferable to troubleshoot *each type of circuit* separately. If the information available from the built-in indicators does not make the nature of the malfunction immediately obvious, then every attempt should be made to isolate the combinational and sequential circuits for separate troubleshooting. Figure 7–10 should be used first on the combinational circuits to ensure that they are capable of furnishing the correct inputs to the sequential circuits. The sequential circuits should then be approached using the modified flowchart (Fig. 10−5).

A starting condition, such as the PRESET condition of the circuit in Fig. 10−1, is established in Block 150. The condition of the built-in indicators is recorded in Block 151 to serve as a reference point. If all possible states of the circuit have not yet been examined (Block 152), an input signal is applied to cause the circuit to assume another state (Block 153). As soon as all possible states of the circuit have been examined and the data recorded (Block 151), operations return to Fig. 7−10. These concepts will be applied in Sec. 10−4, where an example of malfunction location in the diagram of Fig. 10−1 is shown.

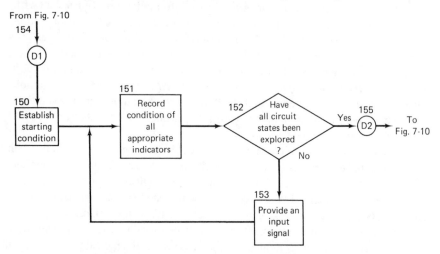

Figure 10−5   Addition to flow chart for troubleshooting logic circuits (Figure 7−10) − sequential circuits

Using External Indicators

When insufficient information is furnished by built-in indicators, the use of *external indicators*, discussed in Sec. 7−3, is a useful troubleshooting technique. The use of external indicators allows a more detailed investigation of the logic circuits. Whereas built-in indicators allow isolation of malfunctions only to a *general* area of circuitry, external indicators allow investigation of *individual* circuit elements. Combinational and sequential sections of the circuit should be separated so that the time dependent operations may be more efficiently investigated. Figure 10−6 shows the modifications required to the flowchart in Fig. 7−12 so that sequential circuits may be examined.

The only new concept used in Fig. 10−6 is that of *opening feedback loops* to investigate circuit operation (Block 267). This allows the sequential circuit to operate independent of the feedback, thus reducing the complexity of the troubleshooting. The remainder of the blocks should be self-explanatory.

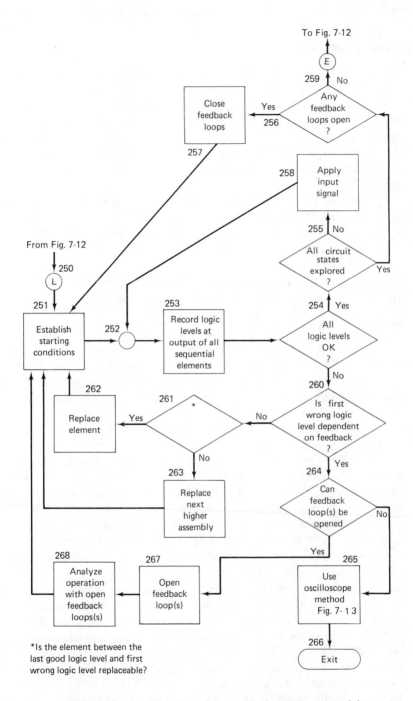

Figure 10–6   Flowchart for troubleshooting sequential circuits using external indicators – use with Figure 7–12

Using Waveforms

The use of *waveforms* as a troubleshooting aid finds considerable application in mixed function circuits — especially in the sequential portions. Oscilloscope presentations correlate directly with the waveform diagrams and allow quick identification of malfunctions. The *cause* of the malfunction may not be immediately apparent, but the reaction of the sequential circuit to inputs and the changing of circuit elements rapidly detects the malfunctioning component. The flowchart in Fig. 7−13 (as modified in Fig. 10−7) is applicable to sequential circuits.

Note that troubleshooting sequential circuits using Fig. 10−7 reduces the number of steps required to locate malfunctioning components. The oscilloscope's ability to show information on a time scale makes its applicability to analysis and troubleshooting of sequential circuits obvious.

10−4   A TYPICAL TROUBLESHOOTING PROBLEM

Preliminary Information

The circuit used in this troubleshooting example is the control unit of Fig. 10−1. To simplify the troubleshooting effort, all the information obtained during the circuit analysis in Sec. 10−2 is available, including the initial description of the unit's requirements, the algebraic expressions, tables, the logic diagram, and the waveforms.

Physically, the control unit consists of a printed circuit card with the logic elements plugged into sockets mounted on the card. Test points are provided on one edge of the card so that they are available when the card is plugged into its socket or into a test fixture. The LEFT and RIGHT lights are also mounted on the edge of the card and are visible either during actual operation of the control unit or during a test. Test points are indicated on the logic diagram in Fig. 10−1.

An equipment malfunction was noted when the worktable ceased to move, although the LEFT light was illuminated. The RIGHT light remained off. Since the maintenance concept for the machine was to remove the apparently defective card and replace it with a known good card, this was done. The machine then worked satisfactorily, and the apparently defective card was returned for repair. Testing and troubleshooting of the control unit card is to be accomplished in a test fixture. The test fixture provides power, individual control of each input signal, an oscilloscope, and a logic probe to detect and display logic levels. These are the given conditions, and it remains to verify the malfunction, locate the cause of difficulty, and repair the unit for future use.

Verification of the malfunction is accomplished by determining if the indicated conditions can be duplicated. The logic diagram shows that it is necessary to obtain at least one I_4 pulse to turn on the LEFT light. An I_4 pulse may be obtained by presetting the sequential counter (Part III) to 1111 at power

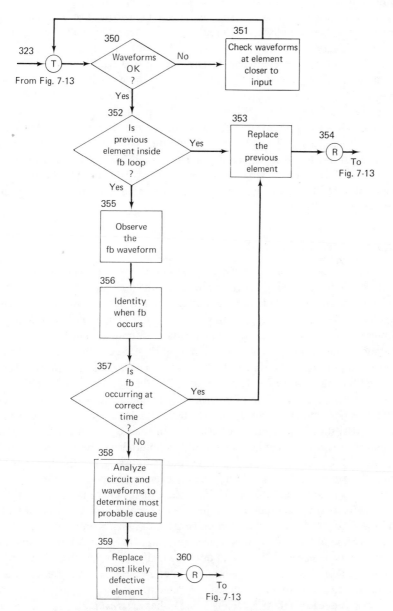

Figure 10–7   Flowchart for troubleshooting sequential circuits using waveforms — use with Figure 7–13

turn-on, receiving an I_n pulse and, subsequently, a delayed I_n pulse. The presetting operation is *automatic*. An I_n pulse is obtained whenever the combinational logic is activated by proper input signals and the clock pulse occurs. It is thus necessary to establish *one* of the input conditions that yields an I_n signal; then allow one clock pulse to occur. The clocked I_n signal CLEARs FF D, which in turn CLEARs C, B, and A. $\overline{Q}_B$ is HIGH under these conditions, and G_{19} is enabled. Arrival of the delayed I_n pulse activates G_{19}; FF E is SET by the resulting pulse; and the LEFT light illuminates. Thus it is necessary merely to set up the required A and B inputs that will generate the I_n signal and then allow one clock pulse to occur. The LEFT light should then illuminate. As long as the A and B inputs are not changed, additional clock pulses may be generated to determine the operation of the sequential counter of Part III. If the initial malfunction is present, the LEFT light will remain illuminated after five or more clock pulses are provided.

The Troubleshooting Task

Following verification of the malfunction, troubleshooting is initiated. It is preferable to use *built-in indicators* to isolate the malfunction. Previous circuit examination should provide sufficient information to decide how to approach the problem. The waveform diagram showed that four I_4 pulses are followed by six I_6 pulses in response to the proper combination of inputs A and B and the presence of clock pulses. Thus the LEFT light should be illuminated for the first four clock pulses and the RIGHT light for the next six. Ten clock pulses then should illuminate the LEFT light 4/10 or 2/5 of the time and the RIGHT light 6/10 or 3/5 of the time. The LEFT light should illuminate first. If the LEFT light does not extinguish and the RIGHT light illuminate upon arrival of the fifth clock pulse (slightly delayed), it probably means that G_{20} has not furnished an I_6 pulse. This could occur because of either a failure in G_{20} or a failure in the counter. Of course, the delayed I_n pulse could be missing, but this is highly unlikely because the LEFT light has already been turned on.

Using built-in indicators and a knowledge of the operation of the control unit has isolated the malfunction to the general vicinity of the counter of Part III and the decoder of Part IV. If we wish to attempt repair without further identification of the malfunctioning elements, G_{20} and the FFs in the counting chain may be replaced (one by one) until the malfunction vanishes. In the interest of rapid repair, such a method is sometimes used when difficulties may be isolated to a relatively few replaceable elements.

A more reasonable approach is to use either *external logic level indicators* or *the oscilloscope*. Use of the external-logic level indicator flowchart (Fig. 7–12) is shown first. The malfunction has already been isolated to a general area, so it is not necessary to apply the flowchart to the complete control-unit circuitry. Parts III and IV may be considered as a separate circuit and examined in detail by using the flowchart.

According to Fig. 7–12, after test-point logic levels have been determined, the test point (TP) nearest the output is checked. In this case, TP_{10} at the output of G_{20} is the first test point to be examined. Use of the built-in indicators and previous investigations have shown that at the fifth clock pulse and through the tenth clock pulse, an output pulse should appear. The procedure to be used is generally the same as the flowchart shows, except that test points are not changed until sufficient input pulses have been generated to allow the counter to go completely through its cycle. For the malfunction under discussion, TP_{10} would remain at a LOW logic level throughout the complete counter cycle time.

Assuming that the delayed I_n pulse is present because of the functioning of the LEFT light, the Q_B output (TP_7) is the next test point closest to the input. The waveform diagram shows that Q_B should be HIGH following preset, go LOW for four input pulses, and then go HIGH for the next six input pulses. With the external logic level indicator attached to TP_7, the control unit is preset by turning power on, and ten clock pulses are generated. The output at TP_7 is investigated following each clock pulse, and it is noted that Q_B presets properly, goes LOW at the first I_n pulse, and remains LOW for the remaining input pulses. This accounts for the lack of I_6 pulses, since G_{20} is continuously disabled and the delayed I_n pulse is not allowed through the gate.

TP_6 is now observed for ten clock pulses to determine if FF C is supplying triggers to FF B. It is noted that FF C also presets properly, but its output (Q_C) goes LOW and remains LOW following the first clock pulse. Thus FF B could not change state because FF C was not furnishing input triggers.

Investigation of TP_5 reveals that FF D responds according to the waveform diagram. FF C is provided with the proper input triggers, yet it does not change state following the initial preset and response to the first clock pulse. When placed into the CLEAR state by the application of a trigger from D, it locks into the CLEAR state. Replacement of FF C and a complete retest of the counter verifies that the malfunction has been cleared and that the counter and decoder circuits are functioning properly.

Before the control-unit card is returned to use again, it should be subjected to a complete test from inputs to outputs. Due to the numerous combinations of A and B that can result in an I_n output and due to the requirement that the counter be actuated ten times to fully prove its operation, it is advisable to check only the combinational logic at this time. The sequential logic has been proven, so all that is necessary to do is to assure that TP_2 shows the correct I_n signal for the proper combinations of A and B. TP_1 and TP_3 are available to check the intermediate signals, C_n and C_{n-1}, respectively.

Once the malfunction has been isolated to the sequential portion of the control unit, the oscilloscope method of troubleshooting might be more expeditious. The clock input could be allowed to operate at its normal speed, and the oscilloscope could be connected to the necessary test points, until the improper waveform is noted. Since the oscilloscope pattern and waveform pattern are very similar, the malfunctioning component would be obvious.

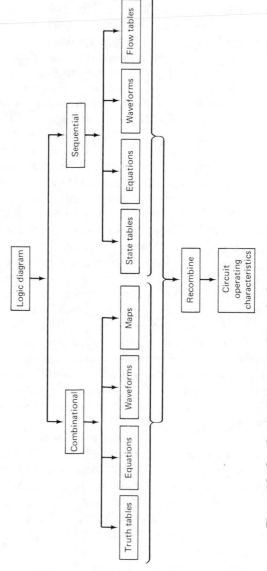

Figure 10–8   Logic circuit analysis methods summary

To generalize, *the oscilloscope method of troubleshooting is usually superior to other methods when sequential logic circuits are under examination.* The oscilloscope may also be used to troubleshoot combinational logic, if input signals are changing rapidly and following a repetitive pattern. Otherwise, the external-logic level indicator is usually more satisfactory.

10—5 SUMMARY

Troubleshooting logic circuits that contain both combinational and sequential elements is adequately summarized in the appropriate flowcharts. However no such visualization of analysis procedures has been shown. Figure 10—8 summarizes the investigative methods for mixed logic circuits.

Since the purpose of analysis is to determine the circuit's operating characteristics, these are shown as the end result of the procedure. As we mentioned earlier, the logic diagram is broken into combinational and sequential parts, and examined separately. The "tools of analysis" for the combinational logic portions (truth tables, Boolean equations, waveforms, and maps) have been used throughout this text. Similarly, the sequential logic tools (state tables, Boolean equations, waveforms, and flowtables) have been employed where applicable.

The similarities between the methods of analysis should be carefully noted. For example, truth tables and state tables perform very similar functions. The only difference is that the *state table depicts circuit states at different times*, while *the truth table shows circuit states with different input combinations at a specific time.* The sequential analysis procedures are thus *time oriented*, whereas the combinational analysis procedures are concerned with *combinations of inputs.*

PROBLEM APPLICATIONS

10—1. Why are combinational and sequential functions separated prior to starting analysis of logic circuits?

10—2. List common identifying characteristics of combinational logic circuits.

10—3. List common identifying characteristics of sequential logic circuits.

10—4. List the common analysis methods used with combinational logic and explain the advantages and disadvantages of each.

10—5. List the common analysis methods used with sequential logic and explain the advantages and disadvantages of each.

10—6. Why is the interface between combinational and sequential circuits sometimes clocked?

10—7. Why must the troubleshooting flowcharts of Ch. 7 be modified for mixed-function logic?

10—8. Discuss the changes necessary to troubleshoot sequential and mixed logic using the built-in indicators flowchart.

10–9. Discuss the changes necessary to troubleshoot sequential and mixed logic using the external indicators flowchart.

10–10. Perform the algebraic and/or map manipulations required to obtain Eq. (10–4) from Eq. (10–2).

10–11. Perform the algebraic and/or map manipulations required to obtain Eq. (10–3) from Eq. (10–1).

10–12. Using algebraic and/or map manipulations, verify that the logic implementation of I_n is the same as Eq. (10–1).

Counters and
Their Applications

A counter is a device, composed of properly connected, sequential (some-times *combinational*) *elements, that performs the function of counting.* Counters are binary in nature, but they may be modified to count in other sequences (e.g., base 7 or base 10). Almost all logic systems require counters. Digital computers use counters to control the sequence and execution of their program steps. Industrial controls use counters to determine the number of steps in machine operation, such as the control unit discussed in Ch. 10. Measuring instruments for the scientific disciplines employ counters to count frequency, voltage, number of events, and so forth. Some typical counter applications will be explored in this chapter, following a discussion of both binary and special-purpose counters. A sample analysis of a typical counter will also be performed — in this case showing the techniques used to investigate multiple flip-flop (FF) circuits.

11–1 THE ASYNCHRONOUS BINARY COUNTER

Basic Concepts

Counters may be divided into two general types: *synchronous* and *asynchronous*. The *asynchronous counter* (sometimes called the *series* or *ripple counter*) is probably the simplest device in the counter family. Since it was discussed in detail as the "logic-signal source" in Chs. 2 and 3, a review of Sec. 2–6 will lay the groundwork for the more detailed discussions to follow.

Simplicity is the major advantage of asynchronous counters. As we can see in Fig. 11–1, simple *T* flip-flops (FFs) with the capability of direct CLEAR may be connected to count in the straight binary sequence. The major disadvantage on these counters is the delay encountered as the changes in each FF in the

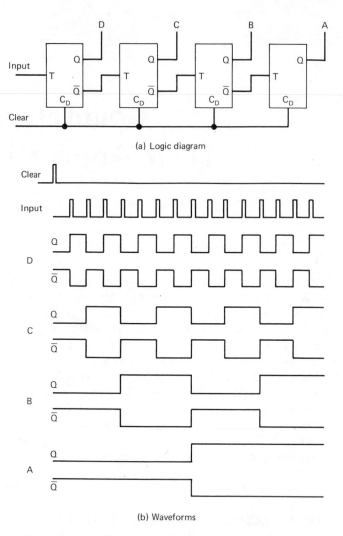

(a) Logic diagram

(b) Waveforms

Figure 11—1   Simple asynchronous binary counter

series string propagate. Cumulative delays may cause errors, when outputs from widely separated stages in the counter are combined in gating circuits for the purpose of decoding specific counts.

The UP Counter

Figure 11—1 is a typical example of an asynchronous, binary UP counter. Its name adequately describes the principle of the counter's operation — it starts counting from some predetermined point (often all FFs CLEARed) in ascending order. The count is accumulated in the binary system; however the ability to convert binary numbers to decimal numbers has already been applied, so no difficulty should exist. Note that the output of each FF represents one binary digit. The first stage, to which the inputs to be counted are applied, represents the number 2^0 or 1. The following stage has a positional value of 2^1 or 2; the next stage has a value of 2^2 or 4; and the final stage represents 2^3 or 8. By adding the values of all stages in the "1" condition, the total count is obtained. Thus n FFs can store any binary number from 0 to $2^n - 1$. The counter in Fig. 11–1 stores the binary equivalents of all decimal numbers from 0 to 15. In the UP counter configuration, these numbers accumulate in the familiar manner shown in Fig. 2–16.

The waveforms should also be closely observed so that the relationship between the input pulses and the outputs of each stage can be seen. The first-stage output, for example, goes through one complete cycle (from LOW to HIGH and back to LOW again) one time for each two input pulses. Effectively, it has an output frequency that is *one-half* the frequency of the incoming signals. The second stage has a frequency that is one-half the frequency of the first stage or one-fourth of the incoming frequency. Due to this characteristic, *counters are sometimes called frequency dividers*, and as frequency dividers, they are extensively used in applications requring submultiples of input frequencies.

Of course, the FFs used to implement the binary UP counter are not restricted to the T type. D, $R-S$, and $J-K$ FFs are all easily worked into ripple counters, merely requiring input connections that allow them to perform as a T FF. Nor is the physical configuration of the counter limited to a single FF per package. *Dual* FFs are commonly available. In addition, complete 4-stage binary counters may be obtained in the 14-pin DIP and other multiple-pin packages. Fig. 11–2 shows the logic diagram for such a counter, which is identified as a

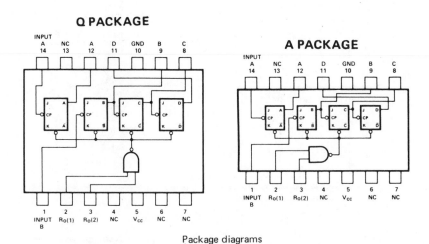

Package diagrams

Figure 11–2   Type 7493 four-bit counter

7493* 4-bit binary counter. Note that the FFs require LOW-going inputs for activation and that a direct RESET capability is available to start the counting at 0000.

Representing Counters in Graphic and Tabular Form

Often, it is advantageous to consider counter circuits in terms of the complete counter rather than individual FFs. We have shown that *sixteen separate states* exist in the simple binary UP counter. But how these states are represented in terms of individual FF outputs *may* not be important. The natural binary sequence (see Fig. 11–3a) is both the most common and most convenient because the arbitrarily assigned state numbers coincide with the decimal equivalents of the binary numbers stored.

However the complete operation of the binary UP counter may be shown by merely drawing a *state flow diagram* (Fig. 11–3b), which displays the sequence of counter states. Each circled number represents one of the states, while the lines and arrows define the transitions taking place. The designation attached to the line shows the input signal necessary to cause the transition from one state to the next. We see that a "1" input will cause the counter to change from state number 0 to state number 1. Likewise, the next "1" input will result in a change from state number 1 to state number 2, and so on. The "1" input

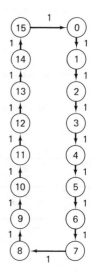

Input pulse number	Present state time = n				Next state time = n + 1			
	A	B	C	D	A	B	C	D
0	0	0	0	0	0	0	0	1
1	0	0	0	1	0	0	1	0
2	0	0	1	0	0	0	1	1
3	0	0	1	1	0	1	0	0
4	0	1	0	0	0	1	0	1
5	0	1	0	1	0	1	1	0
6	0	1	1	0	0	1	1	1
7	0	1	1	1	1	0	0	0
8	1	0	0	0	1	0	0	1
9	1	0	0	1	1	0	1	0
10	1	0	1	0	1	0	1	1
11	1	0	1	1	1	1	0	0
12	1	1	0	0	1	1	0	1
13	1	1	0	1	1	1	1	0
14	1	1	1	0	1	1	1	1
15	1	1	1	1	0	0	0	0
0	0	0	0	0	etc.			

Note: Pins 12 and 1 are connected to
obtain this count sequence

Figure 11–3   State table and flow diagram

* The numbers used to identify specific logic-circuit packages are purposely abbreviated in this text. A certain amount of standardization does exist in industry, but prefixes and suffixes to the actual numbers vary among manufacturers. The *numbers alone* may generally be used to identify a specific type of circuit, regardless of its prefixes and suffixes.

that appears following establishment of state number 15 causes the counter to recycle back to state number 0, and the counter starts over.

The use of 1 instead of HIGH or LOW to define inputs is not out of place. Since a FF is a storage device, its state may be considered as storing either a 1 or a 0. Convention dictates that when a FF is SET it is storing a 1, and when RESET, it is storing a 0. Whether a 1 is HIGH or LOW depends on the implementation of the logic circuit. *The transition from one level to the other is what causes action in the counter.* Thus, regardless of the counts accumulated in the counter, the state flow diagram shows the next state to be encountered. This method of representation is especially valuable when other than natural binary sequences are encountered.

Algebraic expressions may also be used to represent asynchronous binary counters. However, due to the means of triggering each FF from the preceding FF, interpretation of these equations becomes difficult. Therefore waveforms and state tables will be used here to explain the asynchronous operations.

The DOWN Counter

The binary DOWN counter may be implemented in much the same manner as the UP counter. The DOWN counter, however, *decreases* its stored count every time an input pulse is received. One of the easiest ways to obtain the descending count operation is to use the *complement outputs* of the FFs in the UP counter. For example, since the Q outputs were used in the counter in Fig. 11–1 to obtain the ascending count, we need only complement the Q outputs (use the $\overline{Q}$ outputs) to obtain the descending count. Thus, in the waveforms of Fig. 11–1, all of the $\overline{Q}$ outputs are HIGH, following the initial CLEAR operation, and the count represented at that time (using the $\overline{Q}$ outputs) is 1111. Following the first input pulse, the $\overline{Q}$ output of FF D goes LOW, while the other FFs' Q outputs remain HIGH. The number represented is now 1110 or 14_{10}. The first pulse, then, has decreased the stored count by 1. Additional pulses should be examined by the reader.

Further verification of the DOWN counter operation may be made by complementing the outputs shown in the state table in Fig. 11–1. We see that as the output of the UP counter increases, the output of the DOWN counter (its complement) decreases.

Another method of obtaining a binary DOWN counter is to use the Q outputs of each FF to activate the following FF.

Waveforms, a state table, a state flow diagram, and a simple logic diagram of a DOWN counter are all shown in Fig. 11–4.

Bidirectional Counters

The characteristics of both the UP and DOWN counters may be combined into a single UP–DOWN counter. From a practical viewpoint, all that is necessary is to determine whether *the input to each FF in the series string* comes

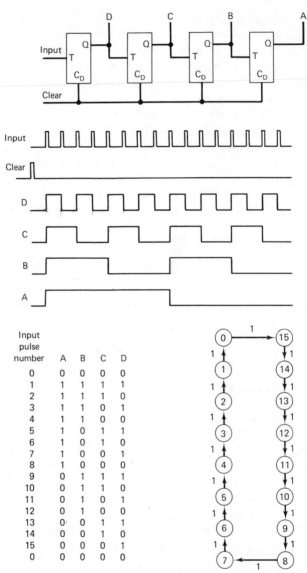

Figure 11—4   An asynchronous binary DOWN-counter

from the Q or the $\bar{Q}$ output. The logic diagram of such a *bidirectional* device is shown in Fig. 11–5. (The dotted lines should be disregarded for the moment.)

The sequence of operation is similar to the other binary counters described in this section. Following an initial CLEAR signal to establish a starting point, input pulses are applied to the T input of FF D. If the UP COUNT input is HIGH, all of the gates connected to the $\bar{Q}$ outputs of the FFs are enabled, and counting will proceed in the UP direction, just as in the UP counter in Fig. 11–1.

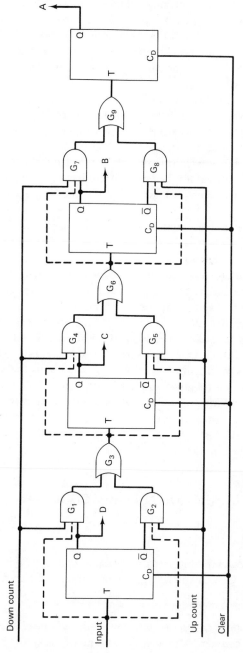

Figure 11—5 Asynchronous binary UP-DOWN counter

Lowering the UP COUNT input and establishing the DOWN COUNT input in the HIGH condition causes the counter to perform in the same manner as the DOWN counter in Fig. 11–4. While both the UP COUNT and the DOWN COUNT inputs may be LOW at the same time (this prevents any FF other than D from changing state), they are *not* allowed to be HIGH simultaneously. The resulting count sequence is somewhat unusual!

Although the restriction on both count-control inputs being HIGH at the same time is observed, other difficulties can arise in the UP–DOWN counter. If the UP COUNT input is HIGH at the time the CLEAR signal is applied, the HIGH-going transition of the $\overline{Q}$ output of FF D will be passed through G_2 and G_3 to the T input of FF C. FF C then attempts to toggle under control of the T input at the *same time* that it is being CLEARed by the C_D input. In most FFs, the direct CLEAR and direct SET inputs override the T inputs, but difficulties can occur if the T input is HIGH when the C_D input returns to the LOW level. Furthermore, under certain count conditions, lowering the UP COUNT input and raising the DOWN COUNT input without *first* CLEARing the counter results in an abnormal count sequence.

Both of these problems may be solved by assuring that all of the AND gates are disabled, *except* during the time an input pulse is being counted. This is easily accomplished by adding the connections shown by the dotted lines in Fig. 11–5. Consider G_1 and G_2. Since the INPUT is at a LOW level except for the short time that the pulses to be counted are present, both G_1 and G_2 are disabled, and they furnish LOW inputs to G_3. G_3 output is then LOW, disabling G_4 and G_5. The LOW output of G_6 disables G_7 and G_8, and no changes on either of the count-control inputs or the FF output may be transferred to following FFs. CLEARing operations or changes in count direction may be established without upsetting the count stored in the counter. The only precaution to be exercised is to assure that *such changes do not occur during the time the clock pulse is present.*

11–2   THE SYNCHRONOUS BINARY COUNTER

Basic Concepts

All FFs are *clocked* (triggered) *at the same time* in the synchronous (parallel) counter. Combinational logic elements (gates) are used to control FF state changes so that the stored count progresses only *one* unit each time a clock pulse is received.

The synchronous counter is better suited to *high-speed* operation than is the asynchronous counter. Although the concept of the synchronous counter indicates that all FFs change state at the same time, this is not strictly true. Delays still exist in the gating structure of the circuit (see Fig. 11–6); thus the top counting speed is considerably below the theoretical speed. Synchronous operation requires the use of relatively high-power clock circuits, since the clock must drive all FFs. Heavily loaded clock circuits may also increase propagation delay and reduce counter speed. Despite these factors, synchronous connters still tend to be superior in speed characteristics when compared to asynchronous counters.

But, as is so often true, speed costs money! The additional gates required to obtain synchronous operation, plus the more complex FF needs, place the synchronous counter in a cost category above the asynchronous counter. And, of course, more complex circuitry usually requires greater power supply capability and overall system cost increases.

The Synchronous Binary UP Counter

An UP counter showing the principles of synchronous operation is seen in Fig. 11–6. Clocked $J–K$ FFs are used, and all FFs are triggered simultaneously. The FFs are interconnected by gating so that the only time a specific FF can change is when all the *least significant* FFs are in the SET state.

Although the waveforms in Fig. 11–6 are almost self-explanatory, the operation of the synchronous binary UP counter is partially discussed to develop an understanding of the use of waveforms with synchronous circuits. Counter operation is based on the already familiar characteristics of the $J–K$ FF. Specifically, the fact that the $J–K$ FF changes state with every clock pulse if both the J and the K inputs are HIGH at the same time is used to allow the FFs to perform the required toggling operation.

Following the initial CLEAR operation, all FFs are in the CLEAR state and all Q outputs are LOW. The Q output of FF D disables FF C. (When both J and K are LOW, no change occurs in FF state upon arrival of a clock pulse.) B is similarly disabled, due to the LOW output of G_1, and A is disabled by G_2. The only FF that can change state with the next clock pulse is D, since its J and K inputs are permanently HIGH. Therefore the first clock pulse changes D from the CLEAR to the SET condition, and Q_D goes HIGH. C is enabled by Q_D, and the next clock pulse changes not only the state of C but also that of D. Investigation of the remainder of the waveforms shows that B can change state only at the time both D and C are SET (G_1) and that A can change state only at the time D, C, and B are SET (G_2). The effect of each clock pulse should be determined, and the operation of the counter should be investigated at least through count 16.

We see that the waveforms for the synchronous binary UP counter are identical to the waveforms for the asynchronous binary UP counter. Thus the circuits of Fig. 11–1 and 11–6 are functionally equivalent. When speed and power parameters are met, either circuit may be used to perform simple binary counting. However, certain limitations exist for each type of circuit; the asynchronous counter may not be able to replace the synchronous counter in all cases.

The state diagram and state table are the same for both asynchronous and synchronous counters. Individual FF equations appear in Fig. 11–6.

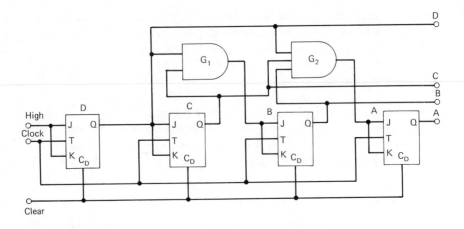

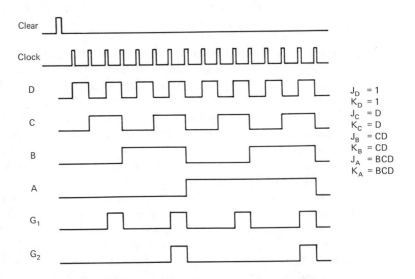

Figure 11—6   The synchronous binary UP-counter

Synchronous Binary DOWN Counter

Like the asynchronous DOWN counter, the only requirement to convert
from an UP to DOWN operation is the use of the *opposite outputs* to feed the
following stages — the UP counter of Fig. 11—6 becomes a DOWN counter by
merely connecting the $\overline{Q}$ outputs to the following gates and/or FFs. Actual outputs continue to be identified with the Q output terminals. Explanatory information in the form of equations and waveform diagrams appears in Fig. 11—7.

The $\overline{Q}$ outputs of the UP counter may also be used as DOWN counter outputs in the same manner as in the asynchronous counter.

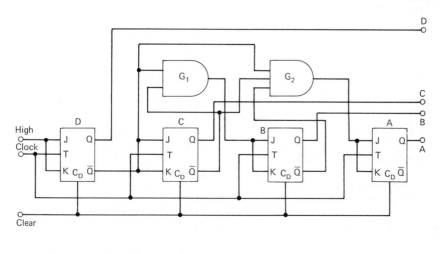

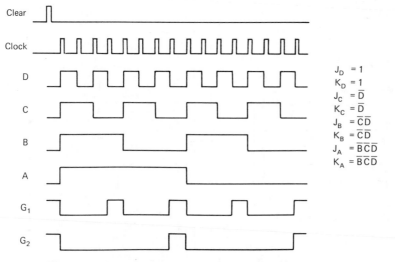

Figure 11—7 The synchronous binary DOWN-counter

Synchronous Binary UP—DOWN Counter

Bidirectional counting may be accomplished in the synchronous mode, just as it was in the asynchronous mode. The logic diagram of a typical synchronous UP—DOWN counter (type 8284) contained in a single integrated circuit (IC) package is shown in Fig. 11—8b.

Because of the complexity of the logic diagram, the "package" format is often used in system diagrams. Fig. 11—8a is the 14-pin DIP form commonly seen in such diagrams. Obviously, it is necessary to have either a complete

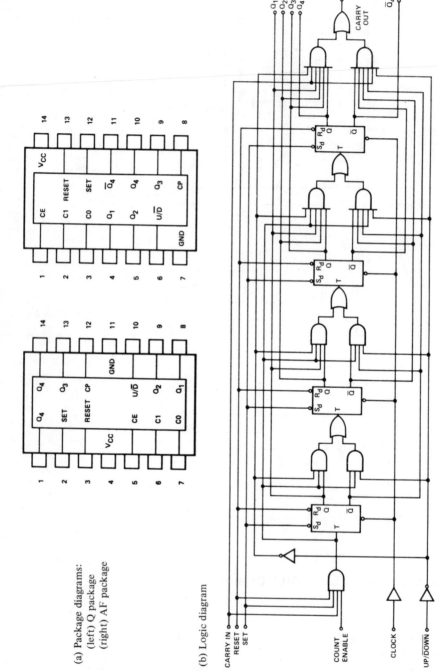

(a) Package diagrams:
(left) Q package
(right) AF package

(b) Logic diagram

Figure 11—8 The synchronous binary UP-DOWN counter (Type 8284) (courtesy of Signetics, a subsidiary of Corning Glass Works, Sunnyvale, CA.)

description of the counter operation (or the logic diagram) available to analyze system diagrams using this form of representation.

The counter in Fig. 11−8 provides the output sequence 0 through 15 in the natural binary code (8−4−2−1), when counting in the UP direction, or 15 through 0 when counting in the DOWN direction. SET and RESET inputs allow a count of either 15 or 0 to be placed in the counter so that a convenient starting point may be established. When these inputs are present, they disable the counting process. LOW-going signals are required to perform the SET and RESET operations.

Input to the counter is via the CARRY IN and the COUNT ENABLE lines. Both must be present (but SET and RESET must *not*) for the counter to operate in either the UP or the DOWN configuration. A LOW level at the UP−$\overline{\text{DOWN}}$ input results in a DOWN count, while a HIGH level results in an UP count. All four Q outputs are available, as is the $\overline{Q}$ output of the final FF. Another output, called CARRY OUT, is provided to indicate that a complete count has occurred.

Although each of the FFs is labeled with a T input, the FFs behave in the same manner as the FFs of Figs. 11−6 and 11−7. For the sake of convenience, the UP−DOWN counter FFs may be considered as "clocked T FFs!" A rather unusual designation, but there is no reason why such a name shouldn't be applied. Just like the J−K FF with the J and K inputs connected together, when the T input is HIGH, the FF changes state with every clock pulse; when it is LOW, no change occurs.

The T input of each FF is gate controlled, so all previous stages must be in the 1 state (for UP counting) or in the 0 state (for DOWN counting) *before* the following FF is allowed to change state on the next clock pulse. (Note the increasing complexity in gating structure as count value increases.)

One more point is worth noting on the logic diagram of the 8284 UP− DOWN counter. The CLOCK and the UP−$\overline{\text{DOWN}}$ inputs have inverting amplifiers inserted prior to use of the signal. These amplifiers are required not only to obtain the correct logic level for circuit operation but also to provide power amplification. The power amplifiers provide sufficient power to operate all of the required circuits, and they also isolate the complete circuit from causing interference with preceding and following circuits. Such isolation is commonly necessary in complex digital systems.

11−3 SPECIAL PURPOSE COUNTERS

Basic Concepts

All of the counters discussed so far have been based directly or indirectly on the binary system or *modulus* 2. *Modulus* is a term used to describe *the maximum number of counts that a counter can achieve.* When using natural binary

counting, such as that shown in Secs. 11–1 and 11–2, the modulus of a counter
may be determined by the equation

$$\text{Modulus} = 2^n \qquad\qquad \text{Eq. (11–1)}$$

where n = the number of FFs.

Thus a single stage counter is modulus 2, a 2-stage counter is modulus 4, a 3-
stage counter is modulus 8, and so forth. The counters discussed in Secs. 11–1
and 11–2 are modulus 16.

The "real" world, however, is not necessarily base 2 oriented. Time of
day uses counts of 6, 10, and 12; distance is measured in inches, feet, and yards
(or millimeters, centimeters, and meters); and quantity is measured in multiples
of 12 (dozen or gross). A need therefore exists for a counting system based on
something other than modulus 2 or multiples of modulus 2. So this section will
investigate methods of counting in numerous bases, using both asynchronous and
synchronous circuits.

Divide by N Counters

Although both asynchronous and synchronous counters may be used to
obtain counters with moduli other than direct multiples of 2, asynchronous
methods will be examined first to develop the general concepts; and then
synchronous counters will be investigated.

The most commonly employed counting element is the J–K FF. Actually,
complete counters such as the 7493 package (see Fig. 11–2) are often used in
place of separate J–K FFs to reduce overall logic element requirements. The
basic package consists of four J–K FFs and a 2-input NAND gate. Internal connections provide a divide-by-2 and a divide-by-8 capability. External connections
allow shortening of the basic count cycle and development of various counters
with moduli from 2 through 16. The basic logic diagram showing the divide-by-2
(A FF) and the divide-by-8 (B, C, and D FFs) connections is seen in Fig. 11–2.
Natural binary counting to base 16 is provided by externally connecting A_{out}
(pin 12) to B_{in} (pin 1). At this time, the counter performs exactly as the asynchronous binary UP counter of Fig. 11–1, except for the LOW-going trigger
requirement.

The NAND gate is used to RESET or CLEAR all stages of the counter —
either for an initial starting count or to return all stages to the reference count
following completion of the count cycle. The same speed and readout restrictions mentioned in Sec. 11–2 apply to the asynchronous counters discussed in
this section.

To operate the J–K FFs in the 7493 counter as T FFs, both the J and K
inputs must be enabled. The method of design in the FFs in this counter allows
both J and K inputs to be left unterminated to accomplish enabling.

In all count applications, the frequency of the incoming triggers is divided
by 2 in the first FF. The natural binary equivalent of the required modulus is

decoded by the reset gate (and any additional gating necessary), and the complete counter is reset to 0000. A divide-by-12 connection that could be used to count inches of measurement or hours of the day is an excellent example of this technique. The logic diagram and waveforms of Fig. 11–9 should be self-explanatory. The count progresses in natural binary sequence from 0_{10} through 11_{10}. When count 12_{10} occurs, the reset gate is enabled and all FFs are returned to the 0 state. The actual count of 12 exists only for as long as it takes to reset all FFs, so the counter can be considered to be counting from 0 to 11 or modulus 12.

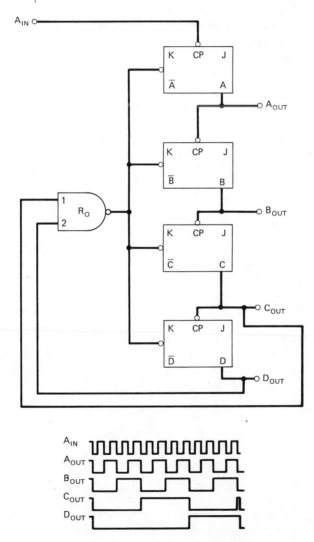

Figure 11–9   Binary divide-by-twelve ripple counter

A divide-by-6 circuit, constructed in the same manner, is shown in Fig. 11–10. The reader is encouraged to verify the waveforms used to explain counter operation; note especially the use of LOW-going logic signals at both the CLEAR (RESET) and CP inputs in both counters.

Since the divide-by-12 counter is used so often in digital systems, another IC example (type 8288) is shown in Fig. 11–11. A divide-by-2 and a divide-by-6 connection is made internally. Divide-by-12 operation is accomplished by externally connecting output A to the clock-2 input. Hence type 8288 and type 7493 circuits are similar.

The 8288 circuit, however, has what is known as *strobed parallel data entry*, which allows the counter to be preset to any desired output state. Considering a 1 as HIGH and a 0 as LOW, a 1 or a 0 at any data (D) input is transferred to the associated output (stored in that FF) when the strobe input goes LOW. The usual common reset is also available to place 0s in all four FFs. Both the strobe and reset inputs are asynchronous with respect to the clock input; that is, they override the clock input.

FF output waveforms for the divide-by-12 portion of the circuit are included in Fig. 11–11c. Strobed parallel data entry is also shown for one of the four FFs (Fig. 11–12) to demonstrate this common technique.

Due to the relative timing of the strobe, reset, and data inputs, a truth-table explanation of circuit operation is shown instead of the usual waveforms. One important point must be emphasized in this analysis: *both the S_D and R_D*

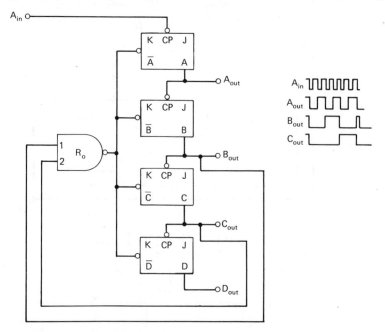

Figure 11–10   Binary divide-by-six ripple counter

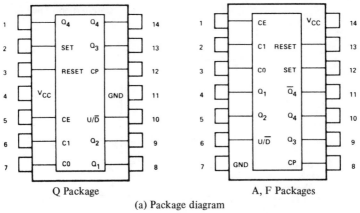

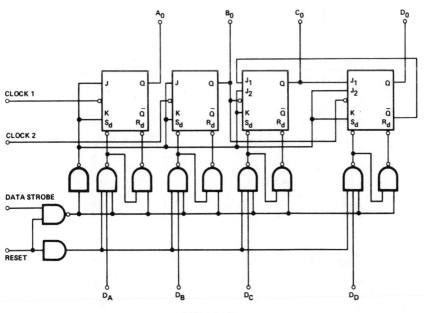

(b) Logic diagram

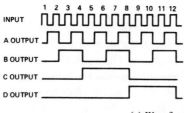

(c) Waveforms

Figure 11–11   Divide-by-twelve counter (Type 8288) (courtesy of Signetics, a subsidiary of Corning Glass Works, Sunnyvale, CA.)

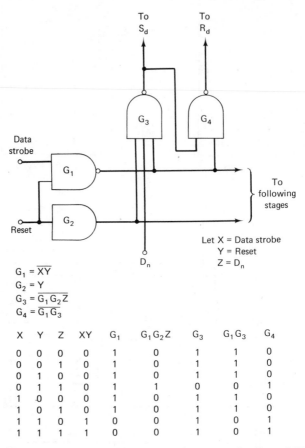

$G_1 = \overline{XY}$
$G_2 = Y$
$G_3 = \overline{G_1 G_2 Z}$
$G_4 = \overline{G_1 G_3}$

X	Y	Z	XY	G_1	$G_1 G_2 Z$	G_3	$G_1 G_3$	G_4
0	0	0	0	1	0	1	1	0
0	0	1	0	1	0	1	1	0
0	1	0	0	1	0	1	1	0
0	1	1	0	1	1	0	0	1
1	0	0	0	1	0	1	1	0
1	0	1	0	1	0	1	1	0
1	1	0	1	0	0	1	0	1
1	1	1	1	0	0	1	0	1

Figure 11—12   Strobed parallel data entry for Type 8288
counter

inputs require LOW-*going activating signals*, as indicated by the FF symbols.
Therefore the outputs of G_3 and G_4 that are important are the 0 outputs — not
the 1 outputs. With this point in mind, we see that S_D is activated *only* when
$Z = 1$, $Y = 1$, and the strobe pulse is LOW. Even if the strobe pulse is LOW and
$Z = 1$, Y must also be 1 in order to SET the FF. If $Y = 0$, RESET takes place,
since the RESET input overrides any attempt to preset the counter. The reset
functions are predictable according to the description of the circuit operation.

One final example of divide-by-N counters illustrates the versatility of
logic circuits. Table 11—1 shows the input equations for binary code counters
from modulus 3 to modulus 31, using J–K FFs and NAND gates. The inputs
necessary for the T, J, and K inputs for each of the FFs are shown in the appropriate
columns for each modulus. NAND gates are used to form the required combinations of inputs. In using the table, those terminals denoted by X should be connected and used as the overall circuit input. A 1 in any column indicates that

Table 11–1 Input Equations for Binary Code Counters

MODULUS	A (1)			B (2)			C (4)			D (8)			E (16)		
	T	J	K	T	J	K	T	J	K	T	J	K	T	J	K
3	X	$\overline{B}$	1	X	A	A									
5	X	$\overline{C}$	1	A	1	1	X	A·B	1						
6	X	1	1	A	$\overline{C}$	1	A	B	1						
7	X	$\overline{B}\cdot\overline{C}$	1	X	1	A+C	B	1	1						
9	X	$\overline{D}$	1	A	$\overline{D}$	1	B	1	1	X	$\overline{A}\cdot\overline{B}\cdot\overline{C}$	1			
10	X	1	1	A	1	1	B	A	1	A	B·C	1			
11	X	$\overline{B}\cdot\overline{D}$	1	X	A	A+D	B	$\overline{D}$	1	B	C	1			
12	X	1	1	A	1	1	B	1	1	B	C	1			
13	X	$\overline{C}\cdot\overline{D}$	1	A	$\overline{C}\cdot D$	1	X	A·B+C·D	A·B+C·D	C	1	1			
14	X	1	1	A	1	1	A	1	B+D	C	1	1			
15	X	$\overline{B}\cdot\overline{C}\cdot\overline{D}$	1	X	1	A+C·D	B	1	1	C	1	1			
17	X	$\overline{E}$	1	A	1	1	B	1	1	C	1	1	X	A·B·C·D	A·B·C·D
18	X	1	1	A	$\overline{E}$	1	B	1	1	C	1	1	A	$\overline{A}\cdot B\cdot\overline{C}$	$A\cdot B\cdot\overline{C}$
19	X	$\overline{B}\cdot\overline{E}$	1	X	1	A+E	B	$\overline{E}$	1	C	1	1	B	A·C·D	A·C·D
20	X	1	1	A	1	1	B	1	1	C	1	1	B	C·D	C·D
21	X	$\overline{C}\cdot\overline{E}$	1	A	1	1	X	A·B	A·B+E	C	1	1	C	D	D
22	X	1	1	A	$\overline{C}\cdot\overline{E}$	1	A	B	B+C	C	1	1	C	D	D
23	X	$\overline{B}\cdot\overline{C}\cdot\overline{E}$	1	X	A	A+C·E	B	1	1	C	1	1	C	D	D
24	X	1	1	A	1	1	B	1	1	C	1	1	C	D	D
25	X	$\overline{D}\cdot\overline{E}$	1	A	1	1	B	1	1	X	A·B·C	A·B·C+D	D	1	1
26	X	1	1	A	$\overline{D}\cdot\overline{E}$	1	B	$\overline{B}\cdot\overline{D}\cdot\overline{E}$	1	A	B·C+D·E	B·C+D·E	D	1	1
27	X	$\overline{B}\cdot\overline{D}\cdot\overline{E}$	1	X	A	A+D·E	B	$\overline{D}\cdot E$	1	B	C	D+E	D	1	1
28	X	1	1	A	1	1	B	1	1	B	1	C+E	D	1	1
29	X	$\overline{C}\cdot\overline{D}\cdot\overline{E}$	1	A	1	1	X	A·B	A·B+D·E	C	1	1	D	1	1
30	X	1	1	A	$\overline{C}\cdot\overline{D}\cdot\overline{E}$	1	A	1	B·C·D	C	1	1	D	1	1
31	X	$\overline{B}\cdot\overline{C}\cdot\overline{D}\cdot\overline{E}$	1	X	A	A+B·C·D·E	B	1	1	C	1	1	D	1	1

FLIP-FLOP INPUTS

the terminal is permanently connected to a source of logic 1. The specification
sheet for the FF used should be consulted to determine whether a logic 1 is
HIGH or LOW and for instructions concerning the termination of unused inputs.
A divide-by-13 counter, which has been constructed using the equations of
Table 11–1, is shown in Fig. 11–13.

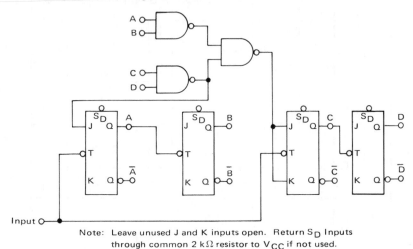

Note: Leave unused J and K inputs open. Return S_D Inputs
through common 2 kΩ resistor to V_{CC} if not used.

Figure 11–13   Divide-by-thirteen counter (courtesy of Motorola
Semiconductor Products Inc., Phoenix, Ariz.)

Binary Coded Decimal (BCD) Counters

The most common of the divide-by-N counters is the *modulus-10 circuit*.
Most measurements (physical and electrical) plus numerous mathematical calculations use base 10 numbers, so it is no wonder that many IC packages contain
binary coded decimal counters in one form or another. These special packages
will be discussed later. Right now, we will present only the simple asynchronous
divide-by-10 connection.

Just as with the divide-by-12 and divide-by-6 circuits, the count representative of the counter modulus is decoded and used to reset all FFs. Figure 11–14
shows the FF interconnections and the associated waveforms of the commonly
used 8–4–2–1 BCD (*Binary Coded Decimal*) counter. Other possible arrangements of FF states are used and have valid applications in special circuits. But
the 8–4–2–1 weights provide simple decoding and excellent correlation with
the already discussed natural binary counter.

The 8290 type IC package is a presettable decade (base 10) counter that is
partially asynchronous and partially synchronous. It may be connected in the
standard BCD counting mode or in the divide-by-5/divide-by-2, biquinary mode
(see Ch. 14). The logic rules concerning strobe, reset, count (clock), and data
inputs are identical with the rules for the 8288 package discussed earlier.

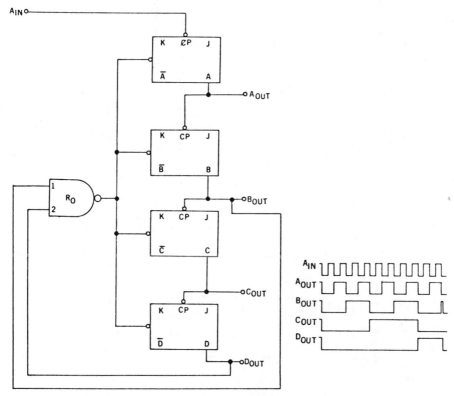

Figure 11–14   Divide-by-ten ripple counter (courtesy Sprague
Electric Company)

Figure 11–15 shows the logic diagram for the 8290 counter, along with
state tables and waveforms for both standard BCD and biquinary counting.
Standard BCD counting is achieved by using Clock 1 as the circuit input and connecting A_{out} to Clock 2. Connection of D_{out} to Clock 1, using Clock 2 as circuit
input, and A_{out} as circuit output results in the biquinary mode of operation.
Only the FF outputs are shown in the waveform diagrams, since the intermediate
waveforms are included as problems at the end of this chapter.

A completely synchronous BCD counter may be constructed using *only*
standard *J–K* FFs. Many standard *J–K* FFs are furnished with multiple *J* and *K*
inputs so that external gating is not required (see Fig. 11–16). By using the
internal gating, the BCD counter of Fig. 11–17 is implemented. From the state
table and waveforms, we see that this counter operates in the standard 8–4–2–1
code, providing the same output sequence as other 8–4–2–1 counters discussed.
As with other synchronous counters, this version is capable of faster counting
speed than asynchronous versions, due to fewer delays in FF operation.

One more counter circuit may be encountered occasionally – a *shift counter,*
which uses the principles of *shift registers.* Shift registers are discussed in Ch. 12.

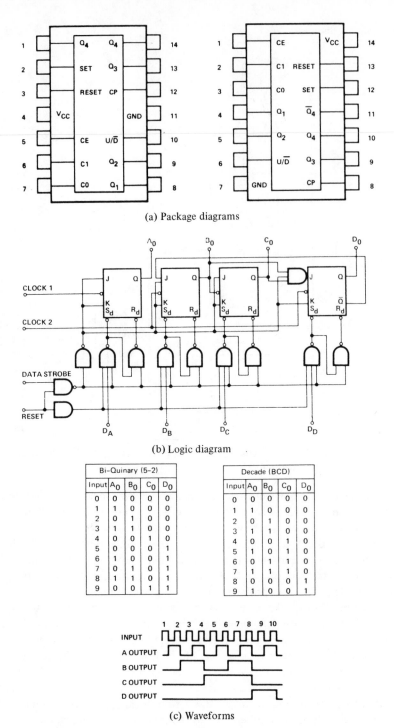

(a) Package diagrams

(b) Logic diagram

Bi-Quinary (5-2)				
Input	A_0	B_0	C_0	D_0
0	0	0	0	0
1	1	0	0	0
2	0	1	0	0
3	1	1	0	0
4	0	0	1	0
5	0	0	0	1
6	1	0	0	1
7	0	1	0	1
8	1	1	0	1
9	0	0	1	1

Decade (BCD)				
Input	A_0	B_0	C_0	D_0
0	0	0	0	0
1	1	0	0	0
2	0	1	0	0
3	1	1	0	0
4	0	0	1	0
5	1	0	1	0
6	0	1	1	0
7	1	1	1	0
8	0	0	0	1
9	1	0	0	1

(c) Waveforms

Figure 11—15   Type 8290 decade counter (courtesy of Signetics, a subsidiary of Corning Glass Works, Sunnyvale, CA.)

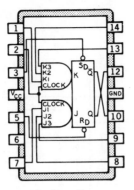

(a) Package diagram

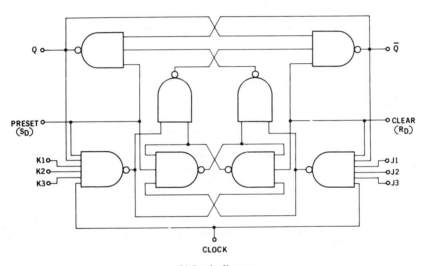

(b) Logic diagram

SYNCHRONOUS					ASYNCHRONOUS			
J	K	Q_{t+1}	$\overline{Q}_{t+1}$		S_D	R_D	Q	Q
0	0	Q_t	$\overline{Q}_t$		0	0	1	1
1	0	1	0		1	0	0	1
0	1	0	1		0	1	1	0
1	1	$\overline{Q}_t$	Q_t		1	1	Q	$\overline{Q}$

NOTES:
1. $J = J_1 \cdot J_2 \cdot J_3$
2. $K = K_1 \cdot K_2 \cdot K_3$
3. Q_t = Bit time before clock pulse.
4. Q_{t+1} = Bit time after clock pulse.

(c) State tables

Figure 11−16   Type 54H72 $J-K$ flip-flop, master-slave (courtesy Sprague Electric Co., Worcester, Mass.)

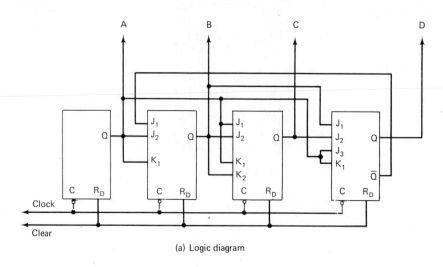

(a) Logic diagram

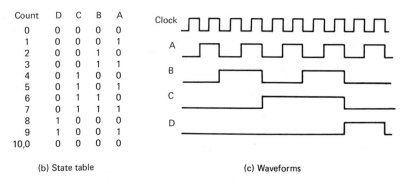

Count	D	C	B	A
0	0	0	0	0
1	0	0	0	1
2	0	0	1	0
3	0	0	1	1
4	0	1	0	0
5	0	1	0	1
6	0	1	1	0
7	0	1	1	1
8	1	0	0	0
9	1	0	0	1
10,0	0	0	0	0

(b) State table   (c) Waveforms

Figure 11—17   Synchronous BCD counter

11—4 COUNTER—CIRCUIT ANALYSIS

A Modulus 5 Counter

A counting circuit with unknown characteristics is shown in Figure 11—18. Sufficient background has been provided in this and previous chapters to determine the count sequence and waveforms, as necessary.

The first step entails a determination of the *characteristics of the logic elements* used in the counter. Assuming MIL-STD-806B symbology, the AND gate follows conventional rules. The J–K FF operating characteristics were detailed in Ch. 8 and are repeated in Table 11—2 (as a limited state table). Manufacturers' specification sheets should be consulted for logic elements that have not previously been encountered.

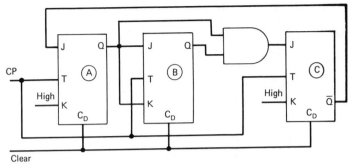

Figure 11−18   Counter circuit to be analyzed

Table 11−2   J−K Flip-Flop Limited State Table

J	K	Q_{n+1}
0	0	Q_n
1	0	1
0	1	0
1	1	$\overline{Q}_n$

Since the clock pulses are made directly available to each of the FFs in the counter, the circuit appears to operate in the synchronous mode. As the equations for each of the FFs are developed, it is assumed that the clock pulse is present. This results in a simpler analysis. Inspection of the logic diagram results in the individual FF equations (J and K inputs) shown here.

$$J_A = \overline{C}_n \qquad \text{Eq. (11–2)}$$
$$K_A = 1 \qquad \text{Eq. (11–3)}$$
$$J_B = A_n \qquad \text{Eq. (11–4)}$$
$$K_B = A_n \qquad \text{Eq. (11–5)}$$
$$J_C = A_n B_n \qquad \text{Eq. (11–6)}$$
$$K_C = 1 \qquad \text{Eq. (11–7)}$$

A complete state table for the counter may now be constructed, using the individual FF equations and the characteristics table for the $J–K$ FF. The present state (time n) of each FF is shown on the left side of Table 11−3, while the state following the next clock pulse (time $n + 1$) is shown on the right side. Note that arbitrary decimal-number state designations are used for reference purposes.

All FFs are established in the CLEAR condition prior to the start of the analysis, resulting in entries $C = 0$, $B = 0$, and $A = 0$ for the present state in row 0. From both the logic diagram and the FF equations, we see that if FF C is in the CLEAR state, its $\overline{Q}$ output is HIGH and the J input of FF A is HIGH. The K input is also HIGH (permanently connected), so FF A changes state with

Table 11–3   State Table for Counter Circuit of Fig. 11–18

State Number		n Present state			$n+1$ Next state	
	C	B	A	C	B	A
0	0	0	0	0	0	1
1	0	0	1	0	1	0
2	0	1	0	0	1	1
3	0	1	1	1	0	0
4	1	0	0	0	0	0
5,0	0	0	0		etc.	

the arrival of every clock pulse as long as C is in the CLEAR state. Thus, in row 0, the A entry for the $n+1$ portion of the state table is 1.

Prior to the arrival of the first clock pulse, FF A was in the 0 state and the Q output was LOW, thus disabling FF B. When the first clock pulse arrives, no change occurs in FF B, and the B_{n+1} entry is 0. Also, since both the A and B FFs are in the 0 state prior to arrival of the first clock pulse, G_1 has a LOW output, which disables the J input of FF C. According to the characteristics table of the $J-K$ FF with $J = 0$ and $K = 1$, the FF goes to or stays in the CLEAR state, depending on its previous condition. It is thus assured that the state of FF C is 0.

The states of all FFs at time $n+1$ are now transferred to the present state side of the state table, and the same process is repeated to obtain the next state, following the arrival of another clock pulse. The analysis continues until repetition of the counting pattern is noted. In the preceding example, this occurs when the fifth clock pulse arrives, recycling the counter to the starting count of 000. Thus the circuit is a modulus 5 counter, operating in the synchronous mode.

Upon completion of the state table, both the state flow diagram (Fig. 11–19a) and complete algebraic descriptions (Fig. 11–19b) may be developed. A waveform diagram may also be developed (Fig. 11–19c) to supplement the other forms of circuit description.

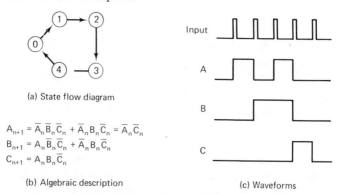

(a) State flow diagram

$$A_{n+1} = \overline{A}_n \overline{B}_n \overline{C}_n + \overline{A}_n B_n \overline{C}_n = \overline{A}_n \overline{C}_n$$
$$B_{n+1} = A_n \overline{B}_n \overline{C}_n + \overline{A}_n B_n \overline{C}_n$$
$$C_{n+1} = A_n B_n \overline{C}_n$$

(b) Algebraic description

Input

A

B

C

(c) Waveforms

Figure 11–19   Description of counter of Figure 11–18

A Modulus 17 Counter

The increased complexity of modern digital equipment has resulted in the availability of IC packages that perform complete system operations. *Multiple-stage counters* have already been discussed at length in this chapter and will be examined further in this section. The logic diagram of Fig. 11−20 shows a device that uses a type 8290 presettable decade counter (discussed in Sec. 11−3), a type 8291 presettable binary counter with natural binary counting sequence and the same presettable inputs as the 8290 unit, and a number of NAND gates.

As with any problem, the first step is to obtain as much information as possible concerning the circuit. The characteristics of the 8290 counter, obtained from manufacturer specification sheets, were used in Sec. 11−3. Type 8291 counter operation is identical, except for the counting sequence. The NAND gates are conventional, except for the unusual connection between G_1 and G_2. This type of symbology is used to indicate an *expansion input*, which allows the output of one gate (prior to inversion) to be connected to the gating structure of another gate. G_3 and G_4 are used in the conventional latch configuration.

The basic circuit input is called *clock* and consists of a sequence of pulses to be counted. *Out* is an indication that the circuit has performed its intended function. The *strobe* is used to preload the counter, just as in previous applications. *Clock* is an output that may be used in other logic circuits, while W is a test point. Some of the individual counter-stage outputs are also available as test points or for use as inputs to other circuits. The only known fact is that the circuit is a counter that counts to a number (which may be preselected) and then recycles to start counting over again.

A bit of constructive thinking about the circuit connections sometimes helps this type of analysis. For example, the specifications for the 8290 and 8291 ICs state that the active condition of the strobe input is LOW-going, so it must be at a HIGH level when the strobe function is not being performed. Since the strobe input is directly connected to the output of G_4, this point must be HIGH most of the time. The latch connection of G_3 and G_4 decrees that G_3 output must be LOW most of the time. The indication of circuit operation then must be a HIGH signal at *out*. Such an indication can occur only if W goes LOW, and this happens only when $A_1, D_1, A_2, B_2, C_2,$ and D_2 are HIGH (along with the normally HIGH strobe input).

Such circumstances exist when the following binary count is in the counter.

D_2	C_2	B_2	A_2	D_1	C_1	B_1	A_1	
1	1	1	1	1	0	0	1	(The state of C_1 and B_1
					1	1		is unimportant.)

Note once again that the physical drawing of the counter and the actual order of the binary number are reversed, as is common in logic diagrams. A_1 is the least-

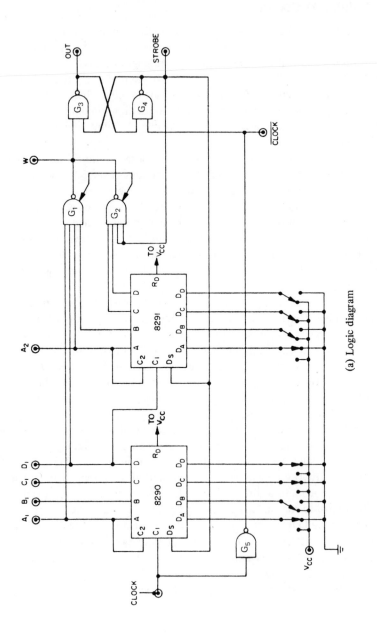

(a) Logic diagram

328

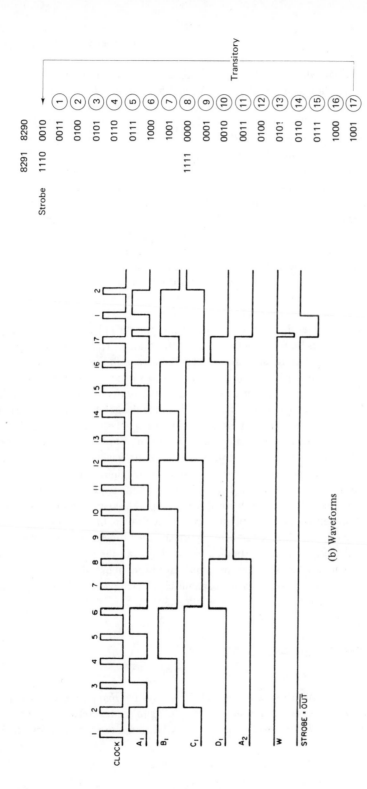

Figure 11–20 Counter circuit using IC packages (courtesy of Signetics, a subsidiary of Corning Glass Works, Sunnyvale, CA.)

329

significant digit, while D_2 is the most-significant digit. Upon occurrence of any of the counts shown above, W goes LOW and the output of G_3 goes HIGH. The strobe output goes LOW, due to the cross-coupled configuration of G_3 and G_4, thus forcing W HIGH and leaving *out* HIGH. Of course, when the NAND-gate FF changes state, the output connected to the strobe terminal goes LOW, and strobing action is set up within the 8290 and 8291 circuits.

Remembering that a LOW input to the strobe terminal of the ICs transfers the levels on the data lines into the respective FFs leads to another interesting discovery. The data inputs may be either HIGH or LOW, depending on the position of the connected switches. If the data terminal is connected to a HIGH potential, a 1 is stored, and if it is connected to a LOW potential, a 0 is stored. Circuit investigation shows that the data inputs D, C, and B on the 8291 and B on the 8290 are connected to the 1 level. Thus, upon occurrence of the strobe pulse, the counter is automatically loaded with the binary number 11100010.

The first clock pulse that arrives following strobing action resets the output latch and advances the counter one count to 11100011. Each subsequent clock pulse advances the counter one more count, until $11111^0_1{}^0_11$ is reached, at which time W goes LOW and activates the output latch. The cycle then repeats, as shown in the counting table and waveform diagrams in Fig. 11–20. Note specifically that the 8290 counter recycles when its maximum count (10) is achieved at clock-pulse number 8, while the 8291 counter never reaches its full count. It recycles and is reloaded as a result of the decoding action of G_1 and G_2.

Analysis has thus shown that the circuit in Fig. 11–20 is a modulus 17 counter, when the control switches are set as shown. Furthermore, the counter decodes the maximum count capability of the circuit, having reached that count from a preloaded count that was equivalent to the maximum count minus the desired count. This type of operation reduces the time required to restart the counting process, since a conventional reset to 00000000 may require the reset signal (under certain circumstances) to ripple the full length of the counter.

The analysis of logic circuits using prepackaged functions such as those in Fig. 11–20 is usually a rather simple process. However the reader should be on the alert for unusual applications of prepackaged functions. Many design engineers take full advantage of the capabilities of such packages, and the specification sheets may not discuss such applications.

11–5 COUNTER APPLICATIONS

Counter applications are manifold — *any system that requires accumulation of a series of events is a candidate for the digitally implemented counter.* Measuring instruments, digital clocks, and digital computers are but a few of the potential uses for the circuits discussed in this chapter. Block diagrams of a few of these applications are presented in this section.

Measuring Instruments

The basis for most direct-reading digital instruments is the simple *binary counter connected in the BCD configuration.* Of course, the count in the individual decades must be decoded and displayed (see Ch. 15) to be useful, but applications may be discussed without actual knowledge of this process.

Precise measurement of the number of events per unit time (whether random or periodic) *may be made with a device called a* "counter." Other names such as totalizer, frequency timer, scaler, time-interval meter, frequency counter, and digital counter are also encountered. Practically all of these devices operate in a similar manner (Fig. 11−21). A source of precision timing is used to open and close a gate that feeds the events to be counted to the counting portion of the instrument. The usual gate open time is 1 second, although most instruments allow the choice of multiples and submultiples of 1 second to improve the instrument's versatility. The events to be counted may be pulses developed from mechanical operations such as switch closures, interruptions of a light beam on a production line, or signals from a radio transmitter.

Digital voltmeters also use counter circuits. The voltage to be measured is converted by some means into a frequency that may be counted. Conventional decade-counting circuits then count and display that frequency in terms of a voltage. Figure 11−22 is a block diagram of a simple digital voltmeter. This voltage-to-time or *ramp* type of voltmeter measures the period of time required for a linear ramp to change from zero to a level that is coincident with the unknown input voltage. During this period, an internal oscillator is gated on and its output is totalized in a counter. With proper selection of ramp slope and oscillator frequency, the total count during the sample period will numerically indicate the unknown voltage.

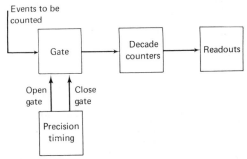

Figure 11−21   Principles of operation for a "Counter"

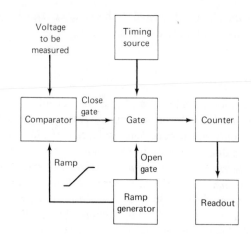

Figure 11—22   Digital voltmeter block diagram

Digital Clocks

Most digital systems used to collect data or record events require some means of indicating *the time at which the data was received or the event occurred.* Such indications may be furnished by *digital clocks.* A block diagram of a typical digital clock is shown in Fig. 11—23.

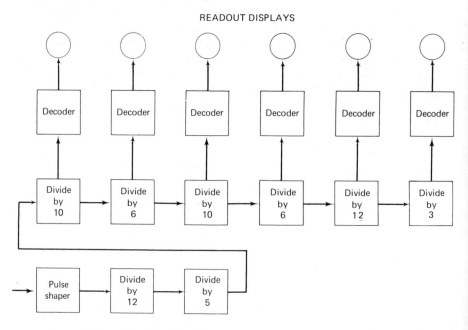

Figure 11—23   Digital clock block diagram

The 60 Hz input signal is derived directly from the power line and is filtered and shaped into a form that will accurately trigger the FF circuits that follow. A divide-by-12 and a divide-by-5 combination is used to obtain one pulse per second. The divide-by-10 circuit that follows develops one pulse every 10 seconds, while the divide-by-6 circuit further divides the output frequency to obtain one pulse every 60 seconds. A similar arrangement is used to develop the minutes and tens of minutes count. Hours and tens of hours are provided by the divide-by-12 and divide-by-3 circuits. A means to preset the clock to any given time is provided. Readout may be visual, in the form of a special code, which is recognized by digital printing devices, or both.

Digital Computers

Although digital computers are not discussed until Ch. 17, counter applications may be discussed without detailed knowledge of the computer. The key fact is that the digital computer operates in an *organized*, *sequential manner* in performing its functions. The basic sequence of operation is controlled from a *system clock*, which is merely a source of precision timing. All of the sequencing within the computer develops from the system clock. The information used within the computer (whether it is data to be operated upon or operations to be performed) is divided into *words*. Words are usually composed of a fixed number of binary digits (*bits*); the computer must keep track of these bits so that it will know when one word is completed and the next one starts. The counter in Fig. 11–20 could do this job quite adequately, for example, if the word contained 17 bits. A change in level at *out* would signify the end of one word and the beginning of another.

The sequential nature of a computer requires that operations be performed in a specific order within given time periods. The counter, together with decoding circuits, provides the necessary information. If, for example, an operation such as addition requires six sequential operations, a modulus 6 counter with decoding circuits is used. A simple block diagram of such a circuit appears in Fig. 11–24.

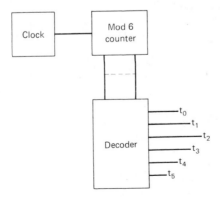

Figure 11–24   Operations sequencer (conceptual)

These examples merely scratch the surface of counter use in digital applications. Since some of these areas (e.g., digital computers) have been pointed out, the reader should experience no difficulty in recognizing many other applications.

PROBLEM APPLICATIONS

11–1.   What is a counter?

11–2.   Define asynchronous counter operation.

11–3.   Define synchronous counter operation.

11–4.   Compare the advantages and disadvantages of asynchronous and synchronous counters.

11–5.   What are the identifying characteristics of an asynchronous binary UP counter? DOWN counter?

11–6.   List three methods of representing asynchronous counters and explain the use of each.

11–7.   Draw waveform diagrams for the UP–DOWN counter of Fig. 11–5 (without the dotted lines) under the following circumstances.
  (a) Initial conditions – all stages CLEARed, UP COUNT HIGH and DOWN COUNT LOW.
  (b) Five input pulses, followed by UP COUNT LOW and DOWN COUNT HIGH.
  (c) Three input pulses, followed by both UP COUNT and DOWN COUNT HIGH.
  (d) Five input pulses, followed by DOWN COUNT LOW, UP COUNT HIGH, AND CLEAR HIGH at the same time.
  (e) Determine the count accumulated at the end of Step (d).

11–8.   Repeat Problem 7 – with the dotted lines connected.

11–9.   Why are the synchronous and asynchronous binary UP counters in Figs. 11–1 and 11–6 equivalent?

11–10.   Using the type 7493 IC package in Fig. 11–2, draw logic diagrams and waveforms of modulus 3, 8, 9, and 11 ripple counters. (External gates may be used if necessary.)

11–11.   Develop the waveforms for the logic diagram of Fig. 11–13.

11–12.   Draw a logic diagram and develop the waveforms for a modulus 17 counter using Table 11–1.

11–13.   Why is the 8–4–2–1 code the most popular of BCD codes?

11–14.   Develop the intermediate waveforms for the type 8290 counter in Fig. 11–15.

11–15.   Make a list of procedural steps required to analyze a counter circuit.

11–16.   List as many applications of counter circuits as possible and explain each of them.

Registers and
Their Applications

A register is a group of logic circuits, usually flip-flops (FFs), *arranged in a manner that allows the storage and processing of information in binary form.* Single FFs *may* be classified as registers, since they are capable of both storing and processing binary information. However registers usually consist of a number of FFs. The number of memory elements in a register is determined by the requirements of the specific application. For example, a register in a small digital computer might have sixteen stages, since a word (the unit of information handled by many minicomputers) contains 16 bits.

One possible classification of registers is functional: *a storage register retains information*, while a *shift register processes information.* Registers can be further classified in terms of the input–output (I/O) methods used. Information is entered into and taken from most registers either *bit-by-bit* (serial form) or *all bits at once* (parallel form). If entered and removed in serial form, the device is categorized as *serial-serial.* If entered and removed in parallel form, the classification is *parallel-parallel.* (Serial-parallel and parallel-serial designations are also commonly encountered.) Thus registers are classified by both their purpose and their method of operation.

As with other digital-logic elements, registers are difficult to identify *physically*, since they appear in the same types of packages as gates, FFs, counters, and other prepacked logic functions. Once again, the manufacturers' specification sheets must be consulted to obtain the characteristics of particular registers. This chapter will illustrate logic diagrams of registers constructed from both individual memory elements (FFs) and complete integrated circuit (IC) packages.

All digital systems that require the *storage and handling of binary information* use registers. The digital computer, of course, is a prime example. Registers are used to perform arithmetic operations, store commands, and transfer data from one portion of the computer to another. In digital measuring equipment, registers store data to be displayed so that the counters may proceed with the next measurement without losing the previous information. The transmission of digital data from one location to another requires conversion from parallel form to serial form — and back to parallel form. Registers perform these functions.

Since the reader will encounter registers in many different forms, the discussions in this chapter should provide sufficient information to recognize and understand the most common implementations of storage, shift, and counting registers.

12–1 THE STORAGE REGISTER

Basic Concepts

Each FF in a register can store *one piece of binary information* (one bit). Since the concept of information storage in FFs was discussed in Ch. 8, it should require little additional coverage here. When storage registers must store more than one bit at a time, additional FFs are used. The interconnections between FFs determine the functional and I/O characteristics of the register.

By the strictest definition, if the FF inputs are connected to the source of data to be stored and the outputs are connected to the circuit that uses the stored information, then a *storage register* exists. Such a definition limits the scope of discussion in this section, but it *does* define a specific, separable class of register applications. Many variations of input and output connections exist and will be discussed in this chapter.

Storage registers are used whenever *temporary storage of information* is needed. A practical example is seen in the block diagram in Fig. 12–1. Many frequency counters have the capability to display the "last" count while accumulating the "next" count. As soon as a counting period is completed, the

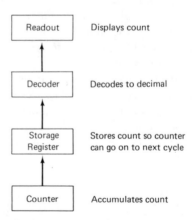

Figure 12–1   A register application

outputs of the counter stages are immediately transferred to the storage register. Once the transfer is complete (which can be accomplished in as little as 50 nanoseconds), the counter can be released to start another counting cycle. The information in the storage register (an exact replica of the information that was in the counter) is decoded into a form acceptable to the display device employed and is displayed for use while the counter is obtaining the latest information to be used in updating the display.

A similar application exists in digital computers, where data from the computer's memory is transferred temporarily to a storage register before it is distributed to other functional parts of the computer.

Implementation

Any device that can (a) *assume a state representative of its input*, (b) *retain that state after the input is removed*, and (c) *make available an output that may be used to represent its state* can function as a storage register. Not only do FFs possess these characteristics, but certain magnetic, electromagnetic, and semiconductor devices may also be so used. FF implementation of the storage register function is the method discussed in this chapter.

$R-S$, $J-K$, and D FFs all see application as storage registers. The clocked (gated) versions are employed so that the input data may be transferred to the FF and then leave the FF unable to respond to any further input changes, until it is time to store new data. Figure 12–2 shows the general concept of data transfer from a FF in a counter to a $J-K$ FF used as a storage register. An $R-S$ FF functions in the same manner as a $J-K$ FF under the input conditions shown in Fig. 12–2a and may be used in place of a $J-K$ FF. Of course, other circuit parameters, such as supply voltage and logic levels, must also be compatible.

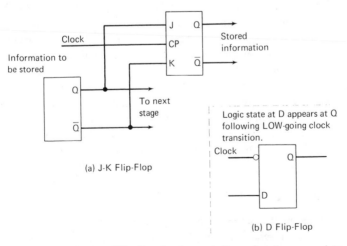

Figure 12–2   Flip-flop implementation of a storage register

D FF implementation of the storage function (Fig. 12–2b) is usually simpler than either the $J–K$ or $R–S$ implementation. Inverted data input is not required due to the characteristics of the D FF (see Sec. 8–7). A single connection between the Q output of the counter FF and the D input of the storage register causes a 1 to be stored in the register, when the counter FF is in the 1 state, and 0 when it is in the 0 state. The clock pulse is assumed to be present.

Whether FFs or other methods are used to furnish the storage register function, the same actions must take place. Whatever data is presented to the input must be stored on command and be made available at the output when needed. If new data is presented at the input, it must not disturb the existing stored data until the next storage cycle takes place. Any medium (electrical, electronic, or mechanical) that performs in this manner is usable in storage register applications.

Storage Registers at Work

FLIP-FLOP STORAGE OF COUNTER OUTPUT

Storage registers, as defined in this section, fit into the *parallel input-parallel output* category. That is, all storage register stages receive their inputs at the same time, and each stage supplies a separate output. Such an application using $J–K$ FFs as storage elements is shown in Fig. 12–3a. A counter composed of four T FFs is operating in the natural binary sequence, as seen in the waveforms of Fig. 12–3b. The Q and $\bar{Q}$ outputs of each counter stage supply the J and K inputs, respectively, of accompanying storage-register stages. A *read pulse* is supplied to the CP inputs of each register stage shortly after the occurrence of every fifth pulse to be counted. The first read pulse samples the output of each counter stage and transfers its logic state to the appropriate register stage. Since Q_A is HIGH at this time, the J input of SR_1 is HIGH, and SR_1 goes to the SET state. Its Q output becomes HIGH, and the state of the A FF is stored in SR_1. Similar actions occur for the other stages, and after the read pulse occurs, the 0101 count of the counter is stored as 0101 in storage register stages SR_4, SR_3, SR_2, and SR_1, respectively. The 1010 count replaces the 0101 count after the second read pulse, and 1111 is stored as a result of the third read pulse.

This example merely demonstrates the operation of storage registers. Without additional FFs, gating, and control, the values placed in the storage register would be representative of *only* the four least-significant digits of the accumulated count. However it *has* been shown that $J–K$ FFs can perform storage register functions.

INTEGRATED CIRCUIT MEMORY BUFFER REGISTER

In the design of equipment requiring storage register functions, *prepackaged* IC *registers* are very common. The type 8200 buffer register (Fig. 12–4a) is an array of ten clocked D FFs that may be used in any parallel input-parallel

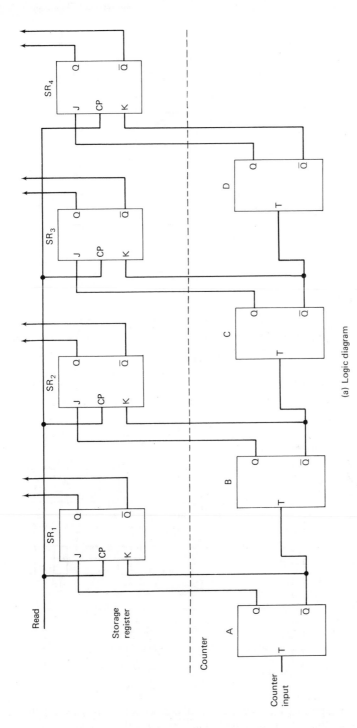

Figure 12—3 J-K flip-flop storage register

(a) Logic diagram

339

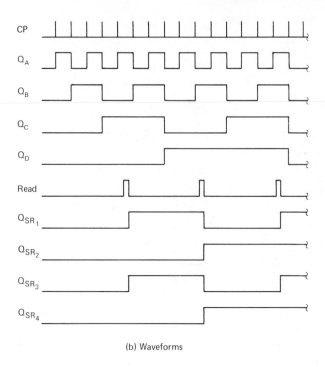

(b) Waveforms

Figure 12–3 (cont'd)

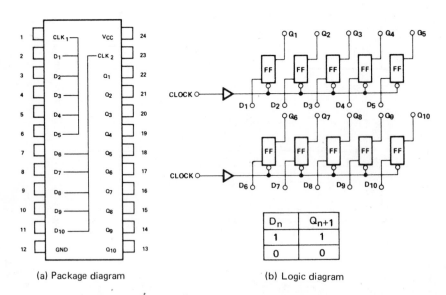

(a) Package diagram

(b) Logic diagram

Figure 12–4   Type 8200 buffer register (courtesy of Signetics, a subsidiary of Corning Glass Works, Sunnyvale, CA)

output application. *Buffer register* is a term often used to describe *a register that serves as a temporary store in transferring information between two units that are asynchronous, operating at different speeds, or performing independent tasks.* The logic diagram of Fig. 12—4b shows that two groups of five D FFs with separated clock inputs are provided to add to the versatility of the package.

A practical (though fictitious) computer application of the type 8200 buffer register is shown in Fig. 12—5. All of the instructions for the solution of a problem and the data required for the problem are placed in the computer memory prior to actually performing the steps of the problem. The computer works toward the problem solution in an automatic sequential manner (unless told to change the sequence), selecting one step after the other and performing the operations required. Information required to perform the operations is con-tained in groups of bits called *words*. One type of computer word consists of 16 bits, divided so the first 6 bits define what action is to be taken (the opera-tion) and the other 10 bits define the address of the data that is to be operated on (the operand).

Each word is removed from memory and placed in a storage register until it is needed. This allows the memory to go on with its other business without waiting for computer operations to be performed.

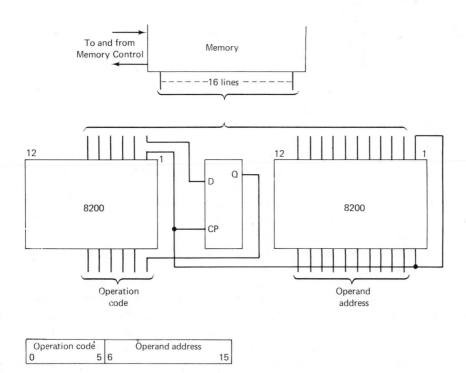

Operation code		Operand address	
0	5	6	15

Figure 12—5   Memory buffer register

Since the type 8200 buffer register is available only in a configuration that results in storing groups of five bits, an extra D FF is used (Fig. 12–5) to provide storage for the sixteenth bit of information. When the 16 bits of a computer word are to be loaded into the storage register (sometimes called the *memory buffer register*), the memory supplies the logic levels to the 16 lines feeding the D input of each of the register stages. A clock pulse is generated by the memory (or memory control) to tell the register to accept the data; and the complete word is "dropped" in parallel into the memory buffer register. The outputs of the register may now be fed elsewhere in the computer to perform various functions, and the memory may proceed with its next step. The memory buffer register has provided the parallel in-parallel out storage register function, showing the practical application of IC prepackaged registers.

12–2 THE SHIFT REGISTER

Basic Concepts

The shift register fits into the overall class of registers, since it may be constructed of FFs and is used to operate on binary data. *Shift registers move data*, usually from one stage of the register to an adjacent stage. The movement (shift) of data may be from left to right, right to left, or in both directions. Hence both *unidirectional* and *bidirectional registers* exist.

Shift registers differ from storage registers in that *adjacent stages are connected* to allow movement of data from stage to stage. Actually, the shift register also may serve as a storage register when proper controls exist. Data may be loaded into the shift register in parallel and read out of the register in parallel, just as in the storage register. However the same shift register can also accept data one bit at a time (serial), move it from stage to stage until all data is in the register, and then store the data until it is ready for readout. The data may be read out one bit at a time (serial) or all bits at once (parallel). Properly designed, then, the shift register can perform all four I/O operations: *serial-serial*, *serial-parallel*, *parallel-parallel*, and *parallel-serial.*

A common application of the shift register is seen in the block diagram in Fig. 12–6. An information display device, such as the keyboard control of a digital computer (similar to an electric typewriter in form and operation), uses a multibit representation of each character to be displayed or entered. One particular representation, often used for remote computer control, employs an 8-bit code. Each alphabetic, numeric, or control character is represented by a group of 8 bits, all of which must be present at the same time (parallel). Transfer of data from one location to another requires eight separate lines, one for each bit. When sending and receiving locations are physically separated by great distances, however, the cost of multiline transmission becomes prohibitive.

Shift registers are used to allow transmission of the complete 8-bit character with only a single line. The 8 bits that result from actuating a key on the key-

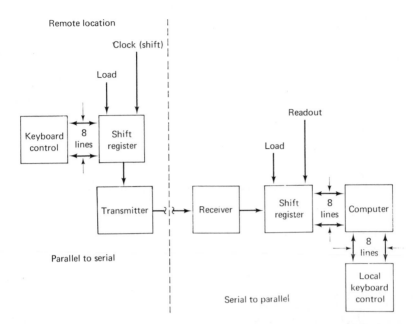

Figure 12—6   A shift register application to data transmission

board are stored in a shift register (8 stages), using parallel loading. As soon as loading is completed, the register is commanded to shift the data. One bit at a time is shifted from the register and converted for transmission. After the eighth bit is shifted out and converted, the register is ready to accept (in parallel) the next character.

At the receiving end, the 8-bit character is converted to logic levels and placed (one bit at a time) into another shift register. When all 8 bits are received, the information is read out to the receiving device and displayed. Thus, at the sacrifice of speed of transmission, the requirement for eight transmission lines has been reduced to one, and the cost of system implementation has been cut drastically.

Implementation

Shift registers are implemented with the same kinds of FFs used in storage registers. Each stage in the shift register may be viewed as an individual storage register, capable of being loaded (in most cases) not only in the parallel mode, as shown in this section, but also in the serial mode. Due to the necessary inter-connections required to move the data from stage to stage, parallel loading is accomplished in a slightly different manner than it is in storage registers. The asynchronous (direct SET and CLEAR) inputs are generally used for parallel loading, while the synchronous inputs provide serial-load capability. Figure 12—7 shows a typical $J-K$ FF used in shift register applications.

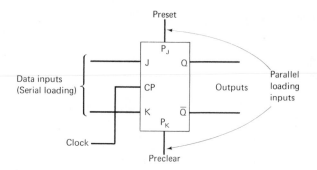

Figure 12—7 $J-K$ flip-flop as a shift register stage

Once again, the $J-K$ and $R-S$ FFs possess the same characteristics under the input conditions of most shift register operations and are functionally interchangeable. Although not as versatile, the D FF used in shift registers will be shown later.

Shift Registers at Work

SERIAL—SERIAL MODE

Perhaps the simplest of shift registers is the *serial in-serial out circuit* shown in Fig. 12—8. (The AND gates and READ OUT input should be disregarded at this time.) It consists of D FFs and requires a logic level that represents only the data to be moved into the register, and a shift input. The CLEAR input is assumed to be activated just prior to arrival of the first shift input so that all stages begin operation in the CLEAR condition.

The output of a D FF is the same as its input *after* the occurrence of the clock or shift input. Thus when the D input to a FF is HIGH, the Q output is HIGH following the next clock pulse. The same is true of the LOW output. Note that the Q output of the A FF is the D input of the B FF. Whatever logic level is present at the Q output of the A FF appears at the Q output of the B FF upon occurrence of the shift pulse. Thus every shift pulse causes the information in the preceding FF to be transferred to the following FF.

The waveform drawing of Fig. 12—8 (disregarding the READ OUT and D, C, B, and A waveforms) does not really show the precise timing requirements that exist between the DATA information and the SHIFT information. Most FFs require that the DATA information precede the SHIFT information by a *specific time interval* (nanoseconds in high-speed FFs). This factor must be taken into consideration during logic design, but it is assumed when using waveform drawings. Although the drawing shows the DATA and SHIFT inputs changing at the same time, the required time differential is considered to be present. Thus when the SHIFT input goes HIGH to activate the FF, it uses the DATA level that

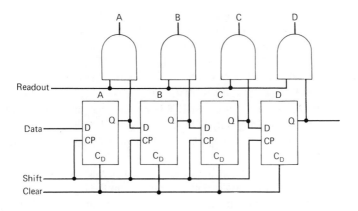

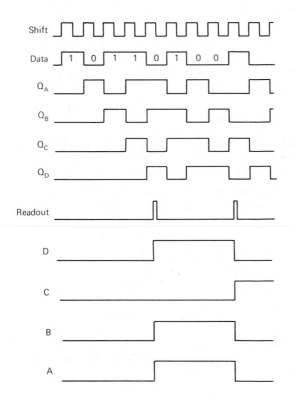

Figure 12-8 Shift register logic diagram and waveforms

was present just before the change in SHIFT input. The first HIGH-going transition of the SHIFT input occurs when DATA is LOW, and since all stages are in the CLEAR state, no change takes place. The second SHIFT input occurs with DATA HIGH, and FF A SETS, making its Q output HIGH. The DATA input has now been transferred to FF A from the input circuit. The third clock pulse occurs when DATA is LOW and FF A CLEARS, making its Q output go LOW. However, at the same time that the SHIFT input was changing the state of FF A, it was also using the HIGH Q output of A to cause FF B to SET. The finite time that it takes to change the state of a FF is sufficient to enable FF B to change state before FF A.

Note that the DATA input has now been transferred from FF A to FF B, and the new piece of information that was present on the input line is now stored in FF A. Evaluation of the remainder of the operations in Fig. 12–8 will show that the incoming data is moved to the right one stage, every time a shift pulse is provided. As soon as all of the data is loaded into the shift register (one bit at a time), it will be shifted out of the rightmost FF and disappear, unless the shift pulses are stopped. The usual mode of operation with serial in-serial out shift registers is to provide enough shift pulses to load the register and then stop shifting, until a time when the output data is needed. The shift pulses are then restarted, and the data feeds from the rightmost FF (one bit at a time), as long as shift pulses are present. Used in this manner, the register functions as a combination shift and storage register.

CIRCULATING REGISTERS

A special case of the serial-serial shift register is the *circulating register*, which is used in certain digital computer applications requiring *repetitive* use of the same information. In the serial-serial shift register, the information is shifted out the last stage and lost, while in the circulating register the last stage supplies its information *back* to the first stage to be recirculated. The informa-

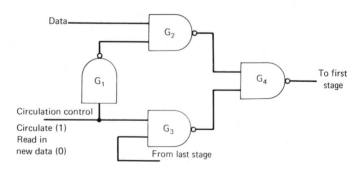

Figure 12–9   Circulating register input gating

tion being recirculated is also available at the output of the register for use in the rest of the circuit.

Conventional serial-serial shift registers may be used as circulating registers by providing input gating similar to that shown in Fig. 12–9. Input to the first stage of the register may be either original data or the data from the last stage of the register, as selected by the circulation control input. If HIGH, the circulation control input enables G_3 and allows data from the last stage to be fed through G_4 to the first stage of the register. G_2 is disabled by the LOW output of G_1. Changing the circulation control input to a LOW level disables G_3 and is inverted by G_1 to enable G_2, thus allowing the original data to be read in.

SERIAL–PARALLEL MODE

The *serial in-parallel out register* may also be implemented by using gates connected to each of the FFs' Q outputs. Data is shifted in, until all desired information is stored. The READ OUT operation is enabled prior to the next clock pulse, and the information in each FF appears at the output of its respective output gate. No information is lost, and the register can continue its shifting operation as if nothing had happened. This operation is shown on the waveform diagrams in Fig. 12–8. It is representative of the type of register that would be used to convert the received serial data in the example in Fig. 12–6 to the required parallel data for machine operation.

PARALLEL–SERIAL MODE

Parallel in-parallel out shift registers are somewhat uncommon — they fit more properly into the storage register category. (Coverage of this configuration was provided in Sec. 12–1.)

$J–K$ FFs are used to demonstrate the *parallel in-serial out* shift register in Fig. 12–10. This type of circuit could also be used in the remote computer control unit of Fig. 12–6 to convert the parallel keyboard data to serial form for transmission. The data to be entered is provided at the A, B, C, and D inputs. When the "enter" pulse arrives, whichever of the two control gates becomes enabled for each FF allows a pulse to be developed, and each FF will be placed in the state represented by the inputs. The FFs do not have to be CLEARed prior to entering information, since both the CLEAR and SET inputs are capable of being activated by the input levels. Following entry of the data, clock pulses may be applied to cause the shifting of the data toward the output. If a 1 has been entered in a FF, its Q output is HIGH, enabling the following FF's J input. In a like manner, if a 0 has been entered, the following FF will have its K input enabled. Due to the interconnections, the J and K inputs cannot both be either 0 or 1, so the FF will respond to the next clock pulse by reproducing the state that was in the previous FF.

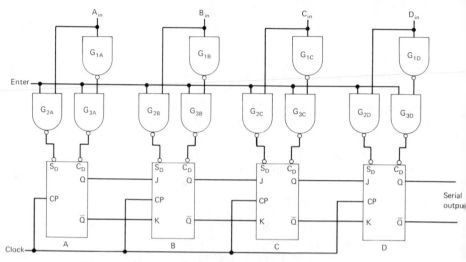

Figure 12—10   Parallel-Serial shift register

SERIAL/PARALLEL ENTRY

Discussion of the four forms of shift registers should prepare us for the presentation of more complex prepackaged IC register circuits. The wide use of shift registers in digital equipment makes it feasible to provide packages that are usable in *more* than one operating mode. Figure 12—11a shows the type 9300 IC, which is a 4-bit shift register with both serial and parallel data entry capability.

The operation of the 9300 shift register is indicated by the logic diagram of Fig. 12—11c. The element is composed of four clocked master-slave FFs with D inputs. The D input of every stage can be switched between two logic sources by the *parallel enable* (*PE*) input. When the *PE* input is LOW, the D inputs of the four stages are connected to the parallel inputs P_0, P_1, P_2, and P_3. When the *PE* input is HIGH, the D inputs of the second, third, and fourth stages are connected to the outputs of the first, second, and third stages, respectively, thus forming a 4-bit shift register. The D input to the first stage (with *PE* HIGH) is obtained from the J and $\overline{K}$ inputs (via gating elements) to produce the action of the first stage, as shown in Fig. 12—11b. All stages are set to 0, when the *master reset* (*MR*) input is LOW, thus overriding the effects of any other input.

One additional point is worthy of note. The $J\overline{K}$ input is the same as the more common JK input except that the LOW level activates the $\overline{K}$ input (indicated by the circle at the $\overline{K}$ input in Fig. 12—11a). The HIGH level activates the J input so that connecting the J and $\overline{K}$ inputs together results in a D-type input.

Conventional parallel-serial conversion is implemented with the type 9300 shift register by causing *PE* to be LOW for one clock period (to parallel load the register), then changing *PE* to a HIGH level so that incoming clock pulses can shift the parallel-loaded data out via the Q_3 or $\overline{Q}_3$ outputs. Serial-parallel conver-

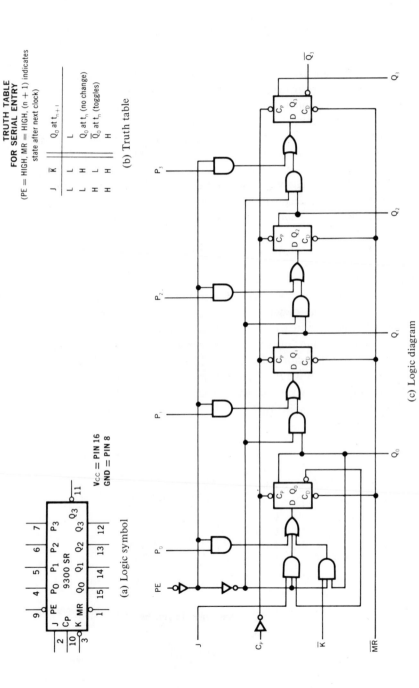

**TRUTH TABLE
FOR SERIAL ENTRY**
(PE = HIGH, MR = HIGH, (n + 1) indicates
state after next clock)

J	$\overline{K}$	Q_0 at t_{n+1}
L	L	L
L	H	Q_0 at t_n (no change)
H	L	$\overline{Q}_0$ at t_n (toggles)
H	H	H

(b) Truth table

V_{CC} = PIN 16
GND = PIN 8

(a) Logic symbol

(c) Logic diagram

Figure 12–11 Type 9300 IC shift register (courtesy of
Fairchild Camera and Instrument Corp., Mountain View, CA)

349

sion is realized by supplying enough clock pulses while *PE* is HIGH to load all information in serial form and by then stopping the clock pulses while the data at Q_0, Q_1, Q_2, and Q_3 is made available to outside circuits for gating.

Shift registers are used not only to convert from serial to parallel form (and vice-versa) but also to *manipulate binary representations of numbers.* Multiplication and division by powers of 2 is performed on binary numbers in a shift register. Such operations may be demonstrated with decimal numbers (Fig. 12–12) so that the concept is more understandable. A movement of all digits in a decimal number 1 position (order) to the right is equivalent to dividing the number by a power of 10. Thus 2000 moved (shifted) one position to the right becomes 200.0. Similarly, left movement is equivalent to multiplication by a power of 10; 20.00 becomes 200.0 when a one-position left shift is accomplished. Since decimal numbers are represented in most digital computers by their binary equivalent, similar operations may be performed. Shift right results in division by a power of 2 instead of 10. Multiplication by a power of 2 results from shift-left operations. Thus a need for registers capable of *either* right or left shifting can be visualized.

Shift-right operations have been discussed in detail in all the shift registers examined, but left shifting of data requires considerably more complexity, especially when it may be necessary to shift *both* right and left upon command. The synchronous parallel inputs of the 9300 are used to produce a register that will shift left or right on each clock. In Fig. 12–13 each 9300 has the Q_1, Q_2, and Q_3 outputs connected to the P_0, P_1, and P_2 inputs, respectively, so that each element now shifts right when the *parallel enable* is HIGH and left when it is LOW. For left shifting, Q_0 is the serial data output and P_3 is the serial data input.

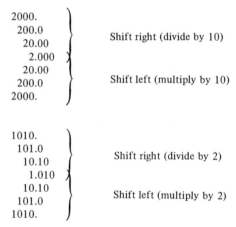

Figure 12–12   Shift multiplication and division

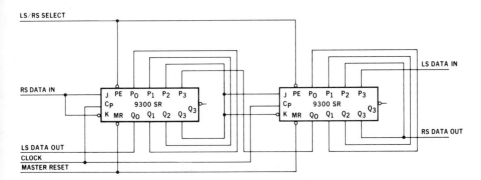

Figure 12–13   Eight-bit Left/Right shift register (courtesy of
Fairchild Camera and Instrument Corp., Mountain View, CA)

12–3 SHIFT COUNTERS

Basic Concepts

Shift counters are a specialized type of *clocked counter.* The name is
derived from the use of shift registers to perform the counting operation. This
type of counter generally results in outputs that are easily decoded, and gating is
usually not required between stages.

The concept of operation is very similar to the serial-serial shift register.
A 1 is loaded into the first stage of the serial connected register and is shifted one
stage to the right upon arrival of each clock pulse. No additional 1s are entered
into the register, so the actual number of shift pulses (clock inputs) counted may
be determined by the location of the stage in the register which contains the 1.
Figure 12–14 shows the principles of operation of the basic shift counter.

This conceptual counter is, of course, impractical. It requires one stage for
each count (a decimal counter would require ten stages), and no means is provided to restart the counter at the reference count without stopping the clock
input and applying the CLEAR input.

The Ring Counter

By providing feedback from the last stage to the first stage in the shift
counter (Fig. 12–15), *count recycling* is provided. A 10-stage decimal counter
could then count from 0 through 9 and start the count at 0 again automatically.
Ten stages are needed (*A* through *M*).

Just as in Fig. 12–14, a 1 is preset into the first stage of the register and
shifted through one stage per count, until the final stage is reached. Note that
the *Q* output of the final stage feeds the *J* input of the first stage and that the

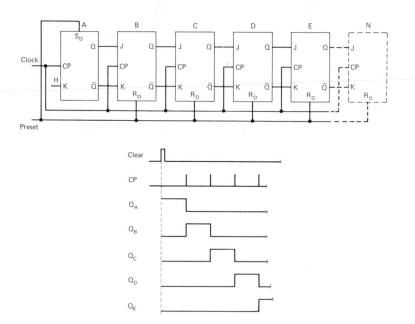

Figure 12—14   Basic shift counter

$\bar{Q}$ output feeds the K input. As long as the final stage is in the CLEARed state, the $\bar{Q}$ output is HIGH and the K input to the first stage is HIGH. Each time a clock pulse is received the first stage is returned to (or left in) the CLEARed state. When the propagating 1 reaches the final stage, the J input of the first stage is enabled, and the 1 transfers from the final stage of the register to the first stage. Ten counts are required for the propagating 1 to recycle, and the circuit of Fig. 12—15 functions as a decimal counter. Since the propagating 1 effectively travels in a circle, the counter of Fig. 12—15 is sometimes known as a *ring counter*. Waveforms for this circuit resemble those of Fig. 12—14. A count sequence table showing the state of each stage is used in place of waveforms.

The Switch-Tail Ring Counter

The requirement for ten FFs in the decimal ring counter can be reduced to five by the circuit shown in Fig. 12—16a. This counter is sometimes called a *switch-tail ring counter* or a *Johnson Counter*. The output is inverted (taken from the complementary output) of the last stage, before it is fed back to the first stage. In addition, a feedback connection from the next-to-last stage is applied to the first stage to make the counter self-correcting and to eliminate the requirement for initial presetting in most applications. The count sequence table is shown in Fig. 12—16b. Waveform development is left to the reader.

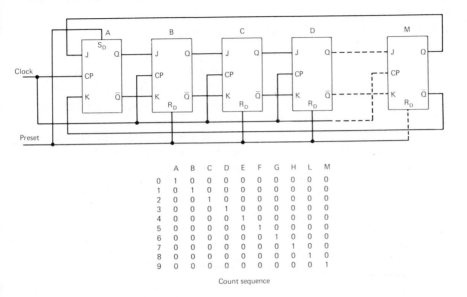

	A	B	C	D	E	F	G	H	L	M
0	1	0	0	0	0	0	0	0	0	0
1	0	1	0	0	0	0	0	0	0	0
2	0	0	1	0	0	0	0	0	0	0
3	0	0	0	1	0	0	0	0	0	0
4	0	0	0	0	1	0	0	0	0	0
5	0	0	0	0	0	1	0	0	0	0
6	0	0	0	0	0	0	1	0	0	0
7	0	0	0	0	0	0	0	1	0	0
8	0	0	0	0	0	0	0	0	1	0
9	0	0	0	0	0	0	0	0	0	1

Count sequence

Figure 12-15   Ring counter

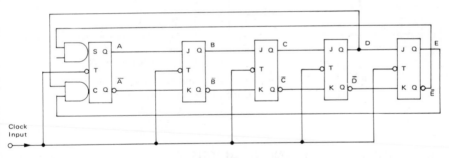

COUNTER SEQUENCE

	A	B	C	D	E
0	0	0	0	0	0
1	1	0	0	0	0
2	1	1	0	0	0
3	1	1	1	0	0
4	1	1	1	1	0
5	1	1	1	1	1
6	0	1	1	1	1
7	0	0	1	1	1
8	0	0	0	1	1
9	0	0	0	0	1

Figure 12-16   Switch-Tail ring counter (courtesy Motorola Semiconductor Products, Inc., Phoenix, AZ)

Shift Counters Using Prepackaged Registers

Prepackaged registers, such as the type 9300 IC, may also be used to produce a wide variety of counting circuits, including simple counters of different moduli, variable modulus counters, and UP–DOWN counters. With the addition of one AND–OR–INVERT gate (another common IC package), the 9300 shift register forms a modulus 10 counter, as shown in Fig. 12–17.

A slightly different form of sequence table is used to describe the counter's operation: no matter what combinations of 1s or 0s are intially in the register, the inner closed sequence of counting will be entered and the ten repetitive states will become the actual counting sequence. The reader is urged to investigate the count sequence of Fig. 12–17 and to verify its accuracy by using the waveform and logic diagram of the 9300 IC (Fig. 12–11).

Many other possible sequences may be used to form a modulus-10 counter with the 9300 shift register. Since the development of logic diagrams for these implementations properly belongs to the design engineer, they are not discussed here.

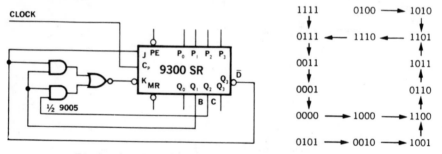

Figure 12–17   Modulus 10 ring counter (courtesy of Fairchild Camera and Instrument Corp., Mountain View, CA)

12–4 APPLICATION AND ANALYSIS

The use of registers in both digital computer and information transfer situations has already been discussed. Numerous other applications in many other types of equipment exist, but one specific use of registers easily demonstrates both multiple register application *and* analysis techniques.

The Block Diagram

A portion of a block diagram of a digital logic system is shown in Fig. 12–18. Such a form is common in large digital systems, since complete logic diagrams tend to be too large. Without some additional information, however, analysis of this register application is very difficult.

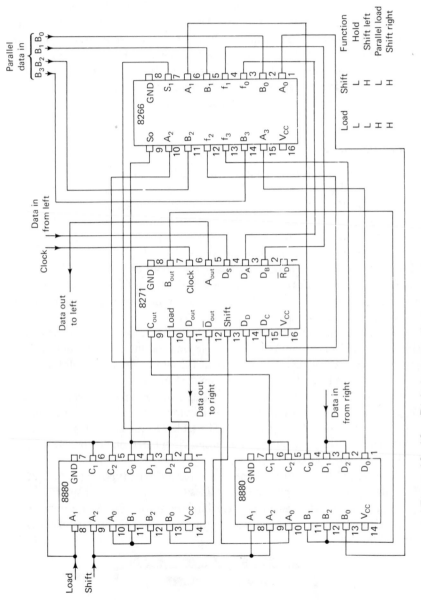

Figure 12–18 Digital system diagram

355

Some clues do exist from the information already included in the logic diagram. The information in the function table implies that the device manipulates data by shifting either right or left. Additional clues to this supposition may be found in the inputs labeled *data in from right*, *data out to right*, *data in from left*, and *data out to left*. The parallel-load function shown in the function table is borne out by the parallel "data-in" inputs. With these clues in mind, we might assume that this device is a register capable of loading information in the parallel mode, storing it (the HOLD function), and shifting it (either right or left).

IC Specifications

The next step requires consultation with specification sheets to determine the nature of each of the ICs. Figure 12–19 is the logic diagram and pin connections for the type 8880 quad 2-input NAND gate package, while Fig. 12–20 furnishes the same information for the type 8271 4-bit shift register. The type 8266 2-input, 4-bit multiplexer is described in Fig. 12–21.

NAND gates, such as those in the type 8880 package, have been adequately discussed in previous chapters. Their use in the circuit of Fig. 12–18 is either as conventional NAND gates or as single-input NAND gates functioning as logic inverters. Unused inputs in the 8880 NAND gates are connected to active inputs or to V_{cc} (the supply voltage).

Type 8266 multiplexer is an IC assembly consisting of four logic switches. Selection of two sets of inputs (A_n or B_n) depends on the control inputs' (S_0 and S_1) logic levels. All switches are actuated at the same time, so all four bits of the parallel input or all four bits of information that are to be shifted left are selected to be loaded into the appropriate stages of the type 8271 shift register. Conventional combinational logic analysis methods are used to determine the operation of each of the four switches. The table accompanying Fig. 12–21 algebraically describes the output of a single switch based on all possible combinations of control inputs.

The capability to enter both serial and parallel data typifies the type 8271 4-bit shift register. Data inputs are single lines that condition their specific register-bit locations after an enabled clocking transition. Since data transfer is synchronous with the clock, data may be transferred in any serial-parallel I/O relationship.

Figure 12–19   Type 8880 quad 2-input NAND gate (courtesy of Signetics, a subsidiary of Corning Glass Works, Sunnyvale, CA)

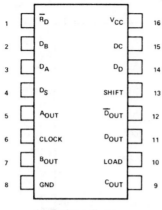

(a) Package diagram

CONTROL STATE	LOAD	SHIFT
Hold	0	0
Parallel Entry	1	0
Shift Right	0	1
Shift Right	1	1

(b) Truth table

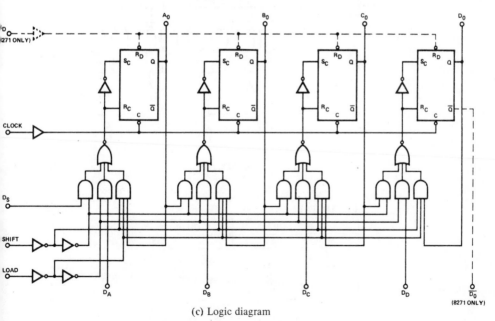

(c) Logic diagram

Figure 12—20   Type 8271 4-bit shift register (courtesy of
Signetics, a subsidiary of Corning Glass Works, Sunnyvale, CA)

357

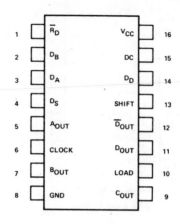

(a) Package diagram

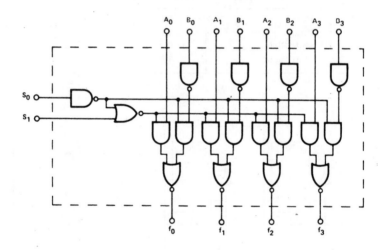

(b) Logic diagram

TRUTH TABLE

SELECT LINES		OUTPUTS
S_0	S_1	f_n (0, 1, 2, 3)
0	0	B_n
0	1	B_n
1	0	$\overline{A}_n$
1	1	1

(c) Truth table

Figure 12—21 Type 8266 2-input, 4-bit multiplexer (courtesy of Signetics, a subsidiary of Corning Glass Works, Sunnyvale, CA)

358

Mode control logic is available to determine three possible control states: *serial shift-right mode, parallel-enter mode,* and *no change (hold) mode.* These states accomplish logical decoding for system control. The truth table for the control modes is shown in Fig. 12—22. For applications not requiring the hold mode, the load input may be tied HIGH and the shift input used as the mode control. The type 8271 shift register also provides a direct reset $(\bar{R}_D)$ and a $\bar{D}$ output line.

Parallel data input is supplied to $D_A, D_B, D_C,$ and D_D, while parallel data output is available at $A_0, B_0, C_0,$ and D_0. Serial input is via D_S, and serial output is available at either D_0 or $\bar{D}_0$. Thus, under proper control conditions, the serial-serial, serial-parallel, and parallel-serial shift registers may be implemented with this universal package.

A register stage consists of a clocked $R-S$ FF with external gating. Each stage operates as a D type FF that can be either enabled or disabled by the logic level applied to the gate input common to both the S gate and the R gate.

Data Flow Diagram

Overall logic circuit understanding may be aided by the use of a *data flow diagram.* Redrawing the diagram in the form of Fig. 12—22 makes the flow of data more apparent. (The actual pin connections are unimportant at this step of the discussion.) Each of the individual truth tables should be verified package by package, if it has not already been done. If the multiplexer and shift register characteristics have already been verified, and it is known that the indicated inputs furnish the indicated outputs, then all that is necessary is to consider the overall circuit inputs in the light of the individual circuits.

Any of the common methods of logic circuit analysis may be applied, but perhaps the easiest is the use of the *supplied truth tables.* Waveforms or algebraic expressions *may* be appropriate in some cases. The truth table method is demonstrated here by determining the inputs to the 8271 and 8266 ICs, as developed by the input control logic (G_1 through G_5). Table 12—1 supplies the

Table 12—1   Functional Analysis of Data Flow Diagram (Figure 12—22)

Control Logic Inputs			8266				8271		
			S_0	S_1			Load	Shift	Function
Function	Load	Shift	G_1	G_2	Output	G_3	G_5	G_4	
Hold	0	0	1	1	1	1	0	0	Hold
Shift Left	0	1	1	0	$\bar{A}_n$	1	1	0	Load
Parallel Load	1	0	0	1	B_n	1	1	0	Load
Shift Right	1	1	0	0	B_n	0	1	1	Shift Right

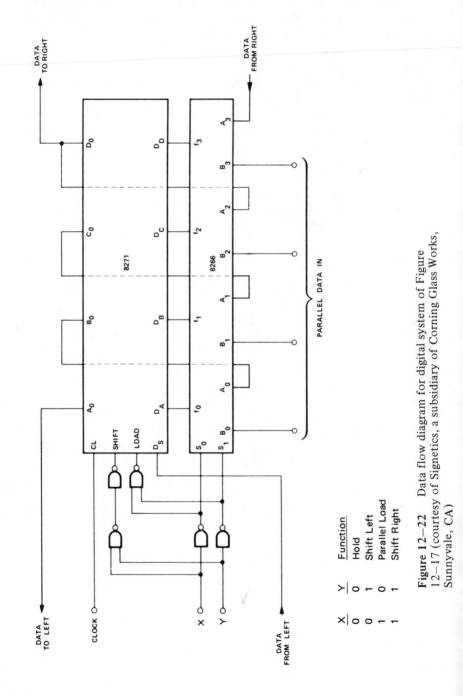

Figure 12–22 Data flow diagram for digital system of Figure 12–17 (courtesy of Signetics, a subsidiary of Corning Glass Works, Sunnyvale, CA)

X	Y	Function
0	0	Hold
0	1	Shift Left
1	0	Parallel Load
1	1	Shift Right

correlations between control-logic input signals and the 8271 and 8266 ICs. Particular note should be made of the SHIFT LEFT and PARALLEL LOAD functions. The 8271 shift register responds with a LOAD operation in both cases. However, in the SHIFT LEFT mode, the LOAD input is supplied by the following FF (gated through the 8266 multiplexer). In the PARALLEL LOAD mode, the "parallel data in" inputs are gated through the multiplexer under the control of the S_0 and S_1 inputs. Data flow should be thoroughly traced through the diagram of Fig. 12–22 to prepare for the more detailed analysis to follow.

Logic Level Analysis

If one stage of register operation (including the multiplexer and the control logic) is analyzed, the complete circuit may be considered to be analyzed. The four stages are identical in their operation, so the first stage (see Fig. 12–23) is used to allow easier visualization of the input gating and control portions of the circuit.

A mixture of analysis methods is used to investigate the operation of the single stage. Algebraic descriptions are developed for the enabling inputs to each of the input selection gates. Logical reasoning is then applied to follow the data from data inputs to the ultimate use of either SETting or CLEARing the flip-flop. Clocked R–S FF characteristics are already known; so are the characteristics of the various gates employed in the total circuit.

The HOLD (storage) mode of operation requires that any change in FF state be inhibited. In the circuit under investigation, one input to both G_{19} and G_{20} is held at the LOW level for this purpose. The output of G_{22} supplies the inhibit and is LOW *only* when both L and S are LOW. [Note that the algebraic expression for the output of G_{22} is $L + S$, which means that when either L is HIGH or S is HIGH (or both are HIGH together), the output of the gate is HIGH. The only remaining combination is both L and S LOW.]

The data that causes the FF to assume a specific state arrives via G_{17}. If the G_{17} output is HIGH, G_{19} is activated, and upon arrival of the clock pulse, the FF SETs and the Q output goes HIGH. If the G_{17} output is LOW, G_{18} inverts the level to HIGH and G_{20} activates to allow CLEARing of the FF with the clock pulse. In the HOLD mode, the output of G_{17} is unimportant, since both G_{19} and G_{20} are inhibited by the output of G_{22}.

However, as soon as either L or S (or both) go HIGH, both G_{19} and G_{20} are enabled; hence the *source* of the logic level at the output of G_{17} determines the mode of operation of the circuit. G_{15} gates data into the register from preceding stages (those to the left of the stage under investigation). G_{16} supplies data from the multiplexer, which has a choice of two different data sources. If G_{10} is enabled, data from following stages (those to the right of the stage under investigation) is used, and if G_{11} is enabled, the information on the parallel data input is selected.

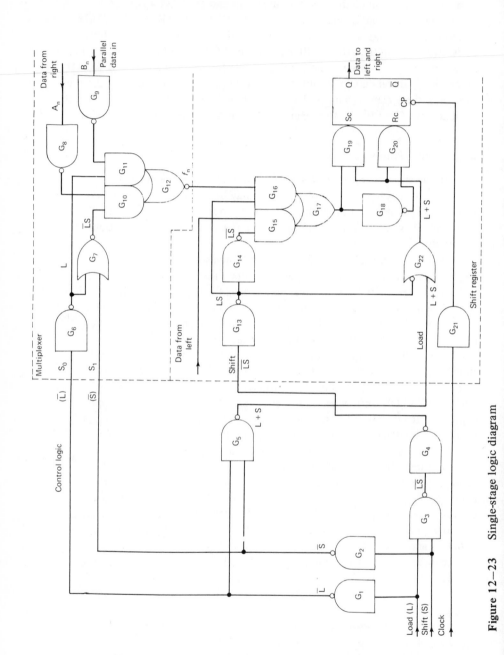

Figure 12–23 Single-stage logic diagram

The SHIFT RIGHT mode requires data from the left (the *preceding* stage), so it is necessary to enable G_{15} and disable G_{16}. If both G_{15} and G_{16} are enabled at the same time, interference between the two data sources results and false information is entered into the FF. G_{15} is enabled by the output of G_{14}, which has a 1 (HIGH) output *only* when both L and S are HIGH. At the same time, G_{16} is disabled by the LOW ($\overline{LS}$) output of G_{13}; the status of G_{10} and G_{11} is unimportant. The information from the preceding stage is thus gated through G_{15} and G_{17} to be stored in the FF when the next clock pulse occurs.

To SHIFT LEFT, the data from the *following* FF must be placed in the FF of the stage being evaluated. G_{10} must be enabled (which happens only when L is LOW and S is HIGH, according to the algebraic expression for the output of G_7); G_{11} disabled; and G_{16} enabled. G_{11} is disabled with a LOW level from G_6; G_{16} is enabled with a HIGH level from G_{13}; and G_{15} is disabled with a LOW level from G_{14}, when G_{10} is enabled by a HIGH level from G_7. The only source of input information to the storage FF under these conditions is from the next FF to the right; SHIFT LEFT is accomplished when the clock pulse arrives.

PARALLEL LOAD operates in a manner quite similar to SHIFT LEFT, except that G_{11} is enabled instead of G_{10}. Other data sources are blocked by disabled gates. Since the four modes of operation of the logic circuit have been discussed for a single stage and since all stages operate in an identical manner, our analysis of the complete logic circuit may be considered complete.

12−5 THE FUTURE

Continuing advancements in semiconductor technology have resulted in great improvements in register design and applications. Now 200 or more bits of static storage and shifting capability using latch-type circuitry are avialable in common IC packages. The use of data-storage techniques that do not depend on FF operation has resulted in decreased complexity and increased capacity. More than 4000 bits of storage and shifting can be accommodated by the use of dynamic (circulating) techniques. This non-FF type of operation requires that the data be constantly circulated − or it will be lost. In contrast, the static type of register retains data without the constant presence of clock pulses to keep it recirculating.

Both of these high-density registers are of the serial-serial type, since parallel operation would require an excessive number of external connections. However the techniques of construction result in operating speeds in the region of 3,000,000 to 5,000,000 operations per second, which makes up for the slower serial type of operation. Technological advances will certainly extend the upper limit in the very near future.

PROBLEM APPLICATIONS

12–1.   Define the term register.

12–2.   What are the characteristics of a storage register?

12–3.   What are the characteristics of a shift register?

12–4.   List the four classifications of register input-output methods and explain the concept of each method.

12–5.   Discuss the applications of $R-S$, $J-K$, and D FFs to storage registers.

12–6.   What is a buffer register?

12–7.   List some advantages and disadvantages of prepackaged IC registers.

12–8.   How do shift registers differ from storage registers?

12–9.   How are shift registers similar to storage registers?

12–10. Discuss the application of $R-S$, $J-K$, and D FFs to shift registers.

12–11. What is a circulating register?

12–12. Draw the waveform diagrams that represent loading of the binary data 1010 into the register of Fig. 12–10. Show the same data being shifted to the right four times.

12–13. Explain the application of shift registers to the performance of arithmetic operations.

12–14. Draw the waveforms that show operation of the ring counter in Fig. 12–15.

12–15. Draw the waveforms that show operation of the switch-tail ring counter of Fig. 12–16.

12–16. Redraw the logic diagram of Fig. 12–8 using $J-K$ FFs.

12–17. Identify the logic circuit shown below and document the steps of the analysis.

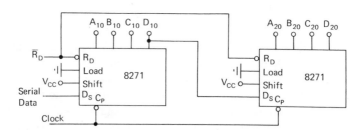

12–18. Identify the logic circuit shown below and document the steps of the analysis.

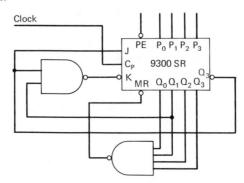

12–19. Identify the logic circuit shown below and document the steps of the
analysis.

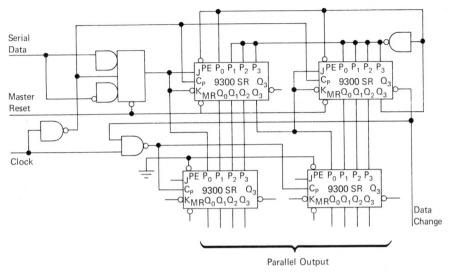

Parallel Output

12–20. Discuss the future developments that may be expected in the design
and application of registers.

13

Miscellaneous Logic Functions

You may have noted that in the two preceding chapters many of the counter and register applications were implemented with *packaged circuits*. It would seem reasonable that such packaged circuits would be used when complete functions may be obtained in sizes much smaller than the flip-flop (FF) implemented functions. The cost of implementation of the hardware by either method is similar. But the reduced manufacturing cost (less components to handle) plus the smaller physical size of the finished product generally leads to a preference for packaged integrated circuits (ICs) whenever possible.

Although the round package of Fig. 13–1a is still encountered, the most popular forms used with the prepackaged logic functions discussed in this text are shown in Figs. 13–1b and c.

Four prepackaged logic functions are explained in this chapter: *Exclusive-OR*s, *full adders*, *multiplexers*, and *memory elements*. Two other common prepackaged functions (decoders and code converters) will be discussed in subsequent chapters, due to their specific applications.

13–1 THE EXCLUSIVE-OR OPERATION

Functional Description

The Exclusive-OR operation is a function that appears so often in digital circuitry that it has been assigned a special symbol. But, since it is not an independent operation, it can be expressed in terms of the basic connectives: AND, OR, and NOT. The expression *A Exclusively-OR B* is written $A \oplus B$. This function is true ($A \oplus B = 1$) when A or B is true but *not* when both are true. Figure 13–2 shows the standard symbol for the Exclusive-OR operation, its truth table, algebraic expression, and a typical IC implementation of the operation. Many other logic implementations of the Exclusive-OR operation are used and will be discussed as they are encountered.

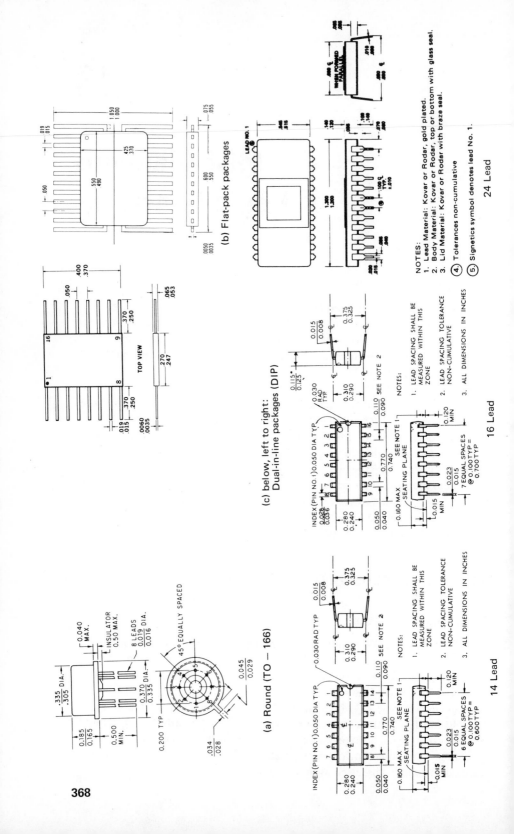

(a) Round (TO — 166)

(b) Flat-pack packages

(c) below, left to right: Dual-in-line packages (DIP)

LEAD NO. 1

NOTES:
1. Lead Material: Kovar or Rodar, gold plated.
2. Body Material: Kovar or Rodar, top or bottom with glass seal.
3. Lid Material: Kovar or Rodar with braze seal.
4. Tolerances non-cumulative
5. Signetics symbol denotes lead No. 1.

24 Lead

16 Lead

14 Lead

NOTES:
1. LEAD SPACING SHALL BE MEASURED WITHIN THIS ZONE
2. LEAD SPACING TOLERANCE NON-CUMULATIVE
3. ALL DIMENSIONS IN INCHES

TOP VIEW

INSULATOR 0.50 MAX.
8 LEADS
45° EQUALLY SPACED

SEE NOTE 1
SEATING PLANE

INDEX (PIN NO. 1) 0.050 DIA TYP
SEE NOTE 1
SEATING PLANE

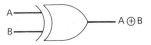

8241   Quad Exclusive-OR

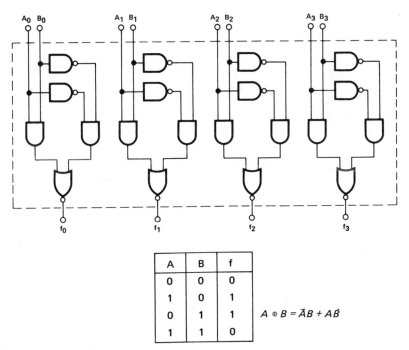

A	B	f
0	0	0
1	0	1
0	1	1
1	1	0

$A \oplus B = \bar{A}B + A\bar{B}$

Figure 13–2   The exclusive-OR operation (courtesy of
Signetics, a subsidiary of Corning Glass Works, Sunnyvale, CA)

The Equality (Exclusive-NOR) Operation

Many useful logic functions are derived from the Exclusive-OR operation.
The expression for the Exclusive-OR circuit $(A\bar{B} + \bar{A}B)$ shows that the nonequal
cases may be determined by noting the logic level at the output. The output is 1,
when A is 1 or B is 1, but not when both are 1 at the same time or when both
are 0 at the same time. The mere addition of a logic inversion following the
Exclusive-OR results in a circuit that can determine whether inputs are equal or
unequal.

Negating the Exclusive-OR expression and simplifying the resulting expres-
sion $\overline{A\bar{B} + \bar{A}B}$ yields $AB + \bar{A}\bar{B}$. When A and B are both 1 or both 0, the logic
value of the negated Exclusive-OR operation is 1. This is sometimes called the
equality (*Exclusive*-NOR) operation. Details of the equality operation are shown
in Fig. 13–3.

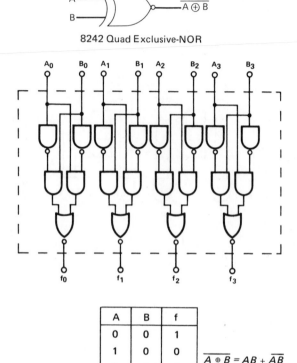

8242 Quad Exclusive-NOR

A	B	f
0	0	1
1	0	0
0	1	0
1	1	1

$$\overline{A \oplus B} = AB + \overline{A}\overline{B}$$

Figure 13–3   The exclusive-NOR (equality) operation (courtesy of Signetics, a subsidiary of Corning Glass Works, Sunnyvale, CA)

Comparators

One of the most common applications of the equality circuit is found in digital computers. The values of variables are often compared, and the results are then used to determine future computer action. For example: if A is less than B, the computer is instructed to perform a specific program step; if A is greater than B, it must perform a different step; if $A = B$, still another "branching instruction" is required. So far, we have discussed the logic circuitry that determines whether $A = B$ or $A \neq B$. But the determination of *relative magnitude* (i.e., $A < B$ or $A > B$) involves a logic circuitry called a *comparator*.

Actually, the Exclusive-NOR circuit is capable of furnishing *all three* of these comparisons, if it is implemented like the logic circuit of Fig. 13–4.* As that truth table shows, $A > B$ is only 1 when $A = 1$ and $B = 0$; $A = B$ is 1 only when A and B are both either 0 or 1; and $A < B$ is 1 only when $A = 0$ and $B = 1$.

---

* This is a relatively simple circuit, comparing only two single-digit binary numbers.

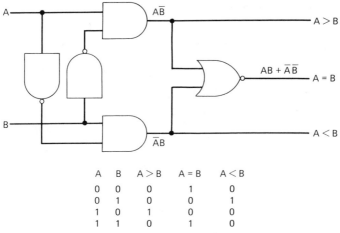

A	B	A > B	A = B	A < B
0	0	0	1	0
0	1	0	0	1
1	0	1	0	0
1	1	0	1	0

Figure 13—4   Single-order comparator

When comparators are used in computers, they must be capable of comparing multi-digit numbers, so the logic circuitry becomes quite complex. A comparator that determines the relative magnitude of the 3-bit binary numbers is shown in Fig. 13—5. The accompanying table lists circuit outputs for the magnitude relationships between A and B.

The function of each gate is listed as follows:

G_1 — determines equality of the most-significant digits in A and B. A HIGH output indicates $A_2 = B_2$.

G_2 — same as G_1, except A_1 and B_1.

G_3 — same as G_1, except A_0 and B_0.

G_4 — determines if $A_2 > B_2$ ($A_2 = 1$ and $B_2 = 0$). A HIGH output means $A_2 > B_2$.

G_5 — determines if $A_1 > B_1$ ($A_1 = 1$ and $B_1 = 0$) and $A_2 = B_2$. A HIGH output results if the gate is activated by these conditions.

G_6 — determines if $A_0 > B_0$ ($A_0 = 1$ and $B_0 = 0$), $A_2 = B_2$, and $A_1 = B_1$. A HIGH output means that the gate has been activated by all of these conditions.

G_7 — ANDs the outputs of all three equality gates to furnish an overall HIGH indication that $A = B$. A LOW output signifies $A \neq B$.

G_8 — ORs the output of comparison gates G_4, G_5, and G_6 so that a HIGH output results if an $A_n B_n$ combination of $A = 1$ and $B = 0$ is present. A LOW output is interpreted as $A < B$.

The circuit operation of the 3-bit comparator is summarized in the following diagram.

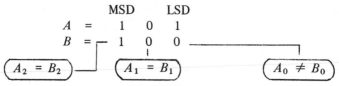

$$A \;=\; 1 \quad 0 \quad 1$$
$$B \;=\; 1 \quad 0 \quad 0$$

$\left(A_2 = B_2 \right)$ $\left(A_1 = B_1 \right)$ $\left(A_0 \neq B_0 \right)$

G_1 output HIGH G_2 output HIGH G_3 output LOW — forces G_7 output LOW

G_4 output LOW G_5 output LOW G_6 output HIGH — allows G_8 output to

be HIGH. Therefore $A > B$ due to $A_0 > B_0$

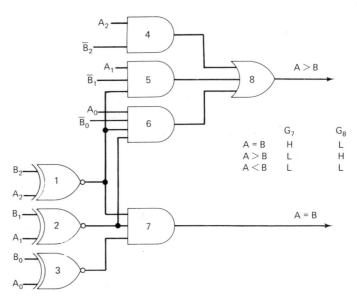

Figure 13—5   3-bit comparator

Parity

The constant movement of data both within digital machines and from machine to machine requires a very *high degree of accuracy*. For example, if the 15-bit computer word 000001000000000 represents the amount of $512, any error in transmission can dramatically affect the actual value of someone's check! (A single 0 changed to a 1 by "noise" during movement of the data — 000001100000000 — results in a value of $768.) The increasing reliability of solid-stage logic devices is helping to attain this high accuracy, but external influences in the transmission path commonly cause errors to develop. Duplication of the system — using a different transmission path and then comparing both sets of data — is another (expensive) way of increasing accuracy.

One of the most common and relatively inexpensive methods of ensuring data-movement accuracy is the addition of a standardizing or *parity bit* at the end of each word.

> Since all data words contain either an even or an odd number of 1s, *if the standard (parity) is an even number of 1s, it is said that* even parity *is being used; if an odd number of 1s is required to meet the standard, the standardizing system is called* odd parity.

Prior to movement of a data word, the *number* of 1s in the word is determined. For a 16-bit word odd parity system, if the number of 1s is even, a 1 is placed in the final position. Thus a 16-bit word is formed, so the data is moved with an odd number of 1s present. When received, the number of 1s is again determined. If an odd number of 1s is present, it is assumed that the data word has been moved correctly. If an even number of 1s is detected, a warning is provided to show that the data is incorrect.

This type of parity system is effective only in detecting an *odd number of* 0s *that have changed to* 1s (i.e., 1, 3, 5, etc.). A simple, *4-bit parity detector/ generator*, with its algebraic expression and truth table, is shown in Fig. 13–6.

Each Exclusive-OR gate is really an even parity generator for the two inputs. Parity generation for a *4-bit word* thus requires one Exclusive-OR gate

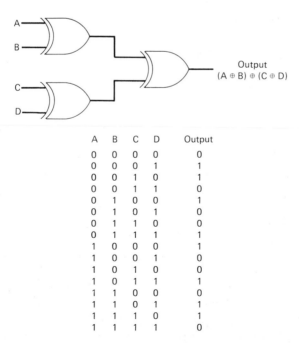

A	B	C	D	Output
0	0	0	0	0
0	0	0	1	1
0	0	1	0	1
0	0	1	1	0
0	1	0	0	1
0	1	0	1	0
0	1	1	0	0
0	1	1	1	1
1	0	0	0	1
1	0	0	1	0
1	0	1	0	0
1	0	1	1	1
1	1	0	0	0
1	1	0	1	1
1	1	1	0	1
1	1	1	1	0

Figure 13–6   4-bit parity detector/generator

for each pair of inputs, plus a third Exclusive-OR gate to compare the outputs of the first two gates.

The 4 bits of the binary word to be transmitted are applied to A, B, C, and D, respectively. Parity is checked, and if an odd number of 1s is detected (as it will be if the input is $A\bar{B} + \bar{A}B$), the output of the parity generator becomes 1. A 1 is then attached as a fifth bit to the transmitted word. If an even number of 1s is detected ($AB + \bar{A}\bar{B}$), a 0 is attached. (The circuit that adds the parity bit to the transmitted message is not shown in Fig. 13–6.)

A *9-bit parity generator/checker* (type 8262) is shown in Fig. 13–7. Eight of the inputs are used to check parity of an 8-bit word. When acting as a parity generator the ninth input of the 8262 controls the operation of the circuit. The parity bit generated at the sending end of the data transmission link is the ninth input when the 8262 functions as a parity checker.

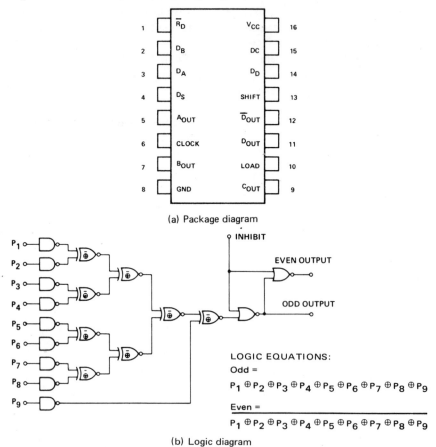

(a) Package diagram

(b) Logic diagram

LOGIC EQUATIONS:

Odd =

$P_1 \oplus P_2 \oplus P_3 \oplus P_4 \oplus P_5 \oplus P_6 \oplus P_7 \oplus P_8 \oplus P_9$

Even =

$\overline{P_1 \oplus P_2 \oplus P_3 \oplus P_4 \oplus P_5 \oplus P_6 \oplus P_7 \oplus P_8 \oplus P_9}$

Figure 13–7   Type 8262 9-bit parity generator and checker (courtesy of Signetics, a subsidiary of Corning Glass Works, Sunnyvale, CA)

Parity systems capable of detecting and even correcting multiple errors in transmitted data are now in use, but they are beyond the scope of this text.

Binary Addition Using the Exclusive-OR Operation

The ability to perform *arithmetic addition* is an inherent requirement in digital computers. Binary addition is quite similar to decimal addition, but the addition operation is greatly simplified. In the decimal system all combinations of *ten* digits must be examined to construct an addition table; in the binary system, only *two* digits are used. The four rules of binary addition are shown in Table 13–1.

By arbitrarily assigning the *logic value* 0 to the *arithmetic value* 0 and the *logic value* 1 to the *arithmetic value* 1, the Carry column of Table 13–1 becomes identical to the logical AND truth table, and the Sum column matches the truth table for the Exclusive-OR operation (Fig. 13–2). The logic circuit that implements the Exclusive-OR operation may then be used to obtain the sum of two single-digit binary numbers, while any AND circuit will develop the required carry value. It should be noted, however, that only *single-digit* numbers may be accommodated by this maneuver. If multi-digit numbers are added and a carry from a previous order of addition is to be included, other means must be provided. Logic circuits that perform this type of operation (Full adders) are discussed shortly.

If a logic circuit designed to provide the sum of two numbers is so arranged that the carry operation may also be derived without additional circuitry, so much the better. Actually, the circuit of Fig. 13–2 is designed to provide both the sum and carry output, although the carry output is not available externally. (The Exclusive-OR output is, of course, the sum.) One of the inputs to the final NOR gate is AB, which is the carry expression. Another implementation of the Exclusive-OR operation is shown in Fig. 13–8. It provides both sum and carry outputs. Proof of the $A \oplus B$ output is left to the reader.

A circuit that performs all the logic operations required to add two binary digits is called a *half-adder.* It must have *two inputs* – one for each digit representation – and *two outputs* – one for the sum and one for the carry. It is called a "half-adder" because two half-adders are required to add two digits *and* the carry that may result from the addition of two less-significant digits.

Table 13–1   Rules of Binary Addition

Digit A		Digit B		Sum	Carry
0	plus	0	=	0	0
0	plus	1	=	1	0
1	plus	0	=	1	0
1	plus	1	=	0	1

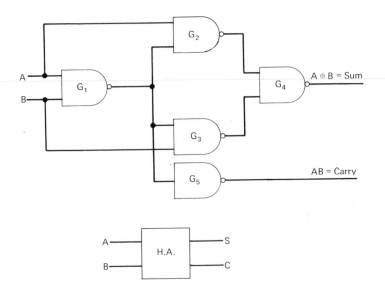

Figure 13—8   Another exclusive-OR circuit (a half-adder)

Both Exclusive-OR circuits discussed so far fit this description and may be used as half-adders. The half-adder as a separate IC package is fast disappearing. In applications where the half-adder is required, it is usually implemented with packaged Exclusive-OR circuits and the necessary separate gates.

13—2 ARITHMETIC / LOGIC CIRCUITS

Although most arithmetic functions, such as addition and subtraction, adapt readily to digital computer applications, many of the circuits designed to implement these functions are also of value in other digital applications. Included in this category are *adders* and *multifunction assemblies*.

Since the adder is the basis for many of the other circuits used in arithmetic functions, we shall discuss it first.

The Full Adder

During the explanation of the half-adder operation, we pointed out that additional circuitry was necessary to accommodate a third variable (the carry from an addition operation in a lower significant position). This requirement dictates a circuit with *three inputs, one sum output*, and *one carry output*. Again, by equating *arithmetic* 0 and 1 to *logic* 0 and 1, we can construct a table that describes the *full adder* function (see Fig. 13—9a).

A	B	C_i	S	C_o
0	0	0	0	0
0	0	1	1	0
0	1	0	1	0
0	1	1	0	1
1	0	0	1	0
1	0	1	0	1
1	1	0	0	1
1	1	1	1	1

$$S = \overline{A}\,\overline{B}C_i + \overline{A}B\overline{C}_i + A\overline{B}\,\overline{C}_i + ABC_i$$
$$C_o = \overline{A}BC_i + A\overline{B}C_i + AB\overline{C}_i + ABC_i$$

(a) Full-adder table

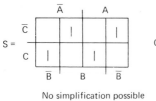

No simplification possible

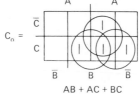

$AB + AC + BC$

(b) Map simplification of full-adder equations

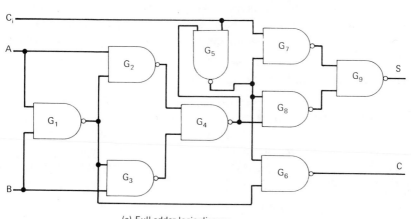

(c) Full-adder logic diagram

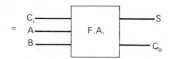

Figure 13–9 Binary full-adder

377

The Boolean expression that results from the tabular description is extracted and subjected to simplification (Fig. 13–9b). A logic diagram (Fig. 13–9c) is then drawn to implement the equations for both the sum and carry terms. All eight of the possible combinations of A, B, and C_i should be applied as inputs to the full adder and logic levels traced through the circuit to assure that the proper logic levels result at the sum and carry outputs.

With proper modification of the numbers to be operated upon, adders may be used to perform other basic arithmetic operations. But since this text is not oriented toward the digital computer, the performance of subtraction, multiplication, and so forth is not discussed here.

In IC packages, the full adder appears in *single*, *dual*, and *quad* configurations. Some versions provide internal interconnections, while others make all (or at least the most important) inputs and outputs available externally. A very versatile, single-bit full adder (type 5480) is shown in Fig. 13–10. Gated complementary inputs are provided, along with both a true sum output (S) and its complement $(\bar{S})$, and an inverted carry output. Input gating and complementing is accomplished by G_1 through G_6, inclusively. The remainder of the circuit functions as a conventional full adder. A 4-bit, binary full adder is also shown implemented with the type 5480 IC package. The complementary inputs allow easy implementation of the subtraction operation.

A 4-bit, binary full adder (type 7483) in a single IC package is shown in Fig. 13–11. While not nearly as versatile as the 5480 circuit, in applications where intermediate outputs are not required, this full adder is quite satisfactory.

Parity Generation and Checking

As pointed out in Sec. 13–2, the Exclusive-OR operation is fundamental to the *parity-checking* concept. Whereas the Exclusive-OR gate is used to detect the existence of an odd number of 1s between two bits, the full adder can be used to detect the same relationship between *three* bits. The sum output of a full adder may be written as $S = A \oplus B \oplus C$, which describes its function as a parity checking device. Expansion of this equation gives us

$$
\begin{aligned}
A \oplus B \oplus C &= (A \oplus B) \oplus C \\
&= (A\bar{B} + \bar{A}B) \oplus C \\
&= (A\bar{B} + \bar{A}B)(\bar{C}) + (\overline{A\bar{B} + \bar{A}B})(C) \\
&= A\bar{B}\bar{C} + \bar{A}B\bar{C} + (AB + \bar{A}\bar{B})(C) \\
\therefore \quad A \oplus B \oplus C &= A\bar{B}\bar{C} + \bar{A}B\bar{C} + ABC + \bar{A}\bar{B}C
\end{aligned}
$$

The diagram of Fig. 13–12 shows how four full adders may be used to generate parity for a 9-bit word. FA_1, FA_2, and FA_3 perform the $A \oplus B \oplus C$ operation on their inputs, and FA_4 then combines the results in its 3-input Exclusive-OR operation. Since each full adder has both a true and a complemented sum output, both odd and even parity are easily obtained — if parity is *not* odd, it must be even!

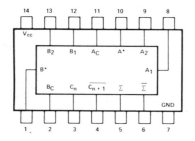

(a) Package diagram

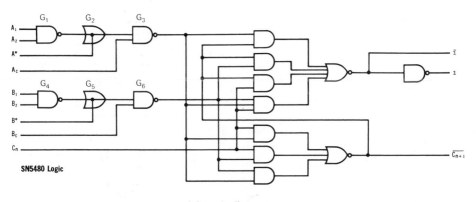

SN5480 Logic

(b) Logic diagram

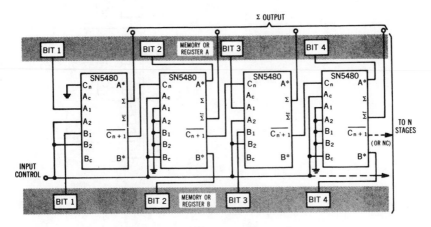

(c) 4-bit binary adder using 5480 packages

Figure 13–10   5480 single-bit full-adder

379

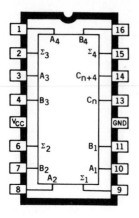

(a) Package diagram

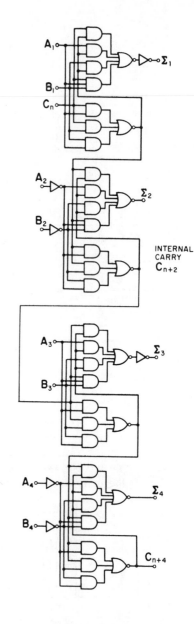

(b) Logic diagram

Figure 13–11 7483 4-bit binary full-adder (courtesy of Sprague Electric Company, Worcester, Mass.)

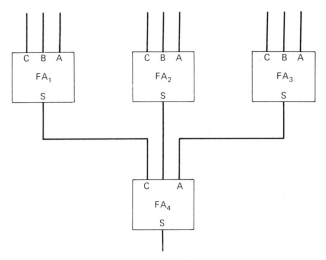

Figure 13–12   9-bit parity generator and checker using full-adders

Multi-Use Circuits

When more than one type of logic operation must be performed on variables in a logic system, a *universal logic element* is helpful. Such elements are not yet common, but two approaches to "universality" are described here.

The type 4610 dual 2-variable function generator (Fig. 13–13a) consists of two separate logic-gate arrays, controlled by common 4-bit inputs. It can select one of sixteen possible Boolean functions performed on the 2-variable inputs. The function generator truth table of Fig. 13–13c may be correlated directly with the logic diagram of Fig. 13–13b.

Circuit flexibility is increased by using S_2, S_3, A, and B inputs as controls and S_0 and S_1 as the input variables. In this mode of operation, two separate output functions are performed simultaneously on a pair of variables, S_0 and S_1, by using both function generators in the package. Figure 13–13d shows the truth table that explains this operation.

In many digital devices, it is necessary to perform operations between *and* with two words of information. These words are usually placed in registers while such operations are being performed, and the results are then placed back in one of the registers. Placing the type 4610 function generator between the two registers and providing 4-bit control on the S_0–S_3 inputs results in the circuit shown in Fig. 13–14. The accompanying table lists the operations that may be performed between the two registers under the control of the 4610.

As IC technology grows, more and more multi function circuits are being designed. In the meantime, however, many of the existing packages are being used for other purposes. Apparently, the application of prepackaged functions is limited only by the imagination of the designer, so we shall describe some of the more interesting applications in sections of this and subsequent chapters.

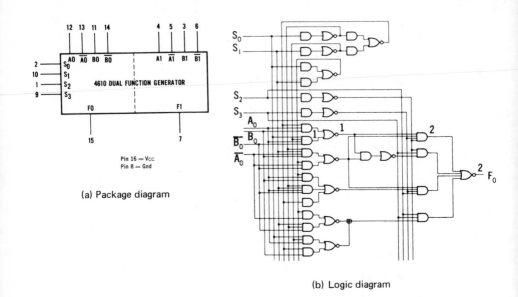

(a) Package diagram

Pin 16 = Vcc
Pin 8 = Gnd

(b) Logic diagram

(c) Normal truth table

S_0	S_1	S_2	S_3	FUNCTION (ACTIVE LOW)	FUNCTION (ACTIVE HIGH)
L	L	L	L	$A + B$	$\bar{A} \cdot \bar{B}$
L	H	L	L	$A + \bar{B}$	$\bar{A} \cdot B$
H	L	L	L	$\bar{A} + B$	$A \cdot \bar{B}$
H	H	L	L	$\bar{A} + \bar{B}$	$A \cdot B$
L	L	L	H	$\bar{A} \cdot \bar{B}$	$A + B$
L	H	L	H	$\bar{A} \cdot B$	$A + \bar{B}$
H	L	L	H	$A \cdot \bar{B}$	$\bar{A} + B$
H	H	L	H	$A \cdot B$	$\bar{A} + \bar{B}$
L	L	H	L	$A \oplus B$	$A \oplus \bar{B}$
L	H	H	L	$A \oplus \bar{B}$	$A \oplus B$
H	L	H	L	1	0
H	H	H	L	0	1
L	L	H	H	A	$\bar{A}$
L	H	H	H	B	$\bar{B}$
H	L	H	H	$\bar{A}$	A
H	H	H	H	$\bar{B}$	B

(d) Truth table using S_0 and S_1 as input variables

S_2	S_3	A	B	FUNCTION (ACTIVE LOW)	FUNCTION (ACTIVE HIGH)
L	L	L	L	$S_0 + S_1$	$\bar{S}_0 \cdot \bar{S}_1$
L	L	L	H	$S_0 + \bar{S}_1$	$\bar{S}_0 \cdot S_1$
L	L	H	L	$\bar{S}_0 + S_1$	$S_0 \cdot \bar{S}_1$
L	L	H	H	$\bar{S}_0 + \bar{S}_1$	$S_0 \cdot S_1$
L	H	L	L	$\bar{S}_0 \cdot \bar{S}_1$	$S_0 + S_1$
L	H	L	H	$\bar{S}_0 \cdot S_1$	$S_0 + \bar{S}_1$
L	H	H	L	$S_0 \cdot \bar{S}_1$	$\bar{S}_0 + S_1$
L	H	H	H	$S_0 \cdot S_1$	$\bar{S}_0 + \bar{S}_1$
H	L	L	L	$S_0 \oplus S_1$	$S_0 \oplus \bar{S}_1$
H	L	L	H	$\bar{S}_1$	S_1
H	L	H	L	$\bar{S}_1$	S_1
H	L	H	H	$S_0 \oplus S_1$	$S_0 \oplus \bar{S}_1$
H	H	L	L	S_0	$\bar{S}_0$
H	H	L	H	$S_0 \oplus S_1$	$S_0 \oplus \bar{S}_1$
H	H	H	L	$S_0 \oplus \bar{S}_1$	$S_0 \oplus S_1$
H	H	H	H	$\bar{S}_0$	S_0

Figure 13–13   Type 4610 dual 2-variable function generator (courtesy Fairchild Camera and Instrument Corporation, Mountain View, CA)

A Typical Processor Application is the transfer and logical operation control between two registers (A and B). The function generators are controlled by a 4-bit operation code field, S_0, S_1, S_2, and S_3. The operation code repertoire includes the 16 operations listed below.

AND $\bar{A}$ and $\bar{B}$ to A	OR A and B to A
AND $\bar{A}$ and B to A	OR A and $\bar{B}$ to A
AND A and $\bar{B}$ to A	OR $\bar{A}$ and B to A
AND A and B to A	OR $\bar{A}$ and $\bar{B}$ to A

Exclusive OR A and $\bar{B}$ to A	Complement A
Exclusive OR A and B to A	Transfer $\bar{B}$ to A
Reset A	No Operation
Set A	Transfer B to A

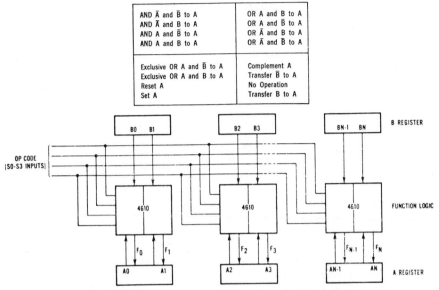

Figure 13–14 Processor Application of 4610 (courtesy
Fairchild Camera and Instrument Corporation, Mountain View, CA)

13–3 MULTIPLEXERS

Another class of logic circuits is the *multiplexer*. The *multiplexer* is
actually *a signal selector gating circuit that is the logic equivalent of a multi-
position selector switch.* A number of input signals are fed into the multiplexer,
and by using control inputs, one of the signals is selected and made available at
the output. Multiplexers can be used in both digital and analog applications.

Switching

The basic concept of digital data selection is shown in Fig. 13–15. A
single-pole, 2-position switch (Fig. 13–15a) is logically implemented with the
increasingly popular *AND–OR–INVERT* (AOI) *gate* (Fig. 13–15b) and an
inverter. AOI gates are available in all the common packaging forms, and in
some cases, they may actually implement the standard logic operations with a
reduction in cost and package count. The accompanying truth table defines the
operation of the *2-input multiplexer.** The output of the multiplexer is either

_____
* As a matter of interest, the AOI multiplexer should be compared with the type
8266 multiplexer used in Ch. 12 (Sec. 12–4) as a data selector for the bidirectional shift
register. The 8266 provides the function of a 4-pole, 2-position switch and may be used in
any digital application requiring this capability.

(a) Single-pole, 2-position swtich

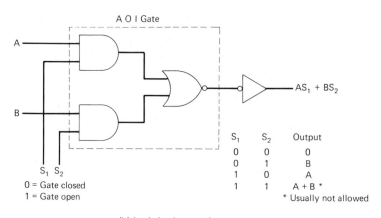

(b) Logic implementation

Figure 13—15   Multiplexer switching concept

A or B, depending on the logic level at the control inputs, S_1 and S_2. When both controls are HIGH at the same time, the conventional OR operation is performed — this is usually not considered a multiplexer function.

As it becomes necessary to accommodate more inputs, the requirements for control signals can become restrictive. For example, if eight inputs must be switched by a multiplexer, eight control signals are necessary, and the number of connections for inputs and outputs exceeds the capability of a standard 16-pin IC package. This problem is solved by using *different combinations of control inputs* to perform selection. Eight different control signals may be derived from only three control inputs by using all possible combinations of 1 and 0 that exist with three variables.

The type 8232 8-input digital multiplexer (Fig. 13—16) uses this concept. Selection of I_2, for example, is made by an address of $A_2 = 0, A_1 = 1$, and $A_0 = 0$. G_2 is then enabled so that the information on input line I_2 is fed to the output NOR gate G_8. Both true and complemented outputs are provided and may be inhibited to provide a constant 0 at both outputs. The truth table accompanying Fig. 13—16 defines the operation of the 8232.

Many other multiplexer I/O combinations exist, but all function under the general switching concepts just described.

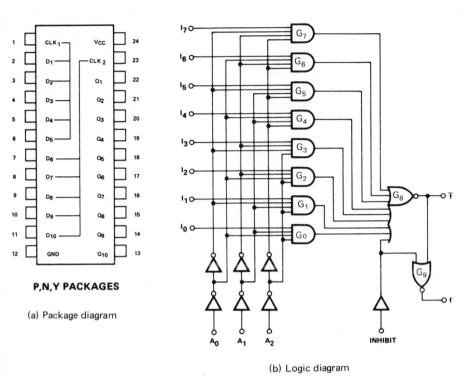

P,N,Y PACKAGES

(a) Package diagram

(b) Logic diagram

ADDRESS			DATA INPUTS									OUTPUT		
A_2	A_1	A_0	I_7	I_6	I_5	I_4	I_3	I_2	I_1	I_0	INH	f	8230 8231 $\bar{f}$	8232 $\bar{f}$
0	0	0	x	x	x	x	x	x	x	1	0	1	0	0
0	0	1	x	x	x	x	x	x	1	x	0	1	0	0
0	1	0	x	x	x	x	x	1	x	x	0	1	0	0
0	1	1	x	x	x	x	1	x	x	x	0	1	0	0
1	0	0	x	x	x	1	x	x	x	x	0	1	0	0
1	0	1	x	x	1	x	x	x	x	x	0	1	0	0
1	1	0	x	1	x	x	x	x	x	x	0	1	0	0
1	1	1	1	x	x	x	x	x	x	x	0	1	0	0
0	0	0	x	x	x	x	x	x	x	0	0	0	1	1
0	0	1	x	x	x	x	x	x	0	x	0	0	1	1
0	1	0	x	x	x	x	x	0	x	x	0	0	1	1
0	1	1	x	x	x	x	0	x	x	x	0	0	1	1
1	0	0	x	x	x	0	x	x	x	x	0	0	1	1
1	0	1	x	x	0	x	x	x	x	x	0	0	1	1
1	1	0	x	0	x	x	x	x	x	x	0	0	1	1
1	1	1	0	x	x	x	x	x	x	x	0	0	1	1
x	x	x	x	x	x	x	x	x	x	x	1	0	1	0

x = don't care

(c) Truth table

Figure 13–16   8232 8-input digital multiplexer (courtesy of Signetics, a subsidiary of Corning Glass Works, Sunnyvale, CA)

An interesting application of multiplexers is shown in Fig. 13—17. A need often arises in both digital test equipment and computers to select *one of a number of different frequencies* that are related to a fundamental clock or timing frequency. Recalling from Ch. 11 that the simple binary counter divides its inputs by 2 is the clue to the operation of the *frequency selector* in Fig. 13—17. Two type 8281 4-bit binary counters (similar to the 7493 counters discussed in Sec. 11—1) are cascaded, thus providing a division ranging from a factor of 2 at

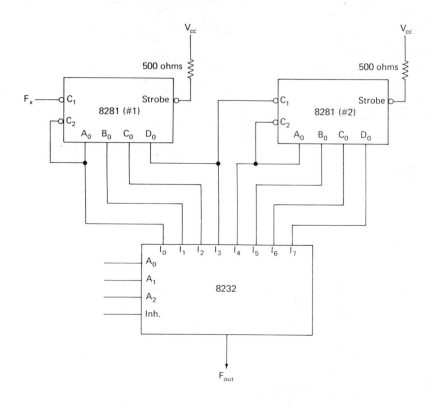

Inh	A_2	A_1	A_0	F_{out}
0	0	0	0	$F_x/2$
0	0	0	1	$F_x/4$
0	0	1	0	$F_x/8$
0	0	1	1	$F_x/16$
0	1	0	0	$F_x/32$
0	1	0	1	$F_x/64$
0	1	1	0	$F_x/128$
0	1	1	1	$F_x/256$
1	X	X	X	0

Figure 13—17   The multiplexer as a frequency selector

the A_0 output of the first counter to a factor of 256 at the D_0 output of the second counter. The eight different frequencies are supplied to the inputs of the 8232 Multiplexer. Selection of the required frequencies is made by proper addressing of the A_0, A_1, and A_2 inputs, as shown in the table accompanying Fig. 13–17. Although only a simple switching application, this example shows the versatility of the multiplexer circuit and the imagination of the designer in applying the principles of logic.

Nondigital Applications

Multiplexer applications are not limited to digital circuits. *Analog information* may also be switched with a multiplexer, if the gates are properly designed. The enabling inputs are merely used to allow the analog information to pass. A specific example is *time-division multiplexing* (TDM). This technique sequentially samples each of the inputs and provides a single output that contains a sample of the amplitude of each input. Voices and music may be transmitted from one location to another in this manner, so long as the sampling rate is at least twice the highest frequency in the signal being sampled.

Once the sample is obtained, the composite output is converted to a form suitable for digital transmission. Numerous transmission forms, such as pulse height, pulse width, pulse spacing, and pulse grouping, may be used. Upon reception of the coded information, it is applied to a *demultiplexer*, which extracts the information from each channel, separates it, and provides an output representative of the information originally encoded at the transmitting end of the circuit. The techniques used for conversion to digital form are many and varied. The basic use of the multiplexer in such a system is seen in the block diagram in Fig. 13–18.

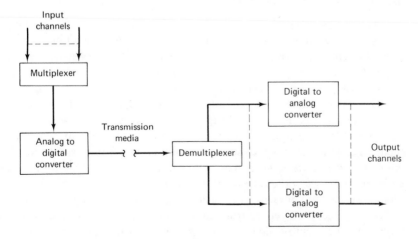

Figure 13–18   Non-digital multiplexer application

13—4 MEMORY ELEMENTS

Applications

Many digital systems require both *temporary* and *semipermanent data-storage capabilities.* Digital computers with their large-scale memory requirements also need smaller memory devices to increase their efficiency. The large memories are relatively slow, so smaller memories act as buffers between the rest of the computer and the large memory. When calculations are being performed, it is most inefficient to move the partial results to main memory. Small-scale memory devices are very useful in this application.

Data communication systems use small-scale memories to temporarily store groups of words, before passing them to the display device or to a computer's main memory. Test equipment used to verify digital system operation uses memory capability to store a sequence of events to be performed or a group of binary digits to exercise the system. Calculators need to store the variables to be operated on. Data display systems must accumulate sufficient data so that a complete line of information may be displayed at one time. Thus applications of *small-scale memory devices* are almost endless.

This section examines some of the common small-scale memory devices used in digital equipment. Discussion will be limited to *semiconductor types of memories,* due to the almost universal application that these devices possess. In addition, they are available in the same manner as the other complex functions discussed in this chapter and fit readily into the prepackaged logic function category. Other modes of storage do exist — but usually at a prohibitive cost and size in terms of the relatively small amount of information that must be stored.

Classification of Small-Scale Memories

The most common of all semiconductor memory elements, the *flip-flop,* has already been examined in great detail. When more than one bit of information must be stored, FFs may be connected together to form registers, as we discussed in Ch. 12. In applications where no more than a few bits of information are to be stored, FF registers adequately fulfill this function.

When more than one group of bits is to be stored, however, FF registers tend to become physically large and expensive. Packages, called *random access memories* (RAMs), are available to fill the gap between single-group storage and computer main memories, which store thousands of groups of bits. *The RAM is a group of storage devices,* which may or may not be FFs, *that store binary data in a digital system when the system is not using that data.* Information is randomly written into or read out of each storage location, as required. The term "random access" is derived from this memory property. Availability of this type of logic circuitry allows temporary memory (with a greater storage capacity than that of a single register) to be distributed throughout the system.

Read-only memories (ROMs) are another form of small-scale information storage devices that are finding unprecedented applications in the digital field. *Anytime a combination of control variables or a fixed value is required from a fixed input, a* ROM *may be used.* Each storage location in the ROM is addressable, just as in the RAM. The information stored may be a complete trigonometric table, a predetermined message or sequence of bits to be used in a data communication system, or numerous other applications. ROMs can supply permanent storage of data that is to be used repetitively, thereby relieving other logic circuits of performing calculations to obtain the data each time that it is to be used.

Random Access Memories (RAMs)

The diagram of a typical IC memory cell is shown in Fig. 13−19. Each of the storage locations is a modified $R-S$ FF, which requires both the X and Y inputs to be enabled before the S or C inputs may become active. The sixteen locations are arranged in an addressable (4 x 4) matrix. The desired bit location is selected by enabling the coincident $X-Y$ address lines (hence the classification of "coincident selection") and by keeping the remainder of the address lines disabled. The data that was stored at the selected location (and its complement) is then read at the output terminals. If the data at the addressed bit location is a 1, the S_1 terminal is LOW. At the same time, S_0 is HIGH, since it is the complement of the S_1 level. Both of the Write inputs, W_0 and W_1 must be LOW during the read operation.

To write a 1 in a specific location, the desired storage location is addressed and the input of the Write 1 amplifier is set to a HIGH level. A 0 is written in a similar manner, using the Write 0 input. This IC is capable then of accepting, storing, and reading out sixteen single-bit binary words.

Expansion to multiple-bit words merely requires an additional IC for each bit in the word. Equivalent address lines are paralleled, e.g., all X_1s connected together, so that when a location is addressed on one IC, the same location is addressed to the other ICs. Of course, the outputs and the Write inputs are not paralleled, since each bit in a word must be available separately, either for output or input. Thus a memory unit capable of storing sixteen 8-bit words could be constructed from 8 of the ICs being discussed.

The use of advanced IC techniques makes possible a 256-bit memory capability in a standard DIP package (Fig. 13−20a and b). Although FFs are not used as storage locations, a semiconductor circuit requiring much less physical space and power provides the memory capability. The 2501 memory is fully decoded, thus permitting the use of the standard DIP package. Read/Write and Chip-select control modes are provided. A LOW logic level applied to the R/W control allows the information from one location in the memory to be made available within about 1 microsecond, following the arrival of the address code. Readout is nondestructive.

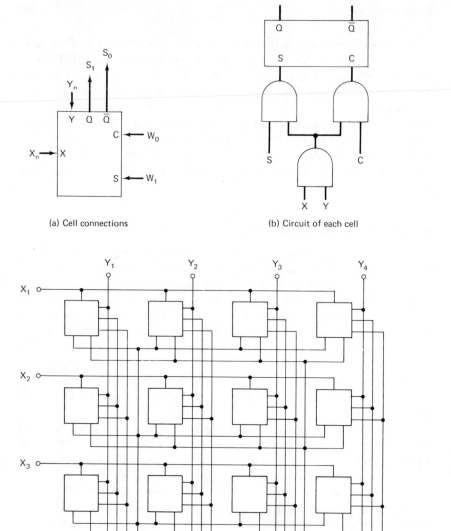

(a) Cell connections

(b) Circuit of each cell

(c) Functional block diagram

Figure 13–19   16-bit memory element concept

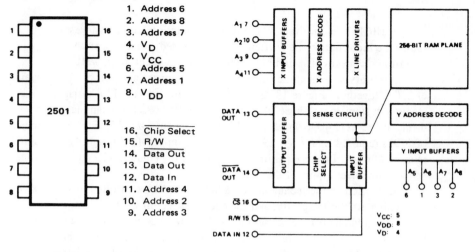

1. Address 6
2. Address 8
3. Address 7
4. V_D
5. V_{CC}
6. Address 5
7. Address 1
8. V_{DD}

16. Chip Select
15. R/W
14. $\overline{\text{Data Out}}$
13. Data Out
12. Data In
11. Address 4
10. Address 2
9. Address 3

(a) Package diagram

(b) Functional block diagram

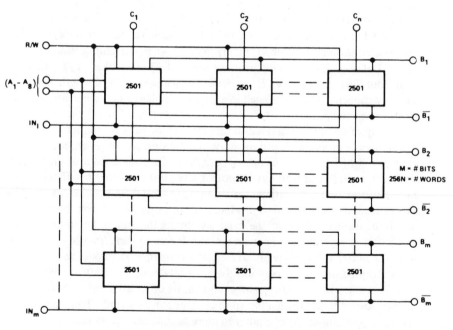

(c) Organization of 2501s into larger memory

Figure 13—20   2501   256 x 1 random access memory (RAM) (courtesy of Signetics, a subsidiary of Corning Glass Works, Sunnyvale, CA)

A HIGH level on the R/W input (in conjunction with an address code) causes the input data to be written into the selected storage location. Complete operation of the circuit is inhibited when the Chip-select input is HIGH.

Arbitrary size memories may be built by connecting appropriate numbers of 2501s together. Figure 13–20c shows a general-concept block diagram of a memory system containing $(256 \times N)$ words by M bits. For example, if the memory size were 4096 words by 12 bits, $N = 16$ and $M = 12$. Thus the number of 2501s required is $M \times N = 192$. The address inputs A_1 through A_8 are common to all rows. Inputs C_1 through C_n provide the column select and are wired to the Chip-select inputs of the 2501s.

Extension of the techniques used in the 2501 have resulted in single packages with a 2000-bit capacity, and the end is not yet in sight.

The coincident type of memory is a relatively common storage device. However other types of circuit organization do exist. A package organized as 16 words of 4 bits each is shown in Fig. 13–21. Selection of the word to be either read or written is accomplished by "decoding" a 4-bit address word. One word is assigned to each decimal-equivalent location of the address word, and no coincident operation is required. The "decoder" activates only one of the sixteen words at a time. This type of addressing is called *linear selection*.

A Chip-select input $(\overline{C}_E)$ is provided to enable the 8225 memory. Words are selected through the 4-input binary decoder when the Chip-select input is LOW. Data is written into the memory when READ ENABLE (R_E) is LOW and read from the memory when R_E is HIGH. Expansion of word length and total memory capacity is accomplished in a manner similar to that described for the single-bit memory ICs.

Any of the units described so far perform one of two functions — a buffer storage device or a scratch-pad memory. *Buffer storage devices are used to compensate for differences in the rate of data flow.* Data is loaded into the memory device at one rate and read out at another rate. For example, in a digital computer, information is sometimes loaded into memory from punched cards. The rate of reading the punched card is hundreds of times slower than the rate at which the memory can store the data. By storing the punched card information in a buffer storage device, the main memory is free to perform other functions, until the buffer storage is ready to feed its information in for storage. It can then be interrupted for a short period of time, rapidly move the contents of buffer storage into its own memory, and go on with its other work, while buffer storage is being loaded with information from another punched card. Thus the buffer storage device allows the computer to more adequately use its time.

Scratch-pad memories are used for intermediate storage of data resulting from arithmetic or other signal-processing operations that have been accomplished in the main memory. Just as one uses a "scratch pad" to work arithmetic problems before transcribing them to a finished form, so does the scratch-pad memory provide a place to "work the problem" without cluttering up main memory.

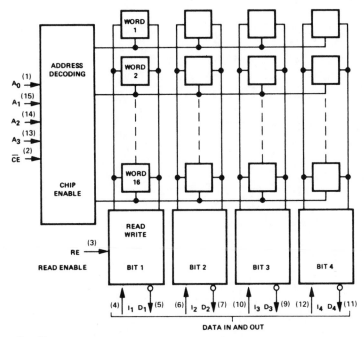

Figure 13—21   Type 8225 64-bit random access memory (RAM) (courtesy of Signetics, a subsidiary of Corning Glass Works, Sunnyvale, CA)

The number of ICs required to implement the buffer storage device in the scratch-pad memory depends on the number of bits per computer word and the maximum number of words that must be stored. Their capacities range from as few as sixteen to thousands of bits, depending entirely on specific applications.

A noncomputer application of a (16 x 4) RAM is shown in the block diagram of Fig. 13—22. Almost all digital systems require some type of *word generator* to test system response. The simple programmable word generator is capable of providing sixteen 4-bit words at its output. Each word is loaded in the RAM by setting the data-entry switches to the desired bit pattern, selecting the location for storage by establishing the required address input combination (using the single-pulse switch), and depressing the R/W control switch. The RAM is changed from the Read to Write mode by the R/W switch action, and the logic data present at the Write inputs is placed in memory. As soon as all sixteen (or as many as are required) words are loaded, the word generator is ready for operation.

The Control Logic is the heart of the generator. Here the sequencing takes place, and clock signals are modified and distributed to perform their various functions. For example, if we desire to start generator operation by using the

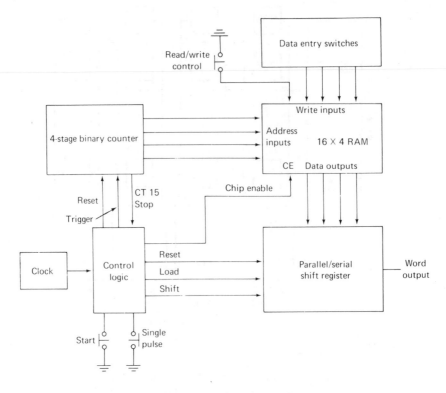

Figure 13—22   Programmable word generator

word stored at location 0, the binary counter must begin operation at the
appropriate count. This can be assured by generating a reset signal to the counter
upon depression of the Start switch. At the same time, the shift register is
cleared. The RAM is enabled (C_E), and data present at the output of the RAM is
made available to the shift register. A Load input is generated, followed by four
shift commands, and the 4 bits of the first word are shifted out to the digital
system being tested.

A trigger pulse is now allowed through the control logic to the counter,
and the counter advances one count. The second word is loaded into the shift
register, shifted out, and the counter once more advanced. Each word is
selected, loaded into the register, and shifted out. The recognition of count 15
in the counter is used to inhibit any further issuance of trigger pulses to the
counter, and following shifting out of the final word, the generator automatically
stops. The digital system has now been exercised with sixteen preselected test
words of 4 bits each, and its response may be evaluated for proper operation.

Except for the Clock and Control Logic, the entire word generator may be
constructed from prepackaged logic functions already discussed. The 4-stage
binary counter is similar to the 8281 binary counter of Sec. 13—3, and the shift

register function may be implemented with an 8271 shift register (Ch. 12). The (16 x 4) RAM operates similar to the 8225 RAM discussed in this section. Clock and Control Logic implementation is accomplished with standard gates and FFs.

Read-Only Memories (ROMs)

A Read-only memory is a system for storing information in a permanent form. In applications that require fixed data, the Read-only memory offers advantages where speed, cost, and reliability are factors. In gross block-diagram form (Fig. 13–23a), the basic ROM has two types of inputs and one type of output. Operationally, the address of a word is placed on the address line, and the Enable input is activated. The contents of the addressed location then appear at the output.

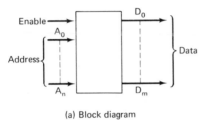

(a) Block diagram

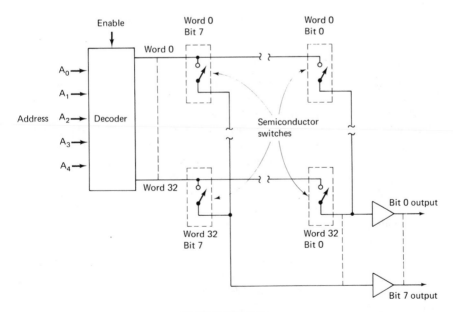

(b) Detailed block diagram

Figure 13–23   Read-only memory (ROM) concept (courtesy of Motorola Semiconductor Products, Inc., Phoenix, AZ)

Most ROMs are not capable of being changed once data has been stored. This means, of course, that loss of power to the equipment or error in operations cannot destroy the contents. However, it also means that if new data is required, another ROM package must be used. Alterable ROMs are not common (but are available), and the future promises increasing use of this type of memory package.

The storage mode in ROMs is somewhat different than that in the RAMs. Whereas some type of semiconductor FF is usually the storage medium in the RAM, a simple semiconductor switch is sufficient for the ROM. An "open" switch may be either a logic 0 or a logic 1, depending on the technique used in design of the memory. The type 8223 ROM shown in the detailed block diagram of Fig. 13–23b uses an open switch as a logic 1 and a closed switch as a logic 0. Linear selection techniques are used to enable one of thirty-two 8-bit words in the RAM. The configuration of the switch at each bit position in the selected word is determined and then applied to individual buffer amplifiers for each bit of the output word.

The 8223 ROM may be obtained with any desired storage pattern, or it may be programmed by the user. Remember, however, that once the device has a program stored, it may not be changed. Mistakes with ROMs are costly.

Word selection techniques are not limited to linear methods. Coincident selection is also commonly used, usually in ROMs with larger storage capacity. Up to 1024 12-bit words are available in some of the larger units, and the top limit is not yet in sight.

ROM applications are seemingly endless. Tables of mathematical data may be stored for use in digital computers, calculators, and computing counters. Repetitive small-scale computer programs may be placed in ROMs to free main memory space for more important tasks. Digital display devices, such as those used with digital test equipment and computers, must generate individual characters (letters, numbers, symbols, etc.). ROMs are used to store the form of each of the characters, so it may be addressed and put to use without having to form it each time it is required. A code used to develop a digital data message may have to be changed to a different form for transmission. ROMs once again tend to be the most efficient method of translation.

Effectively, the ROM is a code converter, since it accepts the input information (the address code) and provides an output (the stored word) that is the input's equivalent in another form. ROM applications are shown in Ch. 15, where code converters are discussed in detail.

PROBLEM APPLICATIONS

13–1.   What are the conditions under which the output of an Exclusive-OR gate is logic 1 or True?

13–2.   Draw the symbol for an Exclusive-OR gate and show the Boolean expression used to describe its output.

13—3. What is an Exclusive-NOR gate? Show its symbol and its output expression.

13—4. What is another name for the Exclusive-NOR gate?

13—5. List the functions that a comparator performs.

13—6. Define odd parity; even parity.

13—7. Determine the output of the 4-bit parity detector/generator in Fig. 13—6 with the following inputs: 0000, 0101, 1010, 1110, and 1111.

13—8. Determine the output of the 9-bit parity generator/checker in Fig. 13—7 with the following inputs: 000000000, 010101010, 101010101, and 111111111.

13—9. What is a half-adder? a full adder?

13—10. Verify algebraically that the logic diagram of Fig. 13—8 performs the half-adder function.

13—11. Verify algebraically,that the logic diagram of Fig. 13—9 performs the full adder function.

13—12. Draw a logic diagram of a full adder using only NOR gates.

13—13. Develop a means of subtracting two numbers using the full adder of Fig. 13—10.

13—14. Verify f_5 and f_9 in Fig. 13—13c for both active LOW and active HIGH inputs.

13—15. Verify f_7 and f_{13} in Fig. 13—13d for both active LOW and active HIGH inputs.

13—16. What is a multiplexer?

13—17. If f_x in Fig. 13—17 is 1 Mhz, what is f_{out} under the following conditions?

$$A_0 = 0, \quad A_1 = 0, \quad A_2 = 0$$
$$A_0 = 0, \quad A_1 = 1, \quad A_2 = 0$$
$$A_0 = 1, \quad A_1 = 1, \quad A_2 = 0$$

13—18. Define random access memory; read-only memory.

13—19. What is the usual storage media in small-scale semiconductor memory elements?

14

Binary
Codes

By the time you reach this chapter, there should be no doubt that the *binary number system* is commonly used in digital devices. Arithmetic operations have been performed using binary numbers. Sequences of operations have been generated. Many of the simple decisions encountered in machine operation were adapted to binary circuits.

One problem remains, however. Although a digital device may conveniently use binary representation, when it becomes necessary for humans to intervene, the information must be converted to a more easily interpretable form. That is, *the input of information to the device* and *the output of information from the device must be in something other than conventional binary form.*

One way of satisfying this requirement is to assign a *unique combination of 1s and 0s* to each number, letter, or symbol that must be represented. Such representations are called *codes*. Many digital devices employ one form of code for input, other forms internally, and even a different form for output. We shall first discuss codes used to represent numbers and then proceed to more complex codes representing letters, symbols, and decimal digits.

14—1 BINARY RELATED RADIX CODES

Three related number systems (*binary*, *octal*, and *hexadecimal*) may be used to represent numbers in digital equipment. Table 14—1 shows the relationship between these systems. The binary number system (with a radix of 2) was discussed in detail in Ch. 2, but positional notation, binary-to-decimal conversion, and decimal-to-binary conversion should be reviewed in preparation for our discussion of the other number systems.

Table 14-1   Binary and related Radix codes

Decimal	Radix 2 Binary	Radix 8 Octal	Radix 16 Hexadecimal
0	0	0	0
1	1	1	1
2	10	2	2
3	11	3	3
4	100	4	4
5	101	5	5
6	110	6	6
7	111	7	7
8	1000	10	8
9	1001	11	9
10	1010	12	A
11	1011	13	B
12	1100	14	C
13	1101	15	D
14	1110	16	E
15	1111	17	F
16	10000	20	10

The Octal System

Numbers are stored within a computer in binary form, but the storage area may be arranged in such a manner that the number can be easily represented in another number system. The *octal* (base 8) *number system*, for example, is closely related to the binary system. Representation of the basic digits of the octal system (0 through 7) is easily accomplished with three binary digits. In fact, if the binary number is separated into groups of three digits, the octal number is immediately obtained. Limiting each group to three binary digits provides equivalent octal digits from 0 to 7, which is the range of the octal system.

Example 14-1   What is the octal equivalent of 101_2?
Solution $101_2 = 1 \times 2^2 + 0 \times 2^1 + 1 \times 2^0 = 4 + 1 = 5_8$

Note the use of positional notation.

Example 14-2   What is the octal equivalent of 110101_2?
Solution   110   101
                6   5   = 65_8

Conversion from octal to binary is the reverse process. The three binary digits that represent each octal digit are written, starting with the least-significant digit first.

Example 14-3   What is the binary equivalent of 65_8?
Solution   6   5
             110   101 = 110101_2

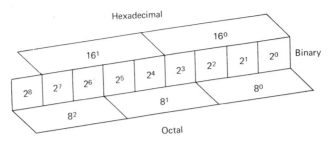

Figure 14—1   Binary, octal, and hexadecimal number system relationships

The relationship between the binary and octal number systems is diagrammed in Fig. 14—1. Octal-decimal and decimal-octal conversion is accomplished in different ways. Although not the most "scientific" way, conversion first to binary, then either to octal or decimal, depending on which direction the conversion is going, may be the easiest. A more organized method for conversion from any radix to radix 10 follows.

CONVERSION FROM ANY RADIX TO RADIX 10

1. Multiply the most-significant digit (MSD) of the chosen number by its radix and add the next most-significant digit.
2. Multiply this sum by the number's radix and add the next most-significant digit.
3. Repeat the process until all digits in the chosen number have been used. The final result is the converted number in radix 10.

Example 14—4   Convert 65_8 to a decimal number.

Solution   6 $\rightarrow 5 = 53_{10}$

$\dfrac{\times 8}{48}$ +

This procedure is the equivalent of the *double-dabble* method used for conversion from binary to decimal in Ch. 2.

The successive division method of Ch. 2 may also be used for whole number conversion from decimal to any other radix, provided the divisor used is the radix of the desired number.

Example 14—5   Convert 53_{10} to an octal number.

LSD

Solution 8$\lfloor$53 $\downarrow$

8 $\lfloor$6   5

0   6 $= 65_8$

$\uparrow$

MSD

Formal mathematical methods for number system conversions exist, but they will not be discussed in this text. However, methods of conversion using logic circuits are examined in Ch. 15.

The octal number system is used for digital data processing input and out-put devices. Ease of encoding and decoding numbers makes the octal system the ideal choice for "minimum hardware" input and output applications. However, the recent advances in semiconductor technology and IC availability are moving octal operations into the background. Converting binary to decimal and vice versa no longer requires extensive hardware and excessive space. As existing digital equipment is updated, octal operations will gradually disappear in favor of the more easily manipulated decimal input and output devices.

The Hexadecimal Code

Some digital computers handle numbers in groups of four binary digits. One code used to represent these sets of four digits is the *hexadecimal* (base 16) *number system.* Sixteen numbers are representable in the hexadecimal system, and additional symbols are required to portray digits greater than 9. Table 14—1 shows a commonly employed set of symbols for the hexadecimal number system.

Conversion from binary to hexadecimal is conveniently performed by group-ing the binary number into sets of four digits and determining the equivalent value using Table 14—1.

Example 14—6   What is the hexadecimal equivalent of 1010_2?

Solution $1010_2 = 1 \times 2^3 + 0 \times 2^2 + 1 \times 2^1 + 0 \times 2^0$

$$= 8 \quad + \quad 0 \quad + \quad 2 \quad + \quad 0$$

$$= A_{16}$$

Example 14—7   What is the hexadecimal equivalent of 10011010_2?

Solution   1001   1010
           9   A $= 9A_{16}$

The process is reversed when converting from hexadecimal to binary. The four binary digits representing each hexadecimal number are written, starting with the least-significant digit first.

Example 14—8   What is the binary equivalent of $9A_{16}$?

Solution   9   A
           1001   1010 $= 10011010_2$

The relationship between the binary, octal, and hexadecimal number systems is diagrammed in Fig. 14—1.

Hexadecimal-decimal and decimal-hexadecimal conversion follows the same procedures as those for octal-decimal and vice versa.

Example 14–9   Convert $9A_{16}$ to a decimal number.

Solution   9 $\rightarrow A^* = 154_{10}$

$\underline{\times 16}$ $+$

144 ⌐ ↰

$^*A_{16} = 10_{10}$

Example 14–10   Convert 154_{10} to a hexadecimal number.

Solution   LSD

$16\underline{|154}$ ↓

16 $\underline{\hphantom{0}|9\hphantom{0}}$ 10^*

$\hphantom{16}\hphantom{1}0$ 9 $= 9A_{16}$

$\hphantom{16000}$↑

$\hphantom{160}$MSD

$^*10_{10} = A_{16}$

Note again that positional structure has been implicit *throughout* the discussion of all number system conversions.

The hexadecimal system will continue to find widespread use in the so-called character-organized digital computers. Engineers, technicians, and programmers who work closely with the computer find no difficulty in using this nondecimal number system. Conversion hardware is reduced in these cases without loss of utility. However, when information must be displayed for the non-computer-oriented user, the decimal system is employed.

14–2 BINARY CODED DECIMAL (BCD) CODES

General Characteristics of BCD Codes

One of the major problems facing the engineer, technician, programmer or any user of digital equipment (whether it is computer, control, or measuring equipment) is *the conversion of a code readily handled by man to a code readily handled by the equipment.* Most people can best work with the decimal code, while digital machines work most effectively with binary codes.

The best solution to this problem is found in the *binary coded decimal* (BCD) *codes.* Ten unique symbols (the decimal number system) must somehow be represented by only two unique symbols (the binary number system). In other words, *ten unique combinations of two symbols* must be found. Previous discussion tells us that any possible combinations of the two binary symbols is an *integer power of 2*, i.e., $2^0, 2^1, 2^2$, and so on. Thus some integer power of 2 must be used that will allow at least ten combinations of binary digits (bits) to exist. Ten is greater than 2^3 but less than 2^4, so four bits must be employed to obtain the ten combinations. This results in sixteen combinations, six of which are not used. Any ten of the combinations may be used, and the specific combinations used will result in a particular type of BCD code.

Each decimal digit, therefore, is represented by a combination of four binary digits. The decimal number 345 may be represented in BCD by 0011 0100 0101. In binary, 345_{10} is represented as 101011001. Note that only nine bits are required for the binary representation, while twelve bits are required for the BCD form. However, since it is easier to recognize the BCD form, the *extra three bits* is the price that must be paid.

BCD codes may be classified as either *weighted* or *unweighted*. The *weighted codes* follow a *positional notation structure*. Positional values do not necessarily have to be ascending powers of 2; they may be arbitrarily assigned. In *unweighted codes*, the digit positions do *not* indicate the relative value of the represented number.

8–4–2–1 Weighted BCD Code

The most common weighted BCD code is the natural 8–4–2–1 code. The most-significant digit possesses a weight of 8, the next most-significant digit a weight of 4, and so forth. Correlation between decimal digits and the 8–4–2–1 BCD representation is shown in Table 14–2. All that is required to convert an 8–4–2–1 BCD number to a decimal number (or vice versa) is a slight familiarity with the binary number system. Remember, each decimal digit is represented by four binary digits, weighted in an 8–4–2–1 order.

Example 14–11 Convert 1001 0101 (BCD 8–4–2–1) to a decimal
number.
Solution   1001 0101
          9   5 $= 95_{10}$

Table 14–2   Weighted Binary-coded-decimal (BCD) codes

Decimal	Natural 8421	2421
0	0000	0000
1	0001	0001
2	0010	0010
3	0011	0011
4	0100	0100
5	0101	1011
6	0110	1100
7	0111	1101
8	1000	1110
9	1001	1111

Example 14–12   Convert 95_{10} to a BCD $(8–4–2–1)$ number.
Solution   9   5
             1001   0101 = 1001 0101

Some digital computers actually operate in a BCD mode, and arithmetic operations may be performed directly in BCD without conversion to binary equivalents. However, since this text is not computer oriented, we shall concentrate on applications commonly encountered in the digital measuring and display fields. A BCD counter was discussed in Ch. 11. Methods of encoding and decoding ing BCD numbers are discussed in Ch. 15.

2–4–2–1 Weighted BCD Code

Another weighted BCD code that is widely used in computers is the 2–4–2–1 code (see Table 14–2). In this case, the most-significant digit of the 4-bit representation of a decimal number has a weight of 2 rather than 8. The major advantage of the 2–4–2–1 codes is that it is *self-complementing*.* Thus the 2–4–2–1 representation of 1 (0001) is the opposite of 8 (1110), and 0 (0000) is the opposite of 9 (1111). All BCD numbers represented in the 2–4–2–1 code have this self-complementing property. And since one method of subtraction in a digital computer uses the minuend *added to* the complement of the subtrahend, the self-complementing feature is quite valuable.

Gray Unweighted BCD Code

Unweighted BCD codes are used in both digital data processing and in data transmission and measuring applications. When it is necessary to convert a physical parameter (e.g., a shaft position on a motor), conventional codes become difficult to work with. Mechanical construction methods place overly restrictive requirements on the encoding device. For example, if the conventional 8–4–2–1 BCD code is used, the encoding device *could* show a change in as many as *four bit positions simultaneously* (0111 to 1000). The *Gray code* alleviates this problem. When transitioning from one number representation to the next, *only one bit position changes*. Table 14–3 shows the Gray code representation of ten decimal digits, while Fig. 14–2 shows a typical mechanical encoder using both 8–4–2–1 and Gray code methods. From these comparisons, it is apparent that the Gray code implementation of BCD is preferable, when minimum possibility of generating improper representation of numbers due to bit change is required. Also, since fewer bits must change for each advance, binary counters implemented with a Gray code tend to have less operating delay.

---

* Any two numbers which add up to their radix minus 1 complement each other, i.e., are opposite.

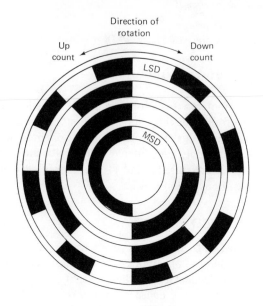

Dark areas represent conductive material and light areas insulative material. Electrical connection to the disc is via wire brushes. A complete circuit (logic 1) exists when a dark area is contacted; an open circuit (logic 0) exists when a light area is contacted.

(a) 8-4-2-1 BCD encoder

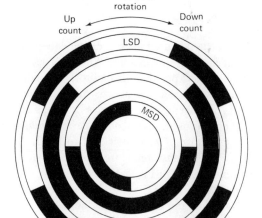

(b) Gray code encoder

Figure 14—2   Encoding binary codes; (a) 8-4-2-1 BCD encoder, (b) Gray code encoder

Table 14–3   Unweighted BCD codes

Decimal	(xs3) Excess 3	Gray	2 out of 5
0	0011	0000	00011
1	0100	0001	00101
2	0101	0011	00110
3	0110	0010	01001
4	0111	0110	01010
5	1000	0111	01100
6	1001	0101	10001
7	1010	0100	10010
8	1011	1100	10100
9	1100	1101	11000

Since no direct relationship exists between binary digits and bit positions in such unweighted codes as the Gray code, conversion is not as easy as it is between weighted codes. So, when converting between Gray and decimal codes, the binary system is used as an intermediate representation of the number. Conversion procedures are detailed in the following lists.

CONVERTING FROM BINARY TO GRAY CODE

1. Write the binary number completely.
2. Starting under the binary number — one place to the right — place the binary number again. The least-significant digit now appears beyond the radix point and should be dropped.
3. Add the two numbers, disregarding any carry terms that then result.

Example 14–13   Convert 93_{10} (1011101_2) to a Gray code number.

Solution   Disregard carry terms generated here

     ⇊

1011101. $= 93_{10}$

  101110.1 (drop)

1110011   Gray code representation

Note: This is equivalent to combining all adjacent digits of the binary number in an Exclusive-OR relationship, as shown in the following diagram.

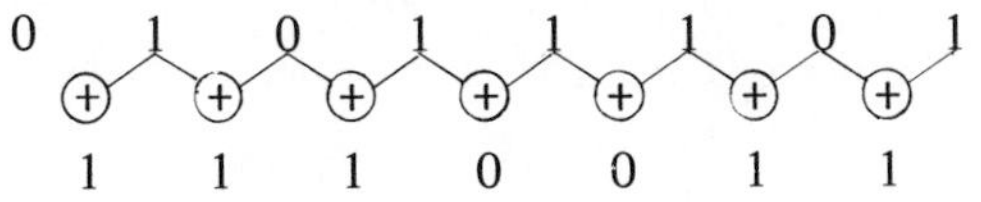

CONVERTING FROM GRAY TO BINARY CODE

1. Write the Gray code number completely.
2. Bring the most-significant digit down and consider it the most-significant digit of the binary number being sought.
3. Add the most-significant digit of the binary number to the Gray code number in the next-significant digit column, neglecting any carry terms that result.
4. Continue the procedure until all Gray code digits have been used. The resulting number is the binary equivalent of the original Gray code number.

Example 14–14   Convert the Gray code number 1110011 to a decimal number, using the binary system.

Solution

$$1 \to 1 \to 1 \to 0 \to 0 \to 1 \to 1$$
$$\downarrow + = + = + = + = + = + =$$
$$1 \nearrow 0 \nearrow 1 \nearrow 1 \nearrow 1 \nearrow 0 \nearrow 1 = 93_{10}$$

Excess 3 (XS3) Unweighted BCD Code

Another unweighted BCD code often encountered in digital computer applications is the *XS3 code*. It is equivalent to the natural $(8-4-2-1)$ BCD representation *plus* 3. Table 14–3 equates the XS3 code with decimal digits. Conversion from decimal to XS3 code may be accomplished either by converting to binary and then adding 3 or by adding 3 to the decimal number and then converting.

Example 14–15   Convert 9_{10} to an XS3 coded number.

Solution $9_{10} = 1001$

$$\underline{+\ 11}$$
$$1100\,(\text{XS3})$$

or $9 + 3 = 12 = 1100\,(\text{XS3})$

Example 14–16   Convert 12_{10} to an XS3 number.

Solution

1	2	Decimal		1	2	Decimal
0001	0010	Binary		+3	+3	Add 3
+ 11	+ 11	Add 3		4	5	
0100	0101	XS3		0100	0101	XS3

While each decimal digit is converted separately, XS3-to-decimal conversion is essentially the *reverse* process. Each digit is converted to its decimal equivalent, and 3 is then subtracted, as shown in Example 14–17. (The subtraction may also be performed at the binary level.)

Example 14–17 Convert 0100 0101 (XS3) to a decimal number.

Solution

0100	0101	XS3		0100	0101	XS3
4	5			− 11	− 11	−3
−3	−3	−3		0001	0010	Binary
1	2	Decimal		1	2	Decimal
	12_{10}				12_{10}	

The XS3 code is *self-complementing*, just like the 2–4–2–1 BCD code. In addition, the proper carry is *always* generated, as though the XS3 number were in decimal form. Both of these characteristics make the XS3 code an excellent choice for arithmetic operations in digital computers, and since many digital measuring instruments interface directly with computers, the XS3 code is also found in this area of digital applications.

2-Out-of-5 Unweighted BCD Code

All of the BCD codes mentioned so far may be used *wherever number representations must be transmitted from one location to another* (either within the digital device or to a remote location). Each of the codes may also be implemented with one or more extra bits to establish a *parity check*, or another way to assure correct transmission is to use an *exact count code*, where the total number of 1s in each number representation is the same. The *2-out-of-5 code* (Table 14–3) is an example. A pseudoweighted BCD method (7–4–2–1) represents the ten decimal digits in the first four bit positions, while the final bit position is reserved for an exact count digit. The added digit in position 5 is selected so that each word has exactly two 1s and three 0s. When decimal numbers are transmitted in this form, all *single errors* are detected. Any character received with less or more than two 1s is in error.

Conversion to and from the 2-out-of-5 code is relatively straightforward due to the pseudoweighted format. However, the decimal digit 0 is represented in a nonweighted manner.

14–3 ALPHAMERIC CODES

The codes that we have discussed so far have represented only *numerical data*. But communicating information from one point to another necessarily includes other forms of data, such as *alphabetical characters* and *symbols*. Hence the total numbers of possible combinations concept of binary digits, called *alphameric representation*, is used to represent numbers, special symbols and alphabetical characters.

As you may have already noted, a direct correlation *may not* exist between a number and its representation in binary digits. The Gray code, for example, adequately represents the decimal digits, but the positions of the binary digits in the representation are unweighted. The most important consideration in its design was the single digit change from one representation to the next, *not* a consistent value for each bit position. Alphameric codes usually fall into the unweighted category, although in some codes a design pattern is noticeable.

5-Level (Baudot) Code

Four binary digits (bits) allow only sixteen different combinations, which are insufficient for alphameric representations. Up to 32 different characters may be represented by 5 bits, thus allowing all the letters of the alphabet to be represented. If 2 of the 32 bits are reserved for the purpose of selecting one of two separate "sets" of information, then a total of sixty-two available combinations is obtained.

A *5-level code*, developed by Jean Baudot (a French engineer) in the late nineteenth century, has been employed in data communication applications, using a typewriter-like device called a *teletypewriter*. The 26 letters of the alphabet are represented on the "lowercase" set of information, while numbers, punctuation marks, and special symbols are represented by the "uppercase" set.

The teletypewriter mechanism contains a group of switches that may be opened and closed by the depression of any key on the keyboard. Switch conditions may be translated into binary representation (e.g., 0 = open switch and 1 = closed switch). Actuation of each key then results in a specific combination of 1s and 0s that is representative of the selected alphameric character. The switches operate in sequence, following a beginning synchronizing indication and followed by a "stop" indication.

The Baudot code is shown in Fig. 14–3, correlated with a drawing of the perforated tape used with many alphameric code devices. The black marks represent holes in the tape (1s), while the unmarked circles represent solid tape (0s). Information that is to be transmitted from one location to another is usually prepared by "punching" a tape; large amounts of data can then be transmitted without interruption. Devices that read the tape sense a hole or a lack of a hole at each of the five hole locations and generate the appropriate binary digit to represent the selected character.

A 5-level code is therefore *capable of representing all alphabetical and numerical characters,* as well as *several punctuation marks and special symbols.* However no means to determine the accuracy of the binary representation is possible with a 5-level code; that is, *no parity is used.* Of course, the code could be expanded to six bits to provide parity, but often it is necessary to communicate more than the minimal number of characters possible with five bits.

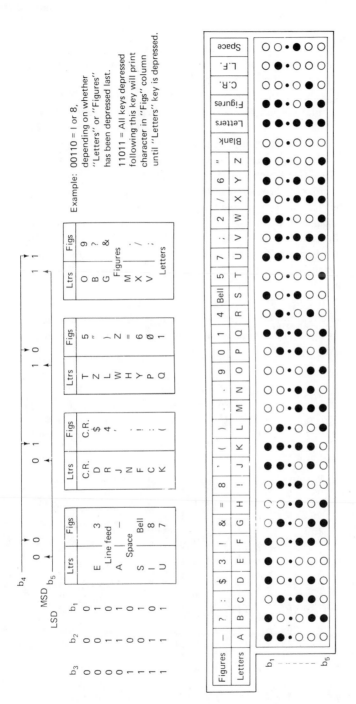

Figure 14–3 Baudot code

411

6-Level Code

A *6-level code*, capable of sixty-four different combinations, easily supplies sufficient variety without the necessity of "uppercase" and "lowercase" control. *One* possible arrangement of a 6-level code is shown in Fig. 14–4. Representation of letters and numbers in this code is of a binary-coded-decimal (BCD) nature. The two most-significant bits (b_6 and b_5) indicate that the next four bits that follow represent either numbers ($b_6 = 0$ and $b_5 = 0$), the first one-third of the alphabet ($b_6 = 1$ and $b_5 = 1$), the second one-third ($b_6 = 1$ and $b_5 = 0$), or the final one-third ($b_6 = 0$ and $b_5 = 1$). Each group is arranged in ascending BCD order (0 through 9) to match either increasing numerical value or letters of the alphabet progressing from the first to the last.

Codes arranged in a sequential manner lend themselves to computing applications. In fact, the 6-level code in Fig. 14–4 is commonly used within computers, and when expanded to seven bits to provide a parity position, it finds application in both magnetic and paper-tape data systems. We should note at this time that codes are described by the number of bit positions, *whether or not* all of the bit positions are used to represent information. Thus, although the data representations are the same in both codes described in Fig. 14–4, the addition of the parity bit resulted in a reclassification from a 6-level to a 7-level code.

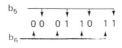

b_5 — 00 01 10 11

b_6

b_4	b_3	b_2	b_1	00	01	10	11
0	0	0	0			−	+
0	0	0	1	1	/	J	A
0	0	1	0	2	S	K	B
0	0	1	1	3	T	L	C
0	1	0	0	4	U	M	D
0	1	0	1	5	V	N	E
0	1	1	0	6	W	O	F
0	1	1	1	7	X	P	G
1	0	0	0	8	Y	Q	H
1	0	0	1	9	Z	R	I
1	0	1	0	Ø	[1]	!	?
1	0	1	1	=		$	.
1	1	0	0	'	(	*	)
1	1	0	1	:	[2]	]	[
1	1	1	0	>	\	;	<
1	1	1	1	√	[3]	[4]	[5]

[1] Record mark
[2] Word separator
[3] Segment mark
[4] Mode change
[5] Group mark

Example: 1110001 = A
0110011 = C

Figure 14–4   6- and 7-level codes (6-level without parity; 7-level with parity)

7-Level (ASCII) Code

The use of 7 data positions results in a code having a total of 128 possible combinations. One such code, commonly used in digital control of industrial machines, is shown in Fig. 14–5. It is compatible with the standard code that we shall discuss shortly, which means that machine operation may easily be integrated with digital computers. Blank areas appear where no characters are used, but this is a result of the compatibility requirement. This code is also expandable to eight bits for parity-type error recognition.

The most common 7-level (expandable to 8-level) code is called the *United States of America Standard Code for Information Interchange* (ASCII) and appears in Fig. 14–6. It is also called the ANSI (American National Standards Institute) code or the Data Interchange code. Any of the 128 locations may be described by a unique combination of high-order (b_7, b_6, and b_5) bit locations and lower order (b_1, b_2, b_3, and b_4) locations. The ASCII is quite versatile and is organized along logical lines. Note, for example, that the device using this code does not have to go beyond the two most-significant bits to determine whether the represented information is a *control* (both 0s) or a *character* (not both 0s).

Numerous variations of the ASCII are in use, but most are compatible. Their differences lie specifically in the area of *names of the control functions* and *the actual symbol represented*, when neither letters, numbers, nor control functions are being used. (Note that all 128 locations are used.)

Expansion to eight bits allows either parity checks or doubling of the number of characters that may be represented.

b_4	b_3	b_2	b_1	$b_7 b_6 b_5$ 0 0 0	$b_7 b_6 b_5$ 0 0 1	$b_7 b_6 b_5$ 0 1 0	$b_7 b_6 b_5$ 0 1 1	$b_7 b_6 b_5$ 1 0 0	$b_7 b_6 b_5$ 1 0 1	$b_7 b_6 b_5$ 1 1 0	$b_7 b_6 b_5$ 1 1 1
0	0	0	0			Space	0		P		
0	0	0	1				1	A	Q		
0	0	1	0				2	B	R		
0	0	1	1				3	C	S		
0	1	0	0				4	D	T		
0	1	0	1			%	5	E	U		
0	1	1	0				6	F	V		
0	1	1	1				7	G	W		
1	0	0	0	BS		(	8	H	X		
1	0	0	1	HT		)	9	I	Y		
1	0	1	0	LF			:	J	Z		
1	0	1	1			+		K			
1	1	0	0					L			
1	1	0	1	CR		–		M			
1	1	1	0					N			
1	1	1	1			/		O			

* Expandable to 8-level if parity bit is used

Figure 14–5   7-level numerical control code (expandable to 8-level if parity bit is used)

b7 b6 b5 →	0 0 0	0 0 1	0 1 0	0 1 1	1 0 0	1 0 1	1 1 0	1 1 1
b4 b3 b2 b1								
0 0 0 0	NUL	DLE	SP	0	@	P	`	p
0 0 0 1	SOH	DC1	!	1	A	Q	a,	q
0 0 1 0	STX	DC2	"	2	B	R	b	r
0 0 1 1	ETX	DC3	#	3	C	S	c	s
0 1 0 0	EOT	DC4	$	4	D	T	d	t
0 1 0 1	ENQ	NAK	%	5	E	U	e	u
0 1 1 0	ACK	SYN	&	6	F	V	f	v
0 1 1 1	BEL	ETB	'	7	G	W	g	w
1 0 0 0	BS	CAN	(	8	H	X	h	x
1 0 0 1	HT	EM	)	9	I	Y	i	y
1 0 1 0	LF	SUB	*	:	J	Z	j	z
1 0 1 1	VT	ESC	+	;	K	[	k	{
1 1 0 0	FF	FS	,	<	L	\	l	|
1 1 0 1	CR	GS	−	=	M	]	m	}
1 1 1 0	SO	RS	.	>	N	^	n	~
1 1 1 1	SI	US	/	?	O	—	o	DEL

(a) Basic 7-level ASCII code

$b_1 \ldots \ldots b_2$ Parity

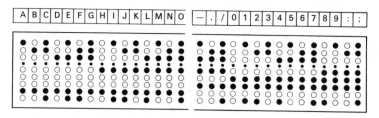

(b) Tape sample showing 7-level code
expanded to 8 level by Parity position

Figure 14-6   ASCII code

8-Level (EBCDIC) Code

One final example of alphameric codes is the *Extended Binary Coded Decimal Interchange Code* (EBCDIC). It is a full *8-level code* (see Fig. 14-7), since each letter, number, symbol, or control function is represented by two BCD numbers — the equivalent of eight binary digits. A number of similarities exist between EBCDIC and the other codes discussed in this chapter. The BCD aspects of the letters and numbers are as explained in the 6-level code. Identification of control functions is made by checking the two most-significant digits, just like the ASCII code. One advantage EBCDIC has over other codes is that the identification of symbols is made easy by investigating only the two most-significant digits.

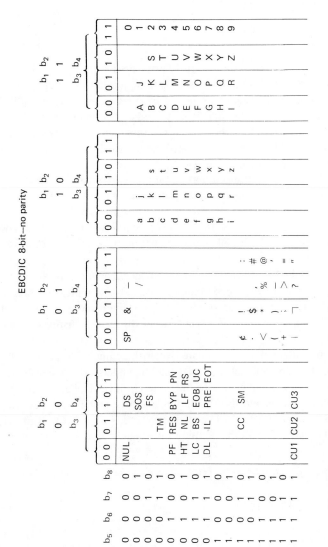

Figure 14—7 Extended binary-coded-decimal interchange code; 8 and 9-level

The EBCDIC is used when the *data transmission and reception devices must interface with a computer that operates predominantly in the BCD mode.* Parity is not provided in EBCDIC; the code must be expanded to nine levels in order to obtain this function.

Hollerith Code

Although the *punched card* is largely associated with digital computers, it should not be overlooked when discussing data representation codes. Letters, numbers, and some symbols are represented by *certain combinations of holes punched in a special card.* The most widely used code in punched cards is the *Hollerith code.* Each of the 80 columns carries one digit, letter, or symbol. Figure 14—8 shows a typical listing of the Hollerith code, which should be correlated with the punched card shown in Fig. 14—9.

Punch positions are arranged horizontally in twelve rows — the bottom ten for digits 0 through 9, plus two rows (numbered 11 and 12) along the top of the card for *zone punching.* Figure 14—8 illustrates how *each digit* (0 through 9) is coded with a *single punch. Alphabetical characters* require a single row punch *plus* a zone punch. Special characters use multiple combinations of row and zone punches.

When a punched card is read, each row in each column is investigated separately. Thus the Hollerith-coded punch card actually represents data with a *12-level code.* Due to the large number of digits, the Hollerith code is not usually approached in its binary form but rather through the listing of the rows and/or zones punched for each character.

It may be enlightening to compare Hollerith coding with the 6-level code in Fig. 14—4. The similarity of the codes may be understood by considering the 01 column in Fig. 14—4 equivalent to the 0 punch of the Hollerith card, the 10 column as the 11 punch, and the 11 column as the 12 punch. Thus conversion from Hollerith code to 6-level BCD code is easy and is a good reason for using BCD coding in digital devices that employ punched-card input.

Number	Punch	Letter	Punch	Letter	Punch	Letter	Punch	Char	Punch
0	0								
1	1	A	12,1	J	11,1			#	3,8
2	2	B	12,2	K	11,2	S	0,2	,	0,3,8
3	3	C	12,3	L	11,3	T	0,3	$	11,3,8
4	4	D	12,4	M	11,4	U	0,4	.	12,3,8
5	5	E	12,5	N	11,5	V	0,5	&	12
6	6	F	12,6	O	11,6	W	0,6	@	4,8
7	7	G	12,7	P	11,7	X	0,7	%	0,4,8
8	8	H	12,8	Q	11,8	Y	0,8	*	11,4,8
9	9	I	12,9	R	11,9	Z	0,9	¤	12,4,8

Figure 14—8   Hollerith code listing

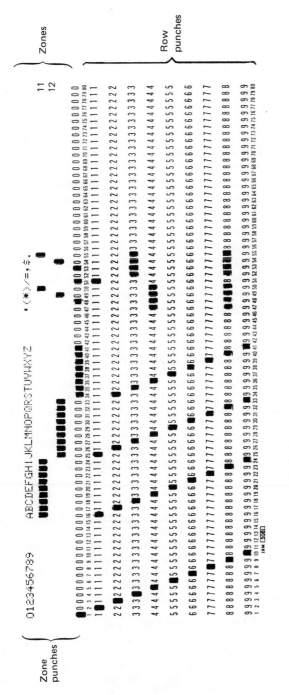

Figure 14–9 Hollerith-coded punched card

417

14-4 SUMMARY

Some of the common binary codes used to represent letters, numbers, and symbols have been discussed in this chapter. More important, however, has been the discovery that *information may be represented in binary form.* With this knowledge, *intelligence* can obviously be encoded, transmitted, decoded, and manipulated with logic circuits. Such capability has provided the impetus for the revolution in machine aids designed to improve our intellectual processes.

Many of the more complex codes used with digital equipment have not been covered in this text. Parity-checking methods may be extended to more than one bit per character — even to complete blocks of information (many groups of characters). All sorts of "exotic" error detection and correction schemes exist, but they are beyond the scope of this text.

In most cases that the digital logic technician will encounter, the codes will be listed in the technical information furnished with the digital device. If such information is not available, the principles discussed in this chapter should enable the technician to systematically approach the equipment and then determine the important characteristics of the codes, before proceeding with troubleshooting and repair activities.

PROBLEM APPLICATIONS

14-1.   Discuss the relationships between the binary, octal, and hexadecimal number systems.

14-2.   Convert the numbers 5_{10}, 76_{10}, and 348_{10} to (a) octal form, (b) hexadecimal form, (c) 8-4-2-1 BCD form, (d) 2-4-2-1 BCD form, (e) Gray code form, and (f) XS3 code form.

14-3.   Why are codes necessary in digital devices?

14-4.   Discuss the advantages and disadvantages of BCD codes versus straight binary codes.

14-5.   What is a weighted BCD code?

14-6.   What is an unweighted BCD code?

14-7.   Discuss the relative advantages and disadvantages of weighted BCD codes versus unweighted BCD codes.

14-8.   Develop a weighted BCD code (other than that used in this chapter) and list its representation of the ten decimal digits.

14-9.   Develop an unweighted BCD code (other than that used in this chapter) and list its representation of the ten decimal digits.

14-10.   Discuss the advantages and disadvantages of the 2-out-of-5 code.

14-11.   What is an alphameric code?

14-12.   What determines the number of bits necessary in an alphameric code?

14-13.   Discuss the advantages and disadvantages of the 5-level Baudot code.

14-14.   Show how the words "6-level alphameric code" would appear in the (a) Baudot code, (b) 6-level BCD code without parity, (c) 6-level BCD code with parity, (d) 7-level ASCII without parity, (e) 7-level ASCII with parity, and (f) EBCDIC without parity.

14-15.   Discuss the relationship between the Hollerith code and the 6-level BCD code.

15

Code Converters
and
Display Devices

This chapter serves a dual purpose. The various codes used to represent alphameric characters in digital equipment were discussed in detail in Ch. 14. The need for conversion between codes was firmly established. A code must function efficiently within the digital equipment, but it must also be easily recognizable to the operator (i.e., it must be capable of being converted to a "human usable" code). To do this, it must be displayed in a manner that is easily recognized; hence *the second major purpose for code conversion is to actuate display devices.* So *both* the conversion between codes and the conversion of codes to actuate display devices will be discussed in this chapter.

The three major *character-display devices* are *cold-cathode direct-character*, *segmented*, and *solid-state units.* Each requires a specific conversion device for actuation, and each displays its information in a different manner. More complex devices capable of displaying complete pages of information are mentioned but are not discussed in detail.

15—1 CODE CONVERTERS

Assuming that most digital devices operate internally in the *binary mode*, numerous requirements for code converters can be identified. Considering only numbers, Fig. 15—1 shows some of the possibilities.

Input to the digital device may assume various physical forms, such as switches, punched paper tape, keyboard devices similar to typewriters, and magnetic tape. The actual process of converting the input information to binary is often accomplished by the input device itself. Of course, if the input device consists of switches (one for each binary order), no need for conversion exists, since the switch is either closed or open (representing a 1 or a 0). Output in binary usually takes the form of individual indicator lights (one for each binary order). Hence *no conversion is necessary*, *if both input and output are of a direct binary nature.*

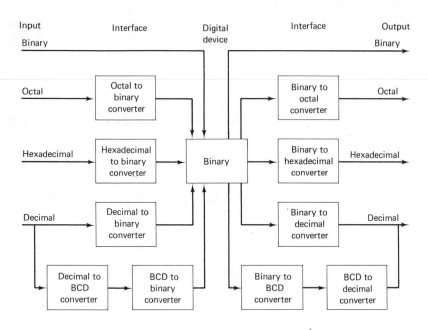

Figure 15—1   Code conversion requirements

Octal to Binary Conversion

Octal input is still used in certain special applications. When the input to a digital device is in octal form, it is necessary to convert to binary form so that the device may operate with the input information. *Mechanical switches*, constructed like those in Fig. 15—2, *provide direct conversion to binary form.*

Conversion of octal quantities in logic-level format requires the use of a *gating arrangement.* Six inputs, representing octal digits 1 through 7, must be combined into three outputs, representing 2^0, 2^1, and 2^2 binary positions. Zero is recognized as the absence of all other digits. A binary to octal conversion table (Fig. 15—3a) provides sufficient information to demonstrate such a circuit (Fig. 15—3b). For example, when the input representing 5_8 is activated and goes HIGH, it will cause the 2^0 and 2^2 outputs to go HIGH. No HIGH inputs are provided to the 2^1 gate, so its output remains LOW. Thus the number represented at the outputs is 101_2, which is equivalent to 5_8. (The remaining seven combinations should be verified by the reader.)

Hexadecimal to Binary Conversion

The octal to binary converter shown in Fig. 15—3b is expandable to hexadecimal capability by furnishing both *input lines* to represent each of the sixteen hexadecimal digits and sufficient *OR gates* to represent the binary equivalents.

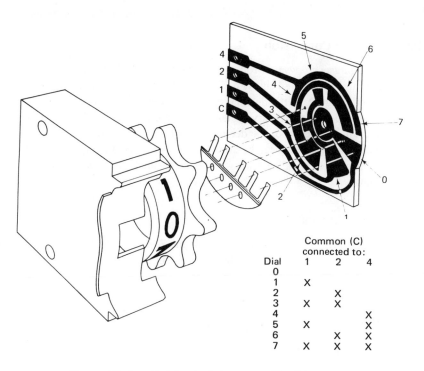

Dial	Common (C) connected to:		
	1	2	4
0			
1	X		
2		X	
3	X	X	
4			X
5	X		X
6		X	X
7	X	X	X

Figure 15—2   Octal switch (courtesy The Digitran Company, Pasadena, CA)

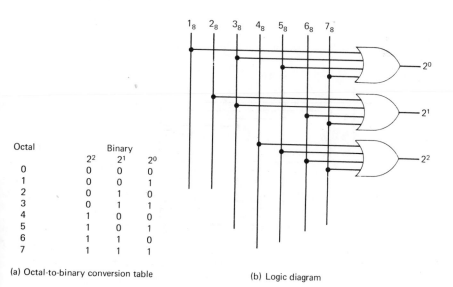

(a) Octal-to-binary conversion table

Octal	Binary		
	2^2	2^1	2^0
0	0	0	0
1	0	0	1
2	0	1	0
3	0	1	1
4	1	0	0
5	1	0	1
6	1	1	0
7	1	1	1

(b) Logic diagram

Figure 15—3   Octal-to-binary conversion

Decimal to Binary Conversion

Decimal to binary conversion follows the same general concepts of octal to binary conversion. Figure 15—4 shows decimal to binary conversion of the decimal digits 0 through 9. In the decimal system, however, there is no *direct grouping relationship* with the binary system, as there is in octal and hexadecimal representation. The full count of the decimal radix (10) is not a multiple of the binary radix (2), so no grouping can be arranged. This results in complex mechanical assemblies or electrical circuits, when large decimal numbers must be converted to binary.

Decimal to BCD Conversion

Many digital devices are designed to accept the *8—4—2—1 BCD codes* — some even perform operations with numbers directly in BCD format. The

		Binary		
Decimal	2^3	2^2	2^1	2^0
0	0	0	0	0
1	0	0	0	1
2	0	0	1	0
3	0	0	1	1
4	0	1	0	0
5	0	1	0	1
6	0	1	1	0
7	0	1	1	1
8	1	0	0	0
9	1	0	0	1

(a) Decimal-to-binary conversion table

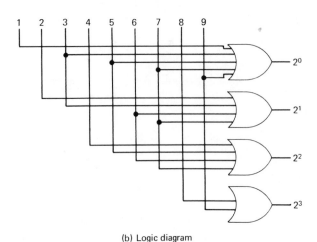

(b) Logic diagram

Figure 15—4   Decimal-to-binary conversion

unwieldy logic circuit that may result from attempting to convert directly
from decimal to binary may be simplified by using an intermediate conversion
to BCD coding. Decimal to BCD conversion is accomplished *one order at a time*
with four bits assigned to each decimal digit, as explained in Ch. 14. The encoding method shown in Fig. 15—4 is sufficient to perform logic-level conversion,
while assemblies similar to the octal encoding switch of Fig. 15—2 are capable of
performing the conversion mechanically.

BCD to Binary Conversion

Conversion from BCD to binary is not quite so easy, but it may be accomplished using a number of different techniques. Perhaps the simplest (and
possibly the slowest) method of BCD to binary conversion is the *count comparison method* shown in the block diagram in Fig. 15—5.

The BCD number to be converted is loaded into a BCD down-counter; at
the same time, a binary up-counter is reset. A Start input is provided to the clock
control, and clock pulses are applied to both the BCD down-counter and the
binary up-counter. Thus, *for each count that the BCD counter decrements, the
binary up-counter increments.* An all-zero detector recognizes when the BCD
counter reaches the zero condition, and clock pulses are terminated. The binary
up-counter now contains the binary equivalent of the BCD number originally
loaded into the BCD down-counter.

This method is quite slow, since an n-bit binary number will require 2^n
clock pulses. Assuming a 10 megahertz clock rate (10,000,000 clock pulses per
second), a count comparison method of conversion could take on the order of
3 milliseconds to convert a BCD number to its equivalent 15-bit binary number.
Table 15—1 shows the actual count sequence for conversion of the BCD number
0010 0001 to its equivalent binary form 10101_2.

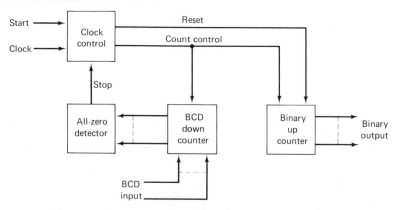

Figure 15—5   BCD-to-binary conversion — count comparison
method

Table 15−1   BCD-to-Binary Conversion − Count Comparison Method

	BCD			Binary	
21	0010	0001	Start	0	00000
20	0010	0000		1	00001
19	0001	1001		2	00010
18	0001	1000		3	00011
17	0001	0111		4	00100
16	0001	0110		5	00101
15	0001	0101		6	00110
14	0001	0100		7	00111
13	0001	0011		8	01000
12	0001	0010		9	01001
11	0001	0001		10	01010
10	0001	0000		11	01011
9	0000	1001		12	01100
8	0000	1000		13	01101
7	0000	0111		14	01110
6	0000	0110		15	01111
5	0000	0101		16	10000
4	0000	0100		17	10001
3	0000	0011		18	10010
2	0000	0010		19	10011
1	0000	0001		20	10100
0	0000	0000	Stop	21	10101

A faster method of BCD to binary conversion uses *full adders* to sum the binary expansions of all the digits in the binary number. The general concept of this type of conversion is shown in Examples 15−1 and 15−2, while implementation in block-diagram form appears in Fig. 15−6.

Example 15−1   Convert BCD number 0010 0001 to its equivalent binary number.

Solution

$$
\begin{array}{ccc}
 & \text{Tens Digit} & \text{Units Digit} \\
\text{Weight} & 80\ 40\ 20\ 10 & 8\ 4\ 2\ 1 \\
\text{BCD} & 0\ \ 0\ \ 1\ \ 0 & 0\ 0\ 0\ 1 \\
 & 1 \times 20 \quad + & 1 \times 1 \quad = 21_{10} \\
 & 10100 \quad + & 00001 \quad = 10101_2
\end{array}
$$

Example 15−2   Convert BCD number 1001 1001 to its equivalent binary number.

Solution

$$
\begin{array}{c}
1\ 0\ 0\ 1 \qquad 1\ 0\ 0\ 1 \\
1 \times 80 + 1 \times 10 \ + 1 \times 8 + 1 \times 1 = 99_{10} \\
1010000 + 1010 + 1000 + 0001 \ = 1100011_2
\end{array}
$$

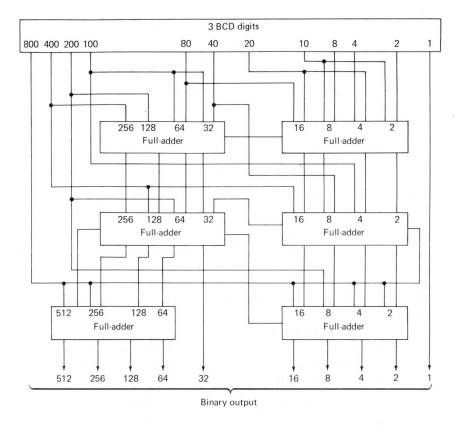

Figure 15—6   BCD-to-binary conversion — full-adder method

Binary to BCD Conversion

Following operations on the information within the digital device, the results must be made available in a form (such as decimal) that is easy to understand. Direct conversion of large binary numbers to decimal becomes just as unwieldy as did decimal to binary conversion. Again, the most common approach is to convert from binary to a BCD code and then to decimal. Both forms of BCD to binary conversion have similar methods for the reverse conversion, but they will not be discussed here.

Perhaps the simplest binary to BCD conversion method uses a *read-only memory* (Fig. 15—7). The binary number serves as the address information to obtain the corresponding BCD code. Each possible binary number to be converted must have a BCD word placed in memory. Therefore this method requires a relatively large amount of storage. For example, a code comprised of ten bits requires 1024 memory words of 13-bit length. Other approaches exist for reducing the amount of read-only memory required. One of the most

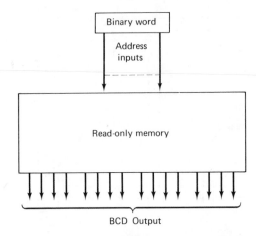

Figure 15—7   Binary-to-BCD conversion — ROM approach

popular is to divide the binary number into parts, using small ROMs to store the equivalents of the now much smaller number, and then recombine the BCD equivalents in BCD adders. This technique can reduce the original storage requirement of 13,312 bits to as little as 576 bits.

One of the most popular binary to BCD conversion methods is the *ADD 3* approach. However this type of conversion from binary to BCD requires further investigation of BCD characteristics. Perhaps some simple additions will reveal the relationships between binary, decimal, and BCD systems.

Example 15—3   Find the sum of BCD numbers 0011 and 0101.

Solution   0011   3
            +0101   +5
            ‾‾‾‾‾‾   ‾‾‾
             1000   8

The addition of 3 + 5 in binary notation yields a valid sum in *either* binary or BCD form, since the BCD form consists of a group of four binary digits.

Example 15—4   Find the sum of BCD numbers 0110 and 0110.

Solution   0110   6
            +0110   +6
            ‾‾‾‾‾‾   ‾‾‾
             1100   12

This sum is valid in binary form, but it is invalid in BCD form because, just as in the decimal system, 9 is the largest digit representable. A carry is required but not generated.

Example 15—5   Find the sum of BCD numbers 1001 and 1000.

Solution   1001   9
            +1000   +8
            ‾‾‾‾‾‾‾   ‾‾‾
            1 0001   17

Just as in Example 15–4, the sum is valid in binary form. In this case, a carry is generated for the BCD form, but the LSD (0001) is incorrect.

The problem in Examples 15–4 and 15–5 is that the decimal carry occurs at 10 and the BCD occurs at 16 – a difference of 6. If the binary equivalent of 6 is added to the invalid sum and any overflow beyond four binary digits is added to the LSD of the next BCD group to the left, the sums will be valid BCD representations. In Example 15–4, 0110 + 0110 = 1 0010, which is the BCD representation of 12_{10}. Also, in Example 15–5, 1 0001 + 0110 = 1 0111, the BCD equivalent of 17_{10}. Thus it is not pure coincidence that 6 must be added to each sum 10_{10} or greater. These examples demonstrate that such action is necessary to maintain *the decimal structure when operating in BCD notation.*

As previously shown, any position (b_x) in a binary number has a decimal equivalent 2^x (i.e., $b_4 = 2^4 = 16_{10}$). The decimal equivalent of a binary number may therefore be obtained by multiplying x times the radix (2) of the binary system by itself.

$$2^4 = 2 \times 2 \times 2 \times 2 = 16$$

The decimal equivalent of a binary number (with the decimal equivalent expressed in BCD form) is the goal of binary to BCD conversion.

Multiplication by 2 is accomplished by shifting the binary number left (toward the MSD) one position.

$$\left. \begin{array}{l} 0011 = 3 \\ 00110 = 6 \end{array} \right\} 3 \times 2 = 6 \qquad \left. \begin{array}{l} 1000 = 8 \\ {\rightarrow}10000 = 16 \end{array} \right\} 8 \times 2 = 16 \qquad \begin{array}{l} 10000 \\ \underline{0110} \\ 1\,0110 = 16_{10} \end{array}$$

Valid   Not valid
                         if ≥ 10, add 6

Note that the 10000 representation of 16_{10} is not a valid BCD number. As shown earlier, 0110 (6_{10}) must be added to the least-significant BCD group to form the proper representation. Actually, the BCD representation of 16_{10} is investigated, and if the least-significant BCD representation is ≥ 10, 6 is added. Another way to perform the same operation is to examine the BCD representation of the number before multiplication (left shift). If it is ≥ 5, add 3 and shift left. This is equivalent to checking for ≥ 10 and adding 6 *after* the left shift. The examination, ADD 3 (if necessary), and shift-left operations are continued until all binary digits are shifted. The binary number has now been multiplied by 2, the number of times required to convert it to its decimal equivalent, although the representation is in BCD form as required. Figure 15–8 shows conversion of two binary numbers to BCD form.

As Fig. 15–9 shows, this method may be implemented with *shift registers*. The binary number is inserted in parallel into a register with serial-output capability. The number is then shifted into a serial/parallel register in a serial manner. Examination of the three most-significant bits is accomplished with a combinational logic circuit. If these bits represent a number ≥ 5, the combinational circuit adds 3 and feeds the modified number back into the serial/parallel register via the parallel inputs. Following the next shift operation, the number is

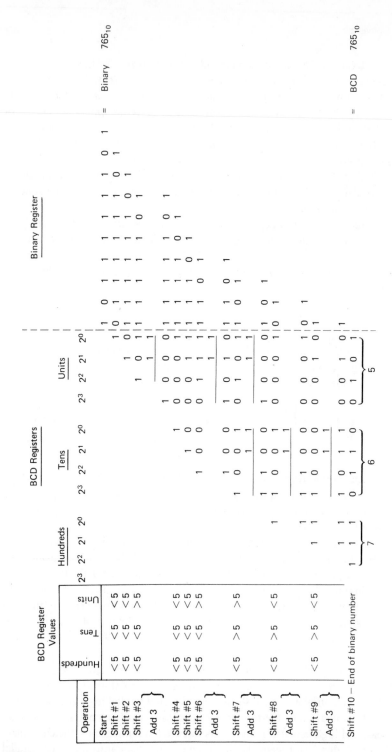

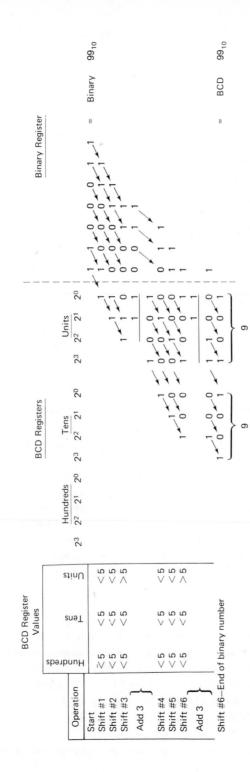

Figure 15–8 Binary-to-BCD conversion – ADD 3 method

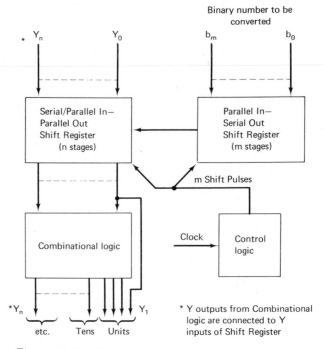

Figure 15—9   Implementation of "ADD 3" method

again examined, 3 is added if necessary, and another shift occurs. The control logic determines when to remove shift pulses, based on the total number of bits in the binary number to be converted.

The ADD 3 method may also be implemented directly with *gates* and/or *ROMs.* Such an approach results in a decreased conversion time but in an increased number of logic elements. Typical speeds for a 15-bit binary word and a 10 megahertz clock rate are 1500 nanoseconds for the sequential shift approach and 540 nanoseconds for the gate approach. Note that both values are well *below* the time required for the count comparison method of conversion.

BCD to Decimal Conversion

The remaining step in the sequence of decimal-in to decimal-out conversion is that from *BCD to decimal.* If the output device displays each of the decimal digits separately, then a means to convert from BCD to a 1-of-10 output is needed. The actual display device* is unimportant from a logic standpoint, just so only *one* of the decimal digits is displayed at a time. As usual, conventional 8—4—2—1 BCD is assumed.

-----

* Different display devices are discussed in subsequent sections of this chapter.

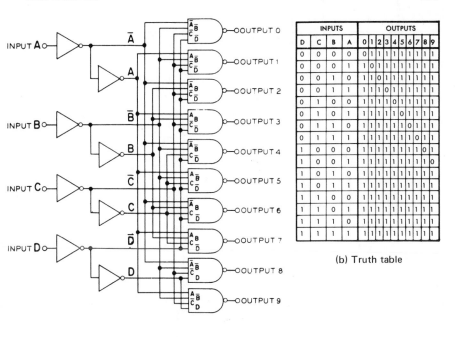

INPUTS				OUTPUTS									
D	C	B	A	0	1	2	3	4	5	6	7	8	9
0	0	0	0	0	1	1	1	1	1	1	1	1	1
0	0	0	1	1	0	1	1	1	1	1	1	1	1
0	0	1	0	1	1	0	1	1	1	1	1	1	1
0	0	1	1	1	1	1	0	1	1	1	1	1	1
0	1	0	0	1	1	1	1	0	1	1	1	1	1
0	1	0	1	1	1	1	1	1	0	1	1	1	1
0	1	1	0	1	1	1	1	1	1	0	1	1	1
0	1	1	1	1	1	1	1	1	1	1	0	1	1
1	0	0	0	1	1	1	1	1	1	1	1	0	1
1	0	0	1	1	1	1	1	1	1	1	1	1	0
1	0	1	0	1	1	1	1	1	1	1	1	1	1
1	0	1	1	1	1	1	1	1	1	1	1	1	1
1	1	0	0	1	1	1	1	1	1	1	1	1	1
1	1	0	1	1	1	1	1	1	1	1	1	1	1
1	1	1	0	1	1	1	1	1	1	1	1	1	1
1	1	1	1	1	1	1	1	1	1	1	1	1	1

(b) Truth table

(a) Logic diagram

Figure 15–10 BCD-to-decimal decoder (Type 7445A) (courtesy Sprague Electric Company, Worcester, Mass.)

Figure 15–10 shows the general approach to 8–4–2–1 BCD to decimal decoding. Investigation of the logic diagram of the 7445A and its truth table shows that the circuit meets the requirements of the preceding paragraph. In addition, all outputs are off for erroneous or not-allowed input combinations.

An increasingly popular display is the *7-segment indicator,* discussed in Sec. 15–2. And, as in more and more digital equipment, the familiar 8–4–2–1 code is used to drive this device. The 8T05 7-segment decoder, which consists of the necessary logic to decode a 4-bit BCD code to 7-segment readout, is shown in Fig. 15–11. To provide an easily readable display, a *ripple-blanking input* is furnished to implement suppression of leading and/or trailing zeros. A lamp-test input is also provided, which forces all segment outputs HIGH. This allows the viewer to check the validity of the display presentation by testing the integrity of the display.

BCD codes other than the 8–4–2–1 version find applications in digital devices. For example, the *XS3 code* is often used when arithmetic operations must be performed. Its self-complementing characteristics greatly simplify some types of arithmetic operations. The XS3 code may be easily decoded to decimal, as shown in Fig. 15–12. No unusual circuitry is employed, and the operation is straightforward.

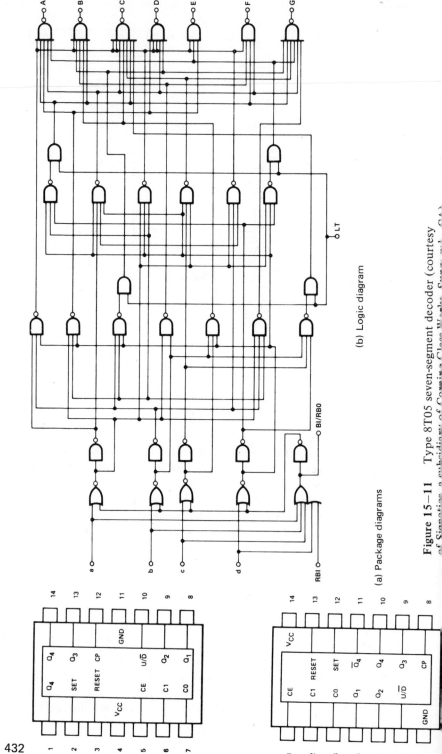

432

(a) Package diagrams

(b) Logic diagram

Figure 15-11 Type 8T05 seven-segment decoder (courtesy of Signetics, a subsidiary of Corning Glass Works, Sunnyvale, CA.)

| | INPUTS | | | | | | OUTPUTS | | | | | | | |
| INPUT CODE | | | | LAMP TEST | RBI | BI/RBO | OUTPUT STATE | | | | | | | DISPLAY CHARACTER |
d	c	b	a	LT			A	B	C	D	E	F	G	
X	X	X	X	0	X	X (Note)	1	1	1	1	1	1	1	8
X	X	X	X	1	X	0 (Note 1 & 2)	0	0	0	0	0	0	0	BLK
0	0	0	0	1	0	0 (Note 2)	0	0	0	0	0	0	0	BLK
0	0	0	0	1	1	1	1	1	1	1	1	1	0	0
0	0	0	1	1	X	1	0	1	1	0	0	0	0	1
0	0	1	0	1	X	1	1	1	0	1	1	0	1	2
0	0	1	1	1	X	1	1	1	1	1	0	0	1	3
0	1	0	0	1	X	1	0	1	1	0	0	1	1	4
0	1	0	1	1	X	1	1	0	1	1	0	1	1	5
0	1	1	0	1	X	1	0	0	1	1	1	1	1	6
0	1	1	1	1	X	1	1	1	1	0	0	0	0	7
1	0	0	0	1	X	1	1	1	1	1	1	1	1	8
1	0	0	1	1	X	1	1	1	1	1	0	1	1	9
1	0	1	0	1	X	1	0	0	0	1	1	0	1	
1	0	1	1	1	X	1	0	0	1	1	0	0	1	
1	1	0	0	1	X	1	0	1	0	0	0	1	1	
1	1	0	1	1	X	1	1	0	0	1	0	1	1	BLK
1	1	1	0	1	X	1	0	0	0	1	1	1	1	L
1	1	1	1	1	X	1	0	0	0	0	0	0	0	BLK

*COMMA

X = Don't care, either "1" or "0".

BI/RBO is an internally wired OR output.

NOTE:
1. BI/RBO used as input.
2. BI/RBO should not be forced high when a, b, c, d, RBI terminals are low, or damage may occur to the unit.

(c) Truth table

Figure 15–11 (cont'd)

433

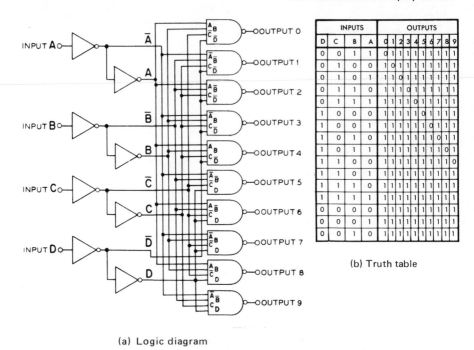

INPUTS				OUTPUTS									
D	C	B	A	0	1	2	3	4	5	6	7	8	9
0	0	1	1	0	1	1	1	1	1	1	1	1	1
0	1	0	0	1	0	1	1	1	1	1	1	1	1
0	1	0	1	1	1	0	1	1	1	1	1	1	1
0	1	1	0	1	1	1	0	1	1	1	1	1	1
0	1	1	1	1	1	1	1	0	1	1	1	1	1
1	0	0	0	1	1	1	1	1	0	1	1	1	1
1	0	0	1	1	1	1	1	1	1	0	1	1	1
1	0	1	0	1	1	1	1	1	1	1	0	1	1
1	0	1	1	1	1	1	1	1	1	1	1	0	1
1	1	0	0	1	1	1	1	1	1	1	1	1	0
1	1	0	1	1	1	1	1	1	1	1	1	1	1
1	1	1	0	1	1	1	1	1	1	1	1	1	1
1	1	1	1	1	1	1	1	1	1	1	1	1	1
0	0	0	0	1	1	1	1	1	1	1	1	1	1
0	0	0	1	1	1	1	1	1	1	1	1	1	1
0	0	1	0	1	1	1	1	1	1	1	1	1	1

(b) Truth table

(a) Logic diagram

Figure 15–12   Excess 3-to-decimal decoder (Type 7443A)
(courtesy Sprague Electric Company, Worcester, Mass.)

Binary to Octal Conversion

If the output is to be in octal form, a *binary to octal converter* is required.
Such a decoder has three lines of binary input, representing the 2^0, 2^1, and 2^2
positions. Eight outputs are provided, only one of which is activated at a time.
A typical prepackaged binary to octal decoder (type 8250) is shown in
Fig. 15–13. The fourth input line (D) is used as an inhibit to allow use in larger
decoding networks. Note that HIGH inputs are required, while the selected output is LOW when activated.

Binary to Hexadecimal Conversion

Hexadecimal conversion of a binary number is accomplished with a logic
circuit *similar* to the binary to octal decoder. In the hexadecimal case, however,
four inputs are provided, and only one of sixteen outputs is activated at a time.
The 8250 one-of-eight decoder may be used, as shown in Fig. 15–14, to provide
the one-of-sixteen operation. Such a scheme is expandable in groups of eight bits
to provide up to a one-of-sixty-four capability.

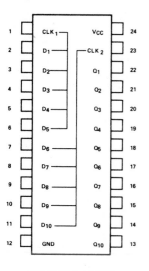

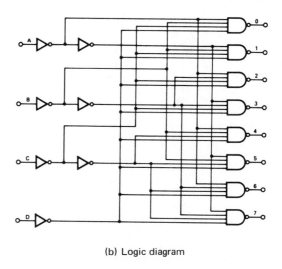

(b) Logic diagram

P,N,Y PACKAGES

(a) Package diagram

| | INPUT STATE | | | | | | OUTPUT STATES | | | | | |
| | | | | 8250 | | | | | | | | |
A	B	C	D	0	1	2	3	4	5	6	7
0	0	0	0	0	1	1	1	1	1	1	1
1	0	0	0	1	0	1	1	1	1	1	1
0	1	0	0	1	1	0	1	1	1	1	1
1	1	0	0	1	1	1	0	1	1	1	1
0	0	1	0	1	1	1	1	0	1	1	1
1	0	1	0	1	1	1	1	1	0	1	1
0	1	1	0	1	1	1	1	1	1	0	1
1	1	1	0	1	1	1	1	1	1	1	0
0	0	0	1	1	1	1	1	1	1	1	1
1	0	0	1	1	1	1	1	1	1	1	1
0	1	0	1	1	1	1	1	1	1	1	1
1	1	0	1	1	1	1	1	1	1	1	1
0	0	1	1	1	1	1	1	1	1	1	1
1	0	1	1	1	1	1	1	1	1	1	1
0	1	1	1	1	1	1	1	1	1	1	1
1	1	1	1	1	1	1	1	1	1	1	1

(c) Truth table

Figure 15–13   Binary-to-octal decoder (Type 8250) (courtesy of Signetics, a subsidiary of Corning Glass Works, Sunnyvale, CA)

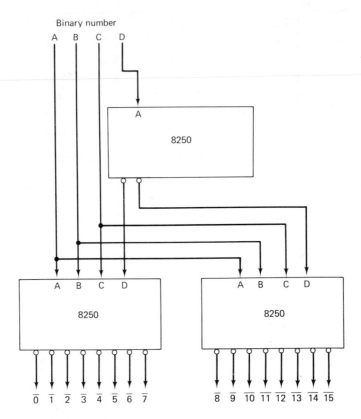

Figure 15—14 Binary-to-hexadecimal decoder using Type 8250 decoders

Conversion Between BCD Codes

At times, it may be necessary to convert from one type of BCD code, such as XS3, to the common 8—4—2—1 code. An interesting application of *logic maps* may then be implemented. The truth tables for both the XS3 and 8—4—2—1 representations are shown correlated with their decimal equivalents in Table 15—2. Examination of each of the B columns supplies the following equations.

$$B_8 = E_4\bar{E}_3 E_2 E_1 + E_4 E_3 \bar{E}_2 \bar{E}_1 \qquad \text{Eq. (15–1)}$$

$$B_4 = \bar{E}_4 E_3 E_2 E_1 + E_4 \bar{E}_3 \bar{E}_2 \bar{E}_1 + E_4 \bar{E}_3 \bar{E}_2 E_1 + E_4 \bar{E}_3 E_2 \bar{E}_1 \quad \text{Eq. (15–2)}$$

$$B_2 = \bar{E}_4 E_3 \bar{E}_2 E_1 + \bar{E}_4 E_3 E_2 \bar{E}_1 + E_4 \bar{E}_3 \bar{E}_2 E_1 + E_4 \bar{E}_3 E_2 \bar{E}_1$$

$$= \bar{E}_2 E_1 + E_2 \bar{E}_1 \qquad \text{Eq. (15–3)}$$

$$B_1 = \bar{E}_1 \quad \text{(Inspection)} \qquad \text{Eq. (15–4)}$$

$$= \bar{E}_4 E_3 \bar{E}_2 \bar{E}_1 + \bar{E}_4 E_3 E_2 \bar{E}_1 + E_4 \bar{E}_3 \bar{E}_2 \bar{E}_1 + E_4 \bar{E}_3 E_2 \bar{E}_1$$

A technique often used with logic maps is the plotting not only of the terms of the equations but also of all terms that can exist with the number of variables *not* included in the truth table. These are called "don't cares," and they often result in maps with greatly simplified terms. An X is used to plot a "don't care" term. The map plots of Eqs. (15–1)–(15–4) appear in Fig. 15–15, and one implementation possibility appears in Fig. 15–16.

Once again, the basic ideas of logic expressions, truth tables, and logic maps have been used to simplify the explanation of a relatively complex logic operation.

Table 15-2   Excess 3 and 8-4-2-1 Truth Tables

| | BCD Input | | | | | 8-4-2-1 Output | | |
E_4	E_3	E_2	E_1		B_8	B_4	B_2	B_1
0	0	1	1	0	0	0	0	0
0	1	0	0	1	0	0	0	1
0	1	0	1	2	0	0	1	0
0	1	1	0	3	0	0	1	1
0	1	1	1	4	0	1	0	1
1	0	0	0	5	0	1	0	1
1	0	0	1	6	0	1	1	0
1	0	1	0	7	0	1	1	1
1	0	1	1	8	1	0	0	0
1	1	0	0	9	1	0	0	1

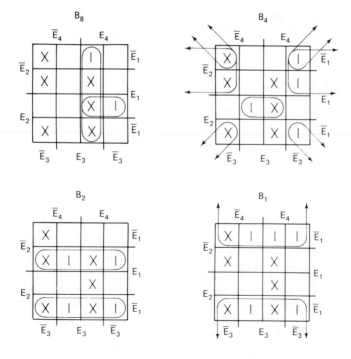

Figure 15–15   Logic map for Table 15–2

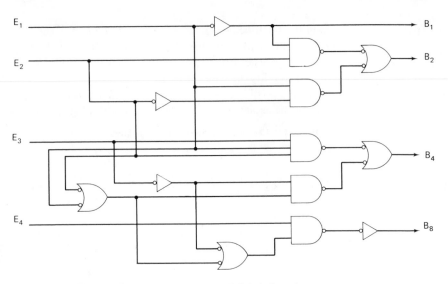

Figure 15—16   Excess 3 to 8-4-2-1 decoder

ASCII to EBCDIC Conversion

An *incompatability* exists between many digital computers and the data-transmission systems seeking to supply inputs to the computer. The computer may use the EBCDIC code, while the data-transmission system uses the ASCII code. However conversion from ASCII to EBCDIC is easily performed using the *type 8224 256-bit read-only memory* (see Fig. 15—17). This device has been programmed to convert the 7-bit ASCII alphabet code to the 8-bit EBCDIC alphabet code. The conversion includes the letters A through Z. Additional gating is required for this ROM to convert both upper and lowercase letters.

15—2 DISPLAY DEVICES

Cold-Cathode, Direct-Character Display

The "old standby" in digital display devices is the *NIXIE* tube* (Fig. 15—18). It is classified as a cold-cathode, gas-filled glow tube. In simple terms, it may be compared to the neon bulbs seen in small orange-colored nightlights. Each of the "neon bulbs" is shaped like one of the ten decimal digits, and when properly activated, only one character glows at a time. The individual characters are so arranged inside the tube that there is a minimum of masking of characters which are in the rear. When one character is illuminated, it is difficult to see any of the nonilluminated numbers.

---

* Registered trademark of The Burroughs Corporation.

	UPPER CASE	LOWER CASE
	0 0	1 1
	0 1	1 0
	1 1	0 0
	1 1	0 0

ASCII
ASCII
CHIP SELECT $\overline{CS} = \overline{b_7}$
EBCDIC #1 OUTPUT $= b_6 \cdot b_7$

To select the ROM only when addressed by an upper or lower case alphabet character, the above truth table applies. Thus, the ASCII to EBCDIC ROM standard product, plus gating as shown, performs the complete conversion.

Applications

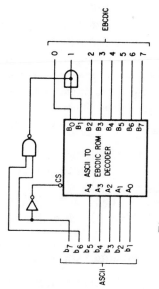

Figure 15–17 ASCII to EBCDIC decoder (courtesy of Signetics, a subsidiary of Corning Glass Works, Sunnyvale, CA)

439

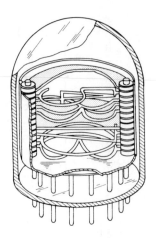

Figure 15—18   NIXIE tube (courtesy Burroughs Corp. Electronic Components Div., Plainfield, NJ)

The major advantages of the NIXIE tube include *the capability to quickly change from one character to another*, *long life*, and *relatively low power consumption.* High voltage is required to operate the NIXIE tube, but recently developed decoder/driver combinations perform the job quite satisfactorily. If there are any major disadvantages, they would fall in the area of decoding complexity. Regardless of the code used within the display device employing NIXIE tubes, it must be converted to a 1-of-10 output. Such a decoding requirement usually results in additional gate requirements.

Assuming BCD number representation, a conventional BCD-to-decimal decoder may be used with supplemental *driver stages*, which are capable of handling the high voltages necessary to operate the NIXIE tubes. A commercially available IC that contains both decoder and driver circuitry (type 7441A) is shown in Fig. 15—19. The symbols between the output gates and the output terminals represent the driver transistors just mentioned. This decoder/driver responds accurately only to the ten combinations shown in the truth table. The six unused combinations in the 8—4—2—1 code, if allowed to occur, will result in more than one output being on at the same time. These combinations should be avoided.

NIXIE tubes are used in many digital applications where *numeric display* is required. Computer displays, automatic test equipment, digital clocks, and even everyday electronic test equipment use cold-cathode display tubes to indicate system conditions. Despite the many advances in display devices, the NIXIE tube will continue to serve a useful purpose for many years.

Segmented Displays

It is unnecessary to use ten separate elements to display the ten decimal digits. A close approximation of digits 0 through 9 may be had by breaking the

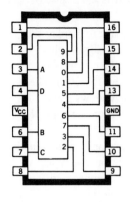

(a) Package diagram

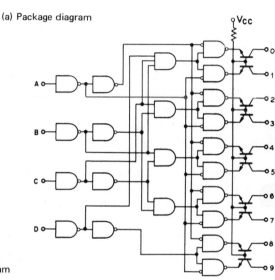

(b) Logic diagram

BCD INPUT				OUTPUT ON*
D	C	B	A	
0	0	0	0	0
0	0	0	1	1
0	0	1	0	2
0	0	1	1	3
0	1	0	0	4
0	1	0	1	5
0	1	1	0	6
0	1	1	1	7
1	0	0	0	8
1	0	0	1	9

*All other outputs are off.

The decoding inputs are coincidental with the BCD
outputs of the US5490A and US7490A Decade
Counters.

(c) Truth table

Figure 15–19   NIXIE decoder (Type 7441A) (courtesy
Sprague Electric Company, Worcester, MA.)

441

digits into *segments* and activating only the segments necessary to display the digit. Seven segments quite adequately synthesize the decimal digits (plus fourteen alphabetic characters), as shown in Fig. 15–20a. When complete alphameric display is required, sixteen segments are used (Fig. 15–20b). The 16-segment display is capable of over 65,000 patterns, which allows great versatility in information display. The physical configuration of one type of segmented display unit is shown in Fig. 15–21a.

The *7-segment display device* is used as the primary example in the remainder of this chapter, since this unit is the least complex of the various types. Illumination of the individual segments is accomplished in a number of different ways. Directly viewed, incandescent filaments are used in the unit in Fig. 15–21a. A small incandescent lamp may be placed at the source end of a "light pipe" with the viewing surface of the display at the receiving end (Fig. 15–21b). Vacuum fluorescent devices (Fig. 15–21c) use individually activated fluorescent segments to form the desired characters. Solid-state, light-emitting diodes (discussed previously) using reflection of the multidirection light produce an easily viewable numeric display. Important characteristics of these and other types of display are summarized in Table 15–3.

Seven-segment displays are activated by decoders similar to the 8T05 discussed in Sec. 15–1. Incandescent and solid-state units generally may be directly operated from this type of decoder. Displays such as the vacuum-fluorescent indicators require isolation amplifiers to safely handle the higher voltages.

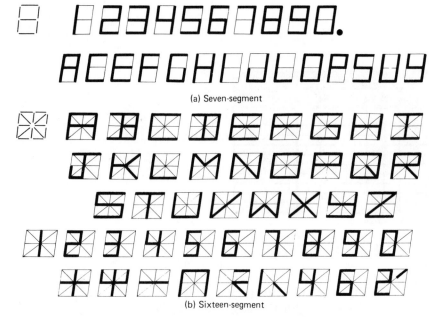

(a) Seven-segment

(b) Sixteen-segment

Figure 15–20   Segmented display capabilities

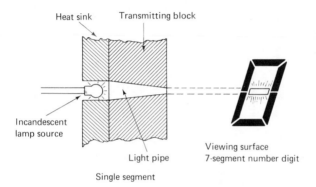

Heat sink   Transmitting block

Incandescent
lamp source

Light pipe

Single segment

Viewing surface
7-segment number digit

Figure 15–21   Segmented displays; (top) incandescent fila-
ment (courtesy Pinlites. Inc., Caldwell, N.J.), (middle) The
"Light Pipe" concept (courtesy Tung-Sol Division, Wagner
Electric Corp., Newark, N.J.); (bottom) vacuum fluorescent
segments (courtesy Tung-Sol Division, Wagner Electric Corp.,
Newark, NJ.)

Table 15-3   Characteristics of common display devices

| | Cold-Cathode, Gas-Filled Displays | | Segmented | | |
	Single Character	Multi-character	Incandescent Wire	Vacuum Fluorescent	Incandescent Lamp
Information displayed	Alphanumeric	Alphanumeric	Alphanumeric	Numeric (Some Alpha)	Alphanumeric
Symbols/Unit displayed simultaneously	1	Up to 256 and growing	1	1	1
Character height	0.3" to 2.5"	0.2" to 0.5" and growing	0.25" to 0.6"	0.3" to 0.6"	0.2" to 0.6"
Operating parameters	170-250 V DC	170-250 V DC	3-5 V DC	20-50 V DC plus filament	3-28 V DC
Brightness	Approx. 200 fL	Approx. 50 fL	1000 fL and easily variable	150-200 fL and easily variable	1000 fL and easily variable
Life (hours)	Approximately 100,000	Same as single character	Up to 100,000	Approximately 10,000	Up to 100,000
Cost range	$2-$30	$18-$1000	$3-$40	$2 upward	$3 upward

Table 15-3 (cont'd)

	Solid State		Miscellaneous		
	Segmented	Dot Matrix	Liquid Crystal	Projected Image	Cathode Ray Tube
Information displayed	Alphanumeric	Alphanumeric	Alphanumeric	Alphanumeric and Special Messages	Alphanumeric + Special Messages
Symbols/Unit displayed simultaneously	1	1	1	1	Unlimited
Character height	0.2" to 0.625"	0.1" to 0.8"	0.4" to 0.7" and growing	0.5" to 3.75"	Variable
Operating parameters	3-5 V DC	3-5 V DC	6-60 V AC	3-28 V DC	1000 V DC
Brightness	100-500 fL	100-500 fL	Depends on ambient light	20-300 fL easily variable	100-300 fL easily variable
Life (hours)	Indefinite	Indefinite	10,000	Approx. 1000 (Lamp life)	Variable
Cost range	$5 upward	$7 upward	$15 upward	$14 upward	$1500 upward

Solid-State Displays

In ever-increasing applications, *solid-state display devices* are beginning to dominate the information-display field. Whether in segmented or dot-matrix form, the basic concept of operation is the same. Semiconductor diodes made of certain materials such as gallium-arsenide-phosphide (only one of many) possess the property of *emitting light when proper potentials are applied.*

For numeric display only, twenty-seven of the light-emitting diodes are arranged in a (5 x 7) rectangular matrix (not all of the thirty-five matrix locations are occupied by diodes); a twenty-eighth diode is offset at the lower left to serve as a decimal point. Figure 15—22a shows a typical solid-state numeric indicator,

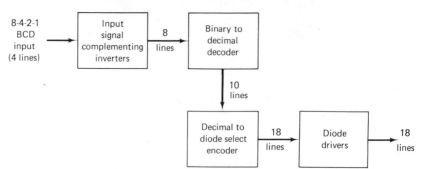

Figure 15—22 Solid-state numeric indicator diagram; (a) LED chip and decoder; (b) block diagram (courtesy Hewlett-Packard Company, Palo Alto, CA)

which includes decoding capability. The block diagram for the complete assembly appears in Fig. 15–22b. BCD input signals are decoded on the IC chip into ten signals, representing digits 0 through 9. These signals are then encoded into signals that drive an encoder that selects the proper combination of light-emitting diodes to display the required digit. The decimal point is activated by a separate input.

When complete alphameric display is required, a full (5 x 7) matrix of light-emitting diodes (LEDs) may be employed (Fig. 15–23a). The (5 x 7) matrix of LEDs, which make up each character, are X–Y addressable. This allows for a simple addressing, decoding, and driving scheme between the display module and the system logic.

To form alphameric characters, a method called *scanning* is used. Information is addressed to the display by selecting one row of LEDs at a time, energizing the appropriate diodes in that row, and then proceeding to the next row. After all rows have been excited (one at a time), the process is repeated. By

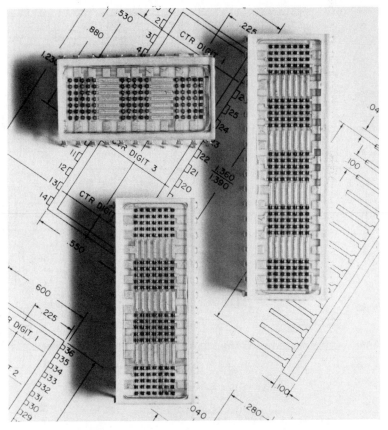

Figure 15–23   Alphameric LED display operation (courtesy Hewlett-Packard Company, Palo Alto, CA)

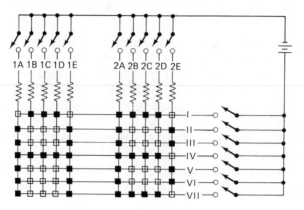

(b) Formation of letters A and B with vertical scanning

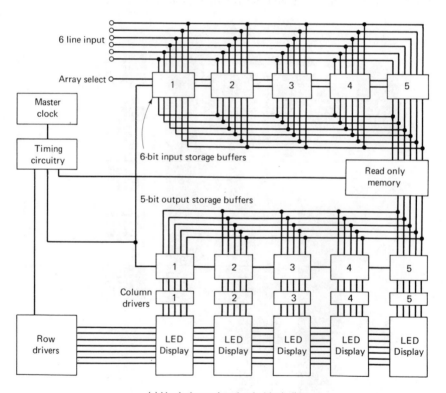

(c) Vertical scanning circuit, block diagram

Figure 15–23 (cont'd)

scanning through all rows at least 100 times a second, a flicker-free character composed of discrete illuminated LEDs can be produced. Information may be moved from row to row of the display (*vertical scanning*) or from left to right (*horizontal scanning*).

Figure 15–23b indicates how (with vertical scanning) letters A and B would be formed by sequentially selecting the rows and energizing the correct diodes in each column. When row I is selected only columns 1B, 1C, 1D, and 2A, 2B, 2C, and 2D are energized. When selecting row II, columns 1A, 1E, and 2A, 2E are energized; the process is continued, as indicated by the solid squares in Fig. 15–23b, until all appropriate LEDs have been lighted. The cycle is repeated at a high rate, so the eye sees a continuous *flicker-free character.*

A typical vertical scanning circuit for addressing the display is shown in the block diagram in Fig. 15–23c. This particular scheme contains five display characters. Operation for other numbers of characters is similar. Circuit operation is as follows.

1. Coded 6-bit alphameric information is sequentially entered and stored in five 6-bit input storage buffers.

2. Next, with the input information stored in the input buffers, timing circuitry enables the ROM and input storage buffer number one, so its stored 6-bit code can be read by the ROM. (All other input storage buffers are disabled.)

3. The 6-bit input is decoded by the ROM, and the first row of character information is stored in the 5-bit output storage buffer number one. In other words, if the character A is being written, diodes 1B, 1C, and 1D of row I will be lighted (see Figure 15–23b); and this information is stored in the first output storage buffer.

4. Having completed the first loading operation, the timing circuitry now activates input storage buffer number two. (All other input storage buffers are disabled.) The coded character is read into the ROM, decoded, and the row information fed into output storage buffer number two. This operation is repeated until all five characters stored in the input storage buffers are sensed by the ROM character generator and the first line of character information is stored in the output storage buffers.

5. Next, the timing circuitry connects the top row driver, so current flows through and lights up all of the appropriate LEDs in the top rows of the five display characters.

6. The complete cycle is then repeated to decode and display row II of the LED matrix characters, then row III, on through row VII. Since the time to decode and load the output storage buffers is short in comparison with the time the display is lit, repeating the character scanning at a rate above 100 times/second gives a flicker-free alphameric character.

Other Displays

We have discussed three display devices and their methods of operation in some detail. The extensive coverage given to these particular devices is because they are presently the ones most commonly used. Since, however, display techniques are constantly changing, we should at least discuss the general characteristics of a number of display devices currently under study and in rather limited use at this time. One — or perhaps all — of these new methods may become *the* display device of tomorrow.

PLASMA PANELS

Cold-cathode, gas-filled display techniques have been modernized, and these improved techniques have resulted in some very versatile display devices. *Segmented displays*, using 7, 9, 13, and 15 bars, have been produced to meet the requirements for alphameric display. *Multicharacter assemblies*, using gas discharge techniques, have also been developed and are being used.

One of the most interesting gas-discharge displays is the so-called *plasma panel*. Characters are displayed in a dot matrix form, one of which is the familiar (5 x 7) form discussed earlier. The physical construction of one version of this type of display (SELF-SCAN* panel) is shown in Fig. 15–24a.

Although the following explanation is grossly simplified, enough of the general concepts are presented to determine its basic principles of operation. When the panel is energized, sufficient potential difference is established between the keep-alive cathode and the keep-alive anode to establish a glow discharge in the keep-alive grooves located in the rear glass cover. The physical construction of the keep-alive cathode and anode are such that charged particles can diffuse into the area of the reset cathode.

The scan anodes are maintained at a high dc voltage, and when the reset cathode is grounded, complete ionization of the partially ionized gas that has diffused from the keep-alive area occurs. A glow is then established in the seven rectangular areas of this single cathode, defined by the intersection of the rear of the cathode strip and the seven scan grooves in the rear glass cover. (This area is not visible from the front of the panel.) The first character-forming cathode is then grounded, immediately following removal of the ground from the reset cathode, and the glow is transferred. Each character-forming cathode has seven tiny holes (.030 inch in diameter) called *glow-priming apertures*. (These holes are small enough that the glow from the rear of the character-forming cathode does not show through, yet charged particles can diffuse through them into the display cavities.) The ground is removed from the first character-forming cathode and transferred to the second cathode. Glow discharge is transferred in this manner until the end of the display is reached. The reset cathode may then be grounded again, and the scan process started over. Thus *the glow discharge is*

---

* Registered trademark of The Burroughs Corporation.

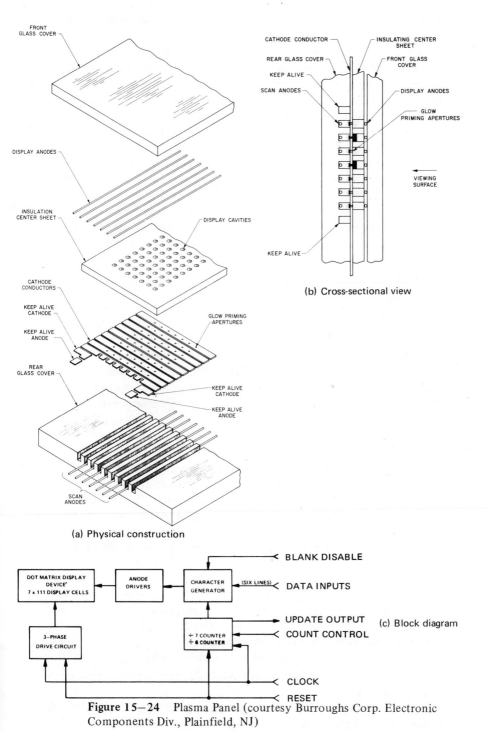

FRONT GLASS COVER

DISPLAY ANODES

INSULATION CENTER SHEET

CATHODE CONDUCTORS

KEEP ALIVE CATHODE

KEEP ALIVE ANODE

REAR GLASS COVER

SCAN ANODES

DISPLAY CAVITIES

GLOW PRIMING APERTURES

KEEP ALIVE CATHODE

KEEP ALIVE ANODE

(a) Physical construction

CATHODE CONDUCTOR — INSULATING CENTER SHEET

REAR GLASS COVER — FRONT GLASS COVER

KEEP ALIVE

SCAN ANODES — DISPLAY ANODES

GLOW PRIMING APERTURES

VIEWING SURFACE

KEEP ALIVE

(b) Cross-sectional view

BLANK DISABLE

| DOT MATRIX DISPLAY DEVICE 7 × 111 DISPLAY CELLS | ← | ANODE DRIVERS | ← | CHARACTER GENERATOR | (SIX LINES) DATA INPUTS |

3-PHASE DRIVE CIRCUIT

÷ 7 COUNTER
÷ 6 COUNTER

UPDATE OUTPUT (c) Block diagram
COUNT CONTROL

CLOCK
RESET

Figure 15—24   Plasma Panel (courtesy Burroughs Corp. Electronic Components Div., Plainfield, NJ)

451

started at the left and transferred one cathode at a time along the entire length of the panel. Actually, the logic circuitry needed to perform the scan may be simplified by using a "3-phase" drive circuit and grounding every fourth cathode. Glow transfer occurs only between adjacent cathodes, so this technique still allows transfer of only one cathode at a time.

As the glow is scanned down the panel, a phenomenon called *glow-priming* occurs. Figure 15–24b shows a cross-sectional view of a panel, depicting one cathode with a glow discharge on the rear (scan side). The scan glow covers the rectangular portion on the scan side of the cathode located beneath each display cavity. The priming apertures have allowed the partially ionized gas to diffuse into the display aperture. If any of the display anodes are placed at a sufficiently high dc voltage, the partially ionized gas will become completely ionized, and a glow will occur in the display cavity. Characters are written on the viewing side of the panel by addressing the display anodes in synchronism with the glow-priming ionization present on the scan side of the cathode that intersects the point where the dot is to appear.

Figure 15–24c is a block diagram showing the general concept of "writing" information into the display panel. The clock input, which occurs at a frequency that allows the overall panel to be scanned approximately 60 times per second, feeds the 3-phase drive circuit to sequentially ground the cathodes that form the characters. A reset input is also provided to return the panel scan to the left-hand side. The character generator converts six bits of primary information and a clock pulse into a dot matrix character format made available column by column, seven bits in parallel. A counter section, using the count control and clock inputs, establishes the spacing between characters. The actual character information is provided by a read-only memory. Anode drivers are provided to interface between the logic level of the character generator and the high-voltage dc levels of the display panel. If information is to remain on the panel for display, external circuits must be provided to refresh the input to the character generator.

Panels using this or similar techniques may be used to display as few as 1 and as many as 256 characters. Display size of characters ranges from 0.25 to 0.5 inch in height. The future will surely see these maximums extended.

LIQUID-CRYSTAL DEVICES

A device called a *liquid-crystal readout* shows great promise in character display applications. The basic principle behind the liquid-crystal readout is simple. It consists of a thin layer of transparent liquid crystal placed between two sheets of glass, which are coated with an electrical conductor. The front electrode is transparent. The rear electrode is shaped in segments to form the desired character.

The display may be operated in either a reflective or a transmissive manner. In *transmissive displays, the rear electrode is transparent*, while in *the reflective unit, the rear electrode is opaque*, having a mirror-surfaced film. When voltage is applied between electrodes of either type, the liquid becomes turbulent and scatters ambient light in the form of the rear-electrode pattern. A source of local light or sufficient ambient light must be present for this type of display to operate. Figure 15—25 shows the general concept of liquid-crystal display devices.

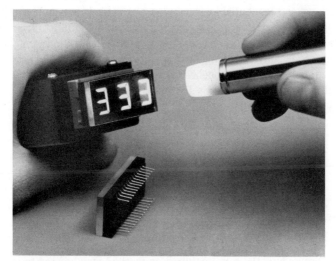

Figure 15—25 A liquid crystal display (courtesy Optel Corporation, Princeton, NJ)

MISCELLANEOUS DEVICES

Two other methods of display, *projected-image* and *edge-lighted panels*, should be mentioned in passing. Both are still in use but are being phased out by the newer, more efficient displays. The projected-image display uses a separate light source and image for each character. The image is usually on film, and activation of the light source causes an enlarged image of the filmed character to appear at the face of the device. Edge-lighted panels have solid characters engraved on acrylic plastic panels, and illumination of the character is accomplished by activating a separate lamp for each panel. Both methods suffer from the decoding complexities and lack of versatility that plague the NIXIE tube, although the relatively low-operating voltage requirements tend to make them more applicable to standard IC decoder/driver packages.

Cathode-ray tubes (CRTs) should not be overlooked as potential display devices. In conjunction with logic circuits called *character generators*, they are finding widespread use as computer terminals and in other data display applications. No limit is placed on the types of characters that may be used, and CRTs promise to fulfill the need of displaying large amounts of data in a relatively small space. The only real disadvantage is cost. CRT displays are perhaps the costliest — although the most versatile — display devices in use today.

15—3 SUMMARY

With the completion of this chapter, the fundamental knowledge required to understand digital logic devices has been presented. The basic concepts of combinational and sequential logic have been expanded to include many of the prepackaged logic functions that are encountered in modern digital devices. Methods of furnishing inputs to digital devices, codes used, timing and storing of digital data, code conversion, logic and arithmetic operations, and presentation of information have all been touched upon. The reader is now prepared to enter the world of digital devices and try his newly acquired skills.

PROBLEM APPLICATIONS

15—1. How many gates are required to convert a 3-digit octal number to binary form? Draw a logic diagram that will perform this operation.

15—2. How many gates are required to convert a 2-digit hexadecimal number to binary form? Draw a logic diagram that will perform this operation.

15—3. Why would it be impractical to convert large decimal numbers directly to binary form?

15—4. Discuss the count comparison method of BCD to binary conversion, noting its advantages and disadvantages.

15—5. Discuss how the count comparison method may be used to convert a binary number to BCD form.

15—6. What are the advantages of the full-adder method of BCD to binary conversion when compared with the count-comparison method?

15—7. Can the full-adder method be used to convert from binary to BCD? Discuss how it must be done.

15—8. Why is the ROM method often used to convert from binary to BCD?

15—9. What are the basic principles of the ADD 3 method of binary to BCD conversion?

15—10. Is the ADD 3 method of binary to BCD conversion applicable to BCD to binary conversion? If so, how is it performed?

15—11. Explain how the 7445A BCD to decimal decoder in Fig. 15—10 guards against improper outputs when erroneous binary input conditions exist.

15—12. What is the purpose of the D input of the 8250 decoder in Fig. 15—13?

15—13. Draw a block diagram showing a 1-out-of-32 decoder, using the 8250 decoder in Fig. 15—13.

15—14. Construct a logic diagram of a converter that will accept 8—4—2—1 input and provide XS3 output. Use the method described in Sec. 15—1.

15—15. Discuss how the method in Sec. 15—1 could be expanded to encompass numbers and special characters when converting from ASCII to EBCDIC.

15—16. What indications would be observed if the BCD input 1100 was applied to the 7441A decoder in Fig. 15—19 while driving a NIXIE indicator?

15—17. Why are 16-segment displays sometimes used in preference to 7-segment displays? What are the advantages and disadvantages of each?

15—18. Discuss "dot-matrix" formation of alphameric characters.

15—19. What is a light-emitting diode?

15—20. Why is scanning used when displaying information with LEDs?

15—21. Discuss the future of plasma-panel and liquid-crystal display devices.

16

Analysis
of a Simple
Digital Subsystem

Before attempting the large-scale digital equipment analysis that will be undertaken in this chapter, we should first review the *systems concept* discussed in Ch. 7. By now, the reader should recognize that a number of different *levels of analysis* exist. The basic system understanding begins at the *systems requirements level*, where the overall functions of the system are described. A *functional block diagram* then follows, thus dividing the system into subsystems. Each of the *subsystems* may be analyzed at its *functional level*. The subsystems may then be divided into *assemblies*, which perform more complex functions. Assemblies may be further broken down into a number of *subassemblies*, which, at an even greater level of diagram complexity, may become a number of separate assemblies.

As the system is broken into smaller and smaller functional components, the diagrams become more detailed, until each diagram may contain numerous discrete components or their equivalents. Since this text is not concerned with discrete-component operations, our analysis stops at the *logic-assembly level*.

The many approaches to understanding digital logic are put to extensive use in this chapter. A *complete digital clock* (which is considered to be a digital subsystem) is examined, starting with the subsystem description, proceeding to the block diagram, examining the logic of each block, and then putting the system back together.

16–1 SUBSYSTEM DESCRIPTION

Requirements

A common requirement in data gathering and recording systems is *time reference*. Conventional (moving-hands) clocks are not adequate, due to difficulty of recording the time information. Not only must some type of film record be made, but it must be coordinated with the data-recording medium.

Digital techniques yield an answer to these problems. Counters may be used to divide some standard frequency. Decoders can convert the counter output to decimal form and drive display devices that can directly display the time. Recording of time information can be performed directly on the data-recording medium by using BCD outputs of counters. With *digital clocks*, no misreading of the position of the clock's hands is possible; no film is required to record time information; and no correlations must be performed with the data record. This adaptability has resulted in their widespread use in numerous other applications.

Basic Functions, Inputs, and Outputs

The basic function of the *digital clock* is simple: *it must display and/or make available in electrical form an indication of time.* Display is directly in digital form, while electrical indication is in BCD form. The input source may be the powerline frequency, a crystal oscillator, or a mechanical tuning fork, depending on the time accuracy required.

A typical digital clock is shown in Fig. 16—1. It displays time in a 24-hour format — *plus* the day of the year.

Figure 16—1   Typical digital clock (courtesy Datatron, Inc., Santa Ana, Ca.)

16—2 THE BLOCK DIAGRAM

The block diagram in Fig. 16—2 displays (in gross terms) the major division of a typical 12-hour (*AM/PM*) digital clock. Although we have not employed a specific manufacturer's clock, the approach used is relatively conventional and will adequately describe the general makeup of most digital clocks.

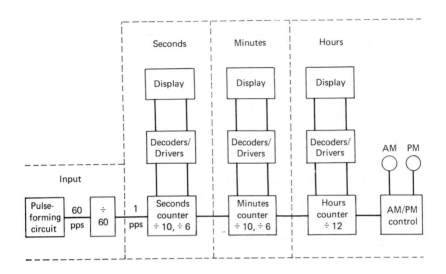

Figure 16—2 Digital clock block diagram

Input Section

In digital clocks using the powerline frequency as the standard for counting, a means must be provided to convert the 60 Hz sine waveform of the line into a pulse form that may be counted. While most digital logic operates on the basis of rapidly changing transitions from one level to another, *the powerline slowly varies from a maximum negative level* (through zero) *to a maximum positive level* and *back* (through zero) *to the maximum negative level*, before repeating its cycle. The *pulse-forming circuit* uses the powerline as an input and converts it to a series of pulses with a repetition rate of 60 pulses per second (pps).

Very little need exists to *display* time in any smaller increments than seconds, so a *divider* (divide-by-60) is provided to convert the 60 pps input into 1 pps. This circuit is peculiar to clocks using the powerline frequency as a standard. If a powerline frequency other than 60 Hz is used, a different division ratio is employed.

When accuracy greater than the powerline is required or when increments smaller than 1 second are necessary, the input to the clock is usually furnished by a high-frequency *crystal oscillator*. For example, some data-recording schemes require time as small as millisecond increments. In such cases, the pulse-forming circuits and the divide-by-60 functions are not required. The proper dividing functions may be included in the external source, and 1 pps is furnished to the clock.

Seconds Section

The 1 pps input is fed to the *seconds counter*, where a division to 1 pulse per minute (ppm) is made. Direct divide-by-60 is not used in this block, due to the display requirements. To separate seconds and tens of seconds, *two counting decades are needed*. The 1 pps input is divided first by 10 to obtain counts 0 through 9. When the tenth count occurs, a divide-by-6 counter activates to begin the tens count. The process continues through count 59, and on the sixtieth count both the units and tens-of-seconds counters reset to start the 0—59 count again. At this time, the 1 ppm signal is developed and fed to the the minutes counter.

Both the units-seconds and the tens-of-seconds counters are decoded separately. The BCD outputs of the counters are supplied to the *decoder/drivers*, and the proper count decoded. External access to the BCD outputs is also provided so that the count may be recorded. The *display device* (cold-cathode, gas-discharge tubes) illuminates one digit (per tube) at a time to display the count in seconds.

Minutes Section

The *minutes counter* (with its associated decoders, drivers, and displays) operates in a manner *identical* to the seconds counter. One pulse per hour is developed and used to activate the hours counter.

Hours Section

Operation of the *hours counter* differs somewhat from the seconds and the minutes counters. The same approach could have been used, but a rather different application of the decoder/driver and counter scheme is employed in the circuit shown. Note that a *divide-by-12 counter* is used rather than a divide-by-10 followed by a divide-by-2 counter. Since the clock displays 12 rather than 24 hours, indication of AM and PM is also provided.

16—3 THE LOGIC DIAGRAM

A complete logic diagram of the digital clock that we are discussing is shown in Fig. 16—3. Note that except for a few simple electronic components, *the complete clock is constructed from IC packages*. The functions of the electronic components are explained only where necessary to complete the logic explanation of the circuitry. Many of the ICs used in the clock have already been discussed, and the figure references are included in the symbol on the logic diagram. Those ICs not previously discussed are shown in this chapter under the referenced figure numbers.

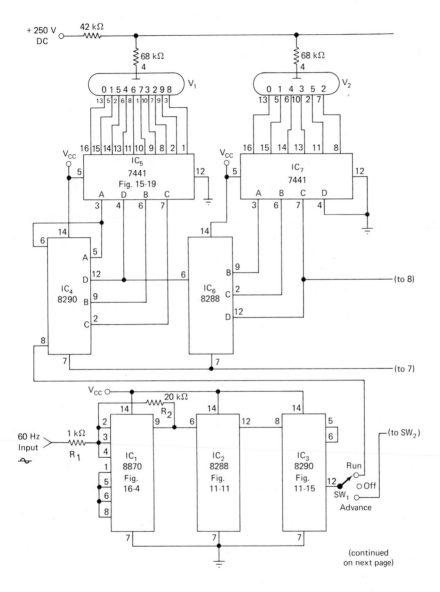

Figure 16—3 Digital clock logic diagram

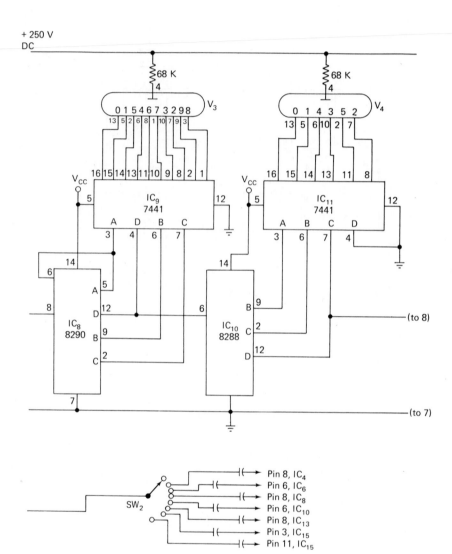

Figure 16-3 (cont'd)

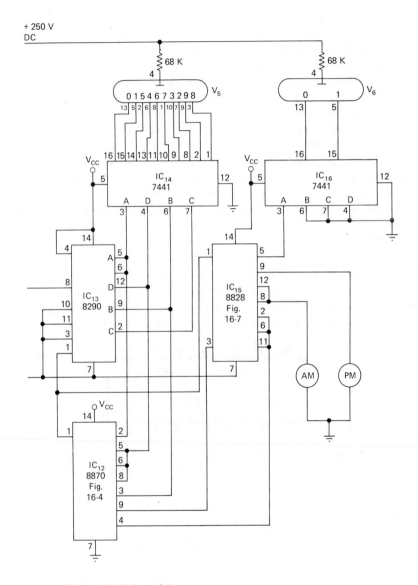

Figure 16–3 (cont'd)

Input Section

The major function of the input section is to *convert* the incoming 60 Hz signal into a 1 pps output signal. The output signal must be in a form that causes proper operation of the seconds counter that follows.

IC$_1$ contains three 3-input NAND gates (Fig. 16—4). Two of the gates are connected to form a circuit called a *Schmitt trigger*, which converts a sine wave input to a rectangular wave output (Fig. 16—5). The leading and trailing edges of the rectangular wave have extremely fast rise and fall times and quite adequately activate the following counter circuit. R$_1$ and R$_2$, in conjunction with the NAND gates, determine the shape and amplitude of the rectangular pulse.

Only the *divide-by-6* option of the 8288 counter (IC$_2$) is used to change the 60 Hz rectangular wave input to 10 pps. The *A* flip-flop (FF) is left unconnected, as are the Parallel-Data inputs and the associated Data-Strobe and Reset inputs. Obviously, a less complex counter IC could have been used in this application. However advantage is taken of quantity purchase prices, and maximum use is made of a minimum number of different ICs. Consideration of only the *B, C,* and *D* waveforms in Fig. 11—11c shows that the *D* output changes state once for every sixth input change, thus resulting in a divide-by-6 operation.

Reduction of the 10 pps output of IC$_2$ to a 1 pps signal is performed by IC$_3$, a type 8290 decade counter. As shown in Fig. 11—15b, the 8290 counter is a divide-by-2, divide-by-5 arrangement, where external connections are made to

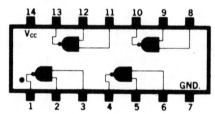

Figure 16—4   Type 8870 NAND gate IC (courtesy Signetics, a subsidiary of Corning Glass Works, Sunnyvale, CA)

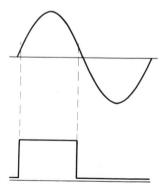

Figure 16—5   Schmitt trigger waveforms

provide decade-counting capability. The waveforms in Fig. $11-15c$ verify the divide-by-10 operation of IC_3.

If the input section is therefore considered as a single circuit, we see that the output (pin 12, IC_3) changes state once for every 60 times that the input wave changes from maximum positive level to zero. IC_1 converts the slow change of the input to a rectangular wave of the same frequency. IC_2 divides by 6, while IC_3 divides by 10. Thus a 60 Hz sine wave input results in a 1 pps output, which meets the requirements for the input section.

Much of the evaluation of this input section was based on previously acquired information. The counter analysis of Ch. 11 drastically reduced the necessary evaluation effort. However, use of different ICs would require a complete investigation of all circuits not previously encountered.

Seconds Counter and Display

Pps signals from the input section are changed to one pulse every 10 seconds signals in IC_4, an 8290 IC operating as a divide-by-10 counter. Further division by 6 in IC_6 (an 8288 IC) results in a 1 ppm output from the seconds counter. Individual IC operation is identical to the applications in the input section.

The *units portion* of the seconds counter is conventional. Correlation between each counter IC and the decoder/drivers requires reference to the state tables for each counter plus the truth table for the 7441 decoder. Examination of the state table for the 8290 shows that it counts in BCD, providing four outputs to the decoder. Since the 7441 requires BCD input, a proper interface occurs without any code conversion. It is necessary merely to assure that *bit positions match*; i.e., the LSD of the counter feeds the LSD input of the decoder. The truth table for the decoder then shows that the proper output is activated for each of the ten input combinations.

The driver portions of the decoder/driver may be viewed as *switches*, which are either open or connected to power common (ground). When a driver is activated by the proper input signal combinations, the switch changes from open to closed, and the appropriate cathode in the display tube is grounded. Sufficient current now flows from the anode to the grounded cathode; the gas around the cathode ionizes; and a numeral illuminates.

Nothing unusual occurs in the *tens section* of the seconds counter. IC_6 divides by 6, just as IC_2 does. Although the count proceeds only to 5, the output of IC_6 is still in BCD form. FFs B, C, and D are used to obtain the count, so B is considered the least-significant FF. The input to decoder IC_7 must be in BCD, and all bit positions must match. Therefore the least-significant output of the counter (B) supplies the least-significant input of the decoder (A); C supplies B; and D supplies C. The slight confusion caused by the intermixing of similar variable names may be cleared up by assigning positional names (e.g., LSD and MSD) and by using pin connections, as shown in Table $16-1$. Note that the D input of the decoder is connected to a permanent logic 0, since the count will

Table 16-1   Tens of Seconds Decoding

Counter output			*supplies*	*Decoder input*			*Displayed count*
LSD	MSD			LSD		MSD	
Pin 9	2	12		3	6	7	
0	0	0		0	0	0	0
0	0	1		0	0	1	1
0	1	0		0	1	0	2
0	1	1		0	1	1	3
1	0	0		1	0	0	4
1	0	1		1	0	1	5

never reach the point where it is required to be logic 1. This helps prevent
inadvertent activation of the decoder by noise or other undesired inputs.

Once again, we see that the analysis of this section of the clock has been
based on previously acquired information. The value of recognizing circuit
similarities and applying such information to current efforts cannot be over-
stressed. Since such circumstances constantly occur in digital logic systems, the
readers' efforts will be greatly reduced by the recognition and use of similar
circuit characteristics.

Minutes Counter and Display

A perfect example of this philosophy may be seen by examining closely
the minutes counter and display section. The four ICs making up this section
are connected *identically* to the four ICs making up the seconds counter and display. Furthermore, recognition that minutes counting requires the *same action*
as seconds counting (except at a slower rate) makes additional analysis unnecessary. All waveforms, tables, and equations derived for the seconds counter and
display apply directly to the minutes counter and display.

Hours Counter and Display

The requirement to count and display 12-hour periods of time tends to
complicate conventional counting arrangements. *Two decades* are obviously
required – *one to count the units-hours* and *another to count the tens-hours*.
This part of the requirement poses no difficulty. However, when the overall
count sequence is investigated, some problems become apparent. In the 12-hour
time system, *zero hour* does not exist. Time progresses directly from 12:59:59
to 01:00:00, skipping a zero hour altogether. The hours counter must therefore
count the following sequence: 1, 2, 3, 4, 5, 6, 7, 8, 9, 10, 11, 12, 1, 2, · · ·, continuously repeating every 12 hours and skipping the zero hour each time.

The complete units-hours and tens-hours counter circuit, less display
elements, is drawn in logic diagram form in Fig. 16–6. The 8290 counter and

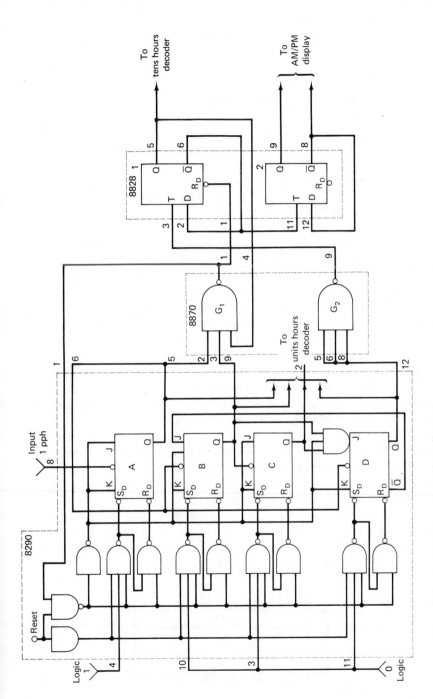

Figure 16–6 Hours counter logic diagram

8870 NAND gate information, which was used previously, is merely repeated here for clarity. A new IC, the 8828 dual D-type binary, is shown in Fig. 16–7. Unit-hours counting is performed by the 8290 and tens-hours counting by FF 1 in the 8828. G_1 in the 8870 recognizes count 13 and recycles the counter back to its starting point (count 1). Logic inversion is provided by G_2 in the 8870, while FF2 in the 8828 is the AM/PM control.

Operation of the 8290 counter in the hours counter differs from regular decade counting only in terms of *the effect of the data-strobe input on pin 1*. It may be recalled that when the data-strobe input goes LOW, the information present at the data inputs (pins 4, 10, 3, and 11) is read into the A, B, C, and D FFs, respectively. Under the circumstances shown in Fig. 16–6, FF A stores a 1, while the remaining FFs B, C, and D all store 0s. Thus the state of the counter following a data-strobe input is 0001 or decimal 1.

As input pulses occur (1 pph), the 8290 counter proceeds to count in the *BCD format* (see Fig. 16–8). The next pulse received after the counter has reached 1001 causes a recycle to count 0000, due to internal connections, and the count starts over again (this time from 0). However, when the D FF transits from 1 to 0, its output is inverted and is supplied to the T input of FF in the 8828. The D FF is connected in the T FF configuration, so any HIGH-going input to the T terminal causes the FF to change state. Assuming that it has been in the RESET state, the recycling of the 8290 counter to 0000 causes it to go to the SET state, resulting in a subsequent 1 indication on the tens-hours indicator. Thus as the 8290 counter continues counting, it is counting 10 *plus* its own count. Eleven and 12 counts occur routinely. When count 13 is reached, all inputs to G_1 are HIGH, and the LOW-going transition at the output of G_1 loads the 8290 counter to the 0001 condition. FF 1 also RESETs, resulting in an overall count indication of 01. The HIGH-going transition of FF1 causes FF 2 (another T type FF) to change state, thus producing an AM/PM display change. The complete cycle repeats at 12-hour intervals, and the hours counter portion of the clock is operating according to requirements.

Display of units-hours is conventional, using the 7441 decoder. Since two digits (0 and 1) must be displayed for the tens-hours counter, only the A input of IC_{16} (7441 decoder) is used. The remaining inputs are connected to a constant logic 0 to prevent undesired indications.

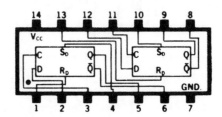

Figure 16–7   Type 8828 dual D-type binary (courtesy Signetics, a subsidiary of Corning Glass Works, Sunnyvale, CA)

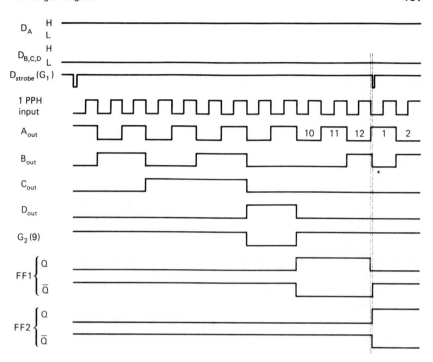

*Counter remains in 13 count for only a few
nanoseconds (long enough for reset to take place)

Figure 16—8   Hours counter waveforms

Switches

Although not a part of the logic of the circuit, *the function of switches* SW₁ and SW₂ should be mentioned. SW₁ selects *the mode of operation.* Position 1 is the normal position with 1 pps inputs provided to the seconds counter from IC₃. No counting operations occur in position 2, since the input to the seconds counter is disconnected. Position 3 connects the 1 pps source to SW₂, which selects the input of one of the counter ICs. SW₂ allows advancing *each of the counter stages to preset and start the counter at some predetermined time.* Position 7 selects the AM/PM FF input: position 8 selects the tens-hours FF input, and so forth. The capacitors at each position of SW₂ isolate the 1 pps source from the counter stages, as they are selected. When each stage has been preset to the selected time, SW₁ is placed in the OFF position until that time is reached. SW₁ is then placed in position 1, and the clock begins counting in normal mode.

For most practical applications, there should be no need to advance or retard the clock, unless a power failure occurs. Except for high-speed data-recording and precision-timing measurements, the powerline frequency is sufficiently accurate.

16—4 SUMMARY

A complete description of the digital clock in Fig. 16—3 may now be provided in English language terms. Detailed analysis has been performed (to the logic level) for all sections of the clock.

The 60 Hz line frequency input is converted to a 60 Hz rectangular wave by IC_1. IC_2, functioning as a divide-by-6 counter, produces a 10 pps output, which is further reduced to 1 pps by IC_3 (a divide-by-10 counter). This basic 1 pps timing signal may be (a) connected to the input of any of the counting stages of the clock via SW_2, (b) disconnected from the clock, or (c) connected to the input of the seconds counter by SW_1. Complete control of the clock's operation is thus provided.

IC_4 and IC_6 are connected as a divide-by-10 and divide-by-6 circuit to count seconds and provide a 1 ppm output to the minutes counter. The BCD output of IC_4 is decoded by the decoder/driver (IC_5), which activates the proper digits on the units-seconds display tube (V_1). A similar function is performed for V_2 (the tens-seconds indicator) by IC_7.

The minutes counter operation is identical to the seconds counter. IC_8 and IC_{10} perform the same operations as IC_4 and IC_6, except at a 1 ppm rather than a 1 pps rate. Units-minutes display by V_3 is controlled by IC_9, while tens-minutes display is decoded by IC_{11} and displayed by V_4. One pph is furnished to the hours counter.

Units-hours counting is performed by IC_{13}, which is connected to count in decade format from 1 through 0, rather than 0 through 9. This unusual count sequence is necessary to accommodate the requirement for the change from 12:59:59 to 01:00:00. IC_{14} decodes and drives the units-hours display V_5. One FF in IC_{15} acts as the tens-hours counter, driving IC_{16} and activating the tens-hours indicator V_6. The other FF in IC_{15} serves as a driver for the AM/PM indications. IC_{12} is used as an inverter (a) to provide the proper level transition to activate the tens-hours FF and (b) to decode count 13 to preset the units-hours and tens-hours counters to 01 hours.

Thus, starting from a description of the functions of the digital clock, a complete analysis (to the logic level) has been performed, and a simplified description of operation has been developed. Any malfunction of the digital clock may be quickly isolated to the malfunctioning IC, and subsequent replace- may be easily accomplished.

Note: All logic operations discussed in this chapter are available as a single Large Scale Integrated (LSI) circuit.

PROBLEM APPLICATIONS

16—1. Assuming a 1 MHz input signal from a precision crystal oscillator, draw a block diagram of the input section that is required to operate the digital clock shown in Fig. 16—2.

Which assembly (or assemblies) should be replaced for each of the following malfunctions?

16—2. V_2 indicator remains at 3, although the remainder of the clock operates satisfactorily.

16—3. V_1 indicates properly, but the remainder of the indicators do not advance.

16—4. V_1, V_2, and V_3 indicate properly, but the remainder of the indicators do not advance.

16—5. All clock stages may be advanced by SW_1 and SW_2, but the clock will not operate in normal mode.

16—6. Units-hours counter continues beyond reset point of count 13.

16—7. AM indicator remains illuminated, even though remainder of the clock operates properly.

16—8. Tens-hours indicator (V_6) does not illuminate.

17

Digital
Applications

As a closing gesture it is only fair that the reader be given a look at the new world of digital logic that he has been preparing for throughout this text. It is in this new world of professional operation that the theorems of Boolean algebra, the logic maps, and the truth tables come together with gates and flip-flops to form real, functional, helpful systems – not just combinations of gates or groups of flip-flops in packages.

Digital techniques are being used practically everywhere. It is difficult to name *any* business-related field where digital techniques have not made their mark. A few of the broad areas of digital applications are covered in this chapter; just enough to encourage the reader to find more.

If the question is *"why* are digital techniques being used?" the answer is easy – almost every application of digital techniques has resulted in faster, more accurate performance of tasks with more easily understood results. The everyday working world is starting to feel the impact of digital methods; the conversion to digital logic is just beginning.

17–1 TELEMETRY

Definitions

Telemetry is a *process that is used to measure data at a location remote from the data source.* The word *telemeter* is of Greek origin, derived from "tele" meaning far off, and "metron" meaning measure. As mentioned in Ch. 7, a telemetering system acquires data at a relatively inaccessible location, encodes, transmits, decodes, and displays the data at the location where it is to be used. Figure 7–7 is expanded in Fig. 17–1, and the general telemetry system is discussed in greater detail next.

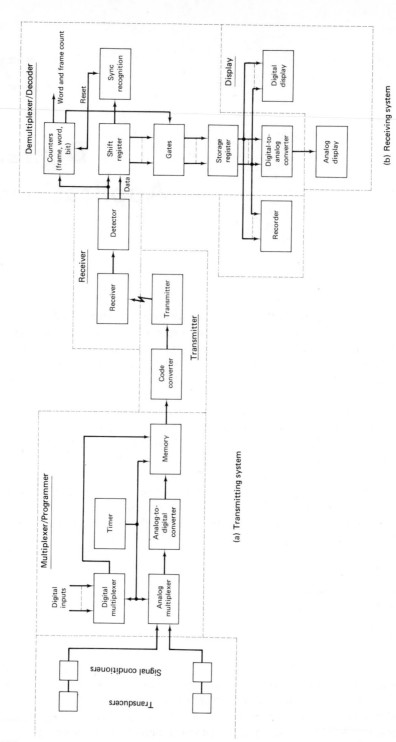

Figure 17–1 A general telemetry system

Transmitting System Description

For the purpose of clarity, the general telemetering system to be described here is divided into two parts: the transmitting system and the receiving system. This section discusses the equipment that is situated at the location where information is gathered, processed, and transmitted.

TRANSDUCER SUBSYSTEM

A typical space vehicle requires hundreds of measurements to determine the operations of the propulsion, airframe, guidance, control, electrical, and perhaps biomedical (if living organisms are aboard) systems. The *physical quantity, property, or condition to be measured is converted to an electrical current or voltage by a transducer*. Transducers take many forms, some of which are:

1. resistive units whose resistance varies with temperature or physical strain;
2. thermocouples whose voltage/current output varies with temperature;
3. moving coil devices whose voltage output and/or phase varies with angular rotation;
4. acceleration sensing devices whose output is a function of the rate of change of motion;
5. pressure-sensing devices whose resistance or voltage output changes with pressure differences; and so on.

Each transducer may have a different basic output range, and it is necessary to *scale* the output of each so that individually or as a group they will be within the range of the remainder of the telemetering system. *Signal conditioners* are used for this purpose. These signal conditioners serve either to reduce the transducer output by attenuators or increase it by amplifiers so that all inputs to the analog multiplexer are in the same range.

MULTIPLEXER-PROGRAMMER SUBSYSTEM

In a typical space vehicle the *multiplexer-programmer subsystem combines, times*, and *encodes analog and digital data into a composite signal which can be transmitted to a receiving location*. A *timer*, consisting of counters (Ch. 11) and gates (Ch. 3 and Ch. 4), is the control point of this subsystem. It develops the timing needed to sample each analog and/or digital input, converts analog inputs to digital values, loads the individual bits into the memory (Ch. 13) in the form of digital *words*, combines the words into *frames*, and supplies the total information as a series of synchronized frames to the *transmitter subsystem*. Digital inputs to the multiplexer-programmer are supplied to the *digital multiplexer*

(Ch. 13), where, under control of the timer, they are sampled periodically and placed in memory.

Switch positions and discrete outputs from guidance and control equipment that can be used to perform space vehicle positioning are examples of digital inputs.

Each of the conditioned transducer outputs is fed to the analog multiplexer, which also samples at the proper time under control of the timer. Analog multiplexers function in a manner similar to digital multiplexers, except that their outputs are nearly exact replicas of the magnitude of their inputs.

The *analog-to-digital converter* changes each analog signal from the analog multiplexer into a group of binary digits whose binary value is equivalent to the analog value. All analog and digital information is stored in memory until a transmittable group (one or more frames) is available. The combined information from the space vehicle is then transferred to the transmitter subsystem for transmission to the display location.

TRANSMITTER SUBSYSTEM

Information presented to the transmitter subsystem is in digital form, which is not easily transmitted via wire lines or radio methods; however binary 1s and 0s can be converted to a change in signal amplitude, phase, frequency, or pulse width, then supplied to the transmitter. Conversion is accomplished by the *code converter.* The *transmitter prepares the data for transmission* so it can be sent out either by wire lines or radio.* The output of the transmitter, then, is a signal that represents, in a given period of time, the combined measured parameters of the space vehicle.

Receiving System Description

At the data display location, transmitted information is received, recovered, separated into its individual measurement channels, and displayed or recorded as required.

RECEIVER SUBSYSTEM

The *receiver* performs the necessary selection and amplification of the transmitted signals. In space vehicle telemetry, the received signals may be very weak, and complex receiving equipment may be required to recover and amplify the information satisfactorily. A *detector* converts the received signal into digital form, separating the transmitted timing and data for use in the demultiplexer/ decoder.

---

* Transmitter operation is non-digital and therefore not discussed in this book.

DEMULTIPLEXER/DECODER SUBSYSTEM

In the demultiplexer/decoder subsystem the received and separated data is reconstituted for storage and display. The extracted timing is used to shift each unit of information into the shift register (Ch. 12), and to operate the various controlling counters (Ch. 11). As data is shifted into the shift register, the synchronizing pattern that accompanies each word and/or frame is recognized, and the counter is synchronized to indicate which frame or word is being received. The word counter gates each word into the storage register, where it is made available to the display devices (Ch. 15) and recorders. The next word may be shifted in while the existing word is being displayed. The display of information varies between the coding systems. Analog data is converted by the digital-to-analog converter and fed to display devices, whereas digital data is directly displayed. Recorders are commonly supplied so that received data may be stored for future use.

DISPLAY SUBSYSTEM

The end result of the data put into the general telemetering system is seen at the display device. For example, analog measurements of physical parameters inside or outside of a vehicle somewhere in space are displayed on gauges and meters so that vehicle operation may be closely monitored, second-by-second. Switching operations and commands to steering rockets are displayed instantly — as they happen. As this occurs, various recorders and plotting devices retain such telemetered information input for future study and incorporation into design improvements. There is no doubt that the ability to measure physical parameters remotely by use of telemetry has been especially influential in our rapid progress in the exploration of space, and many land-based, commonplace but necessary applications hold such promise as well.

Applications of Telemetry

A partial list of telemetry applications must include:

1. protection systems, such as fire and burglar alarms;
2. weather balloons;
3. natural gas pipeline monitoring;
4. highway and air traffic monitoring;
5. remote buoys gathering oceanographic information;
6. biomedical monitoring of patients;
7. industrial control systems, like chemical batching, sewage effluent discharge, electrical power distribution, and so on almost ad infinitum.

Only the user's imagination may limit the use of telemetry in the modern world.

17–2 DATA COMMUNICATION

Definitions

The term *data communication* implies the movement of information in some type of machine language. Data communication is usually concerned with machines that exchange information via some link, whether it be wires, radiowaves, or optical beams. The machines involved may be as simple as a teletypewriter or as complex as a computer, and the data communicated may be simple business records like the water company bills, or complex trigonometric calculations that concern the operation of a manned space vehicle.

Data communication is allied to telemetry in that information is transmitted in both cases; however, in data communication, the volume and speed of transmission are usually much lower. Many data communication devices operate at regular audio frequencies (frequencies that may be heard by the human ear) via regular voice-grade telephone lines. Thus high-speed operation becomes difficult due to the limited frequency response characteristics of such lines.

Data communicated from one machine to another is in the form of 1s and 0s and, as in telemetry, must be converted. The techniques mentioned in the transmitter subsystems section are also used in data communications, but the device used in data communication has a special name – *modem*. The term modem is a contraction of the words *modulator/demodulator, which defines the process that is used to prepare data for transmission.* The 1s and 0s of the message to be transmitted are used to modulate (change) some characteristic – phase, frequency, or amplitude – of a continuous carrier signal furnished by the transmitting device. At the receiving end, the signal is *demodulated* (the modulation is removed) and used to actuate the display device. Detailed discussions follow.

The Modem

A simple modem is shown in Fig. 17–2 in block diagram form. Frequency shift (FS) transmission is used, whereby the frequency of the output signal is different for a mark (1) than it is for a space (0). The FS transmitter of Fig. 17–2 contains an *oscillator* to generate the basic carrier frequency, and a *dc keying switch* to shift the carrier frequency when a mark is to be generated. Remote turn-off of the carrier may be accomplished if desired. The basic carrier frequency in the Fig. 17–2 example is about 30,000 Hz (cycles per second). This basic frequency is supplied to the modulator, where a non-varying frequency is supplied from the RF oscillator and demodulator to translate the 30,000 Hz FS signal to approximately 1700 Hz. Frequency translation is used to reduce distortion at high data rates and to generate a stable transmitted signal. An *output amplifier* increases the power of the FS signal to a level compatible with telephone-line operation, and a *transmitter filter* is used to reduce interference.

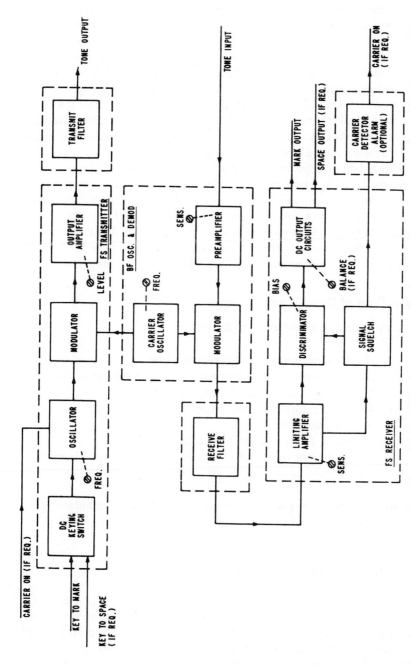

Figure 17—2   Data modem block diagram (courtesy RFL Industries, Inc., Boonton, NJ)

At the receiving end, the incoming signal is fed to a *preamplifier* in the RF oscillator and demodulator. Here it is amplified and translated to a higher frequency in the *modulator* for reasons similar to those previously mentioned. A *receive filter* removes *extraneous components of the signal*, and the signal is sent to a *limiting amplifier* in the *FS receiver*. Variations in signal amplitude are removed in the *limiting amplifier* and *mark-and-space frequencies* are detected in the *discriminator*. If the incoming signal falls below a preset level, *signal squelch* is provided to *disable the discriminator. Loss of carrier signal will actuate the carrier detector alarm*. Next, an output circuit converts the detected marks and spaces into the dc levels needed to operate the display device.

One important point should be made at this time. Note that very little digital logic is used in the modem; other electronic techniques borrowed from the communications field are applied in order to accomplish digital communication. However, the digital technician need not be alarmed for many modems are constructed with replaceable assemblies, just as digital units are. The instruction manual provides techniques, which, used in conjunction with provided test points, usually enable one to repair malfunctioning units quickly and completely.

Applications of Data Communication

Current data communication applications include:

1. the interconnection of law-enforcement data files;
2. computer-assisted instruction, where classroom terminals are used to communicate directly with computers;
3. remote stock-brokerage offices;
4. computer time-sharing;
5. medical data-bank interconnection;
6. confirming travel reservations via remote-to-central computer;
7. verification of credit card accounts.

These are but a few of the present-day applications of data communications. The future leaves much to be discovered and applied in this field.

17–3 DIGITAL TEST EQUIPMENT

Digital testing equipment is different from other types of test equipment in that it directly displays in decimal form the parameter being measured or the signal being supplied. A digital voltmeter shows the voltage being measured in the types of displays discussed in Chapter 15, whereas a conventional moving coil voltmeter shows the voltage being measured by the position of a pointer on a scale.

Both conventional and digital equipment afford high accuracy, ruggedness, convenience, and utility. In addition, the digital measuring and testing equipment offers the special advantages of speed and direct numerical readout. These devices do have a high initial cost, are excessively influenced by ac fields, and are, in some cases, more difficult to maintain. Except for their high initial cost, the other two disadvantages can be bypassed by good design and application practices.

The General Digital Instrument

As indicated in Ch. 11, the basis for most direct-reading digital instruments is the BCD counter. The count in each decade is decoded and displayed in terms of the parameter(s) being measured. Figure 11−21 shows the general concepts of a device that counts events per unit time, such as pulses per second, cycles per second, and the like. Each decade in the BCD counter has an indicator associated with it, and the number of decades in the counter is determined by the number of significant figures to be displayed. Typical decade counters have been shown in Figs. 11−14, 11−15, and 11−17. Decoders are discussed in Ch. 15.

Because a digital counting device is the basis for most digital measuring instruments, it becomes necessary to convert the input parameter(s) being measured to units that may be counted. Time-dependent waveforms, such as the frequency of pulses and ac voltages, are directly measured with the counter. (The counter has internal shaping circuits to convert input waveforms to pulses that may be counted.) The measurement of other parameters, such as *magnitudes* of voltages, currents, and light intensities, requires conversion to time-dependent, counter-compatible parameters. In other words, *some type of analog-to-digital conversion must be performed.*

Analog-to-digital converters are a subject unto themselves, and only the simple ramp-type converter will be discussed here. The concept of digital-measurement of dc voltage magnitudes was shown in Fig. 11−22 and is expanded and modified in Fig. 17−3 to demonstrate the use of digital techniques in combination with analog circuits.

The Digital Voltmeter

Note that the major functions exist in both Fig. 11−22 and Fig. 17−3, although the greater detail of Fig. 17−3 makes the operation more evident. The voltage to be measured is applied to the input terminals, where it is either manually or automatically magnitude-adjusted to fall within a predetermined range that is compatible with the *level* or *zero comparator* devices. The amount of division or amplification needed to establish this compatible range determines the position of the decimal point on the digital display.

The two comparators are used to determine the number of clock pulses that are provided to the counter. Magnitude-adjusted input is compared with an

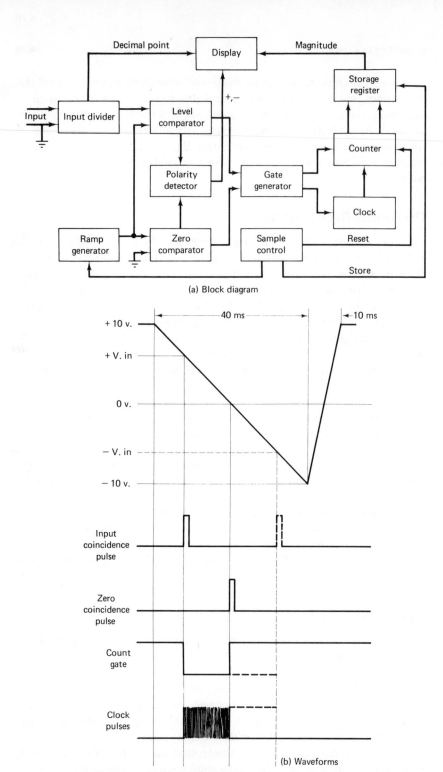

(a) Block diagram

(b) Waveforms

Figure 17–3   A digital voltmeter

internally generated ramp signal in the level comparator; a *reference voltage* (0 volts, or ground) is compared with the ramp in the *zero comparator*. The level comparator pulses when the magnitude-adjusted input and the ramp signal are equal in magnitude; the zero comparator pulses when the ramp is equal to 0 volts. Typical comparator waveforms are shown in Fig. 17–3b.

During the time the ramp is changing from +10 volts through 0 volts to −10 volts, the input (magnitude-adjusted) is being examined. As shown in Fig. 17–3, when coincidence between the ramp and the level comparator input occurs, a pulse is generated which starts the clock and simultaneously enables the counter. The ramp continues downward, and when the zero comparator is enabled by coincidence between ground (0 volts) and the ramp, the generated pulse stops the clock and disables the counter. Thus the counter operates for a period that depends on the difference between the input voltage and zero volts.

The polarity of the input voltage is determined by a circuit, called the *polarity detector*, that determines the relationship between the input coincidence pulse and the zero coincidence pulse. In Fig. 17–3, the positive input causes the input coincidence pulse to occur first, so the polarity detector provides an input to the digital display that illuminates the + sign. Occurrence of the zero comparator pulse first indicates that the input voltage is −, so the − indicator is illuminated.

A *sample control circuit* determines the number of samples to be observed each second by controlling the frequency of ramp occurrence. The counter is reset following each sample period, but the contents have already been transferred to the *storage register* by the *store pulse*, so the counter is free to establish a new count if necessary.

Digital voltmeter resolution (precision) is determined by the relationship between the slope of the ramp and the frequency of the clock. For example, if the ramp changes at the rate of 500 V/sec, and clock frequency is 500 kHz, the total count displayed corresponds to input in millivolts.

Many other techniques for measuring electrical parameters are in use. Resistance, current, phase angle, capacitance, inductance, and the like, all fall easily within the scope of digital measurement. However, one common denominator exists: the parameter *must* be converted from analog to digital form. There are many excellent texts on this subject that may be consulted if more detailed information is desired.

17–4 THE DIGITAL COMPUTER

Introduction

Time and time again the term "digital computer" has appeared in this text, and not without reason. The digital computer is probably the best publicized, least understood, and most over-rated of all examples of digital logic application. "Digital computer" may describe something as simple as a $5 toy that demonstrates the principles of binary arithmetic to a multi-million dollar complex capable of mathematical operations almost beyond human conception. It may refer to equipment that processes and stores information relating to a bank's financial transactions, or to equipment that accepts data from a spacecraft and performs calculations used to correct its flight path toward the outer planets. This section simplifies the digital computer so that some of the mystery is removed and the applications of digital logic become clear and understandable.

Obviously a more restrictive definition than "digital computer" must be developed so that an easily understood system may be examined. The popular "minicomputer" incorporates all of the features of larger digital computers, and provides easily understood explanations of system operation. It is "mini" in physical size, cost, and volume of data that can be manipulated in a reasonable period of time. The basic internal organization of most minicomputers parallels that of larger general-purpose machines, so a knowledge of the mini's operation is a valuable step toward understanding the computing giants.

Functional Description

General-purpose digital computers may be approached in the classical functional analysis manner. The basic requirements of the system are identified, defined, and major functions listed. A block diagram is drawn to portray the equipment's operational sequences. Each functional block is then examined to reveal the necessary subfunctions. When sufficient detail is available, hardware implementation of the operations is shown and explained in a manner that fulfills the original system requirements. A functional block diagram of a general-purpose digital computer is shown in Fig. 17–4.

In very broad terms, the general-purpose digital computer accepts information, arithmetically and logically manipulates that information, and makes the results of those manipulations available. Since intermediate results may occur before final data manipulations are accomplished, storage must be provided. Finally, a means to sequence and control the hardware that performs the various

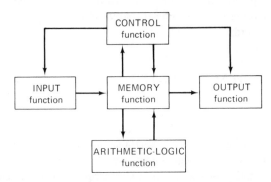

Figure 17–4   Digital computer functional block diagram

operations is needed. Actually the general-purpose digital computer is a collection of digital circuits that, when properly connected, will perform these three key operations. And these operations, when properly sequenced, provide solutions to a myriad of mathematical, logical, and data handling problems.

The collection of digital circuits called a computer will not, however, solve any problem unless it is told, step-by-step (sequentially), what to do. The sequential list of steps to be performed is called a *program*; the program is written and coded by a *programmer* who has analyzed the problem, determined the methods of solution, and proposed a plan for producing a solution to the problem.

In order to obtain any appreciable speed of operation when performing problem solutions, the sequenced steps of the program must be available immediately as needed. In addition, it may be necessary to modify the sequence of steps according to intermediate solutions that might be obtained during problem solution. These two factors, immediate availability of program sequence and the ability to change that sequence, dictate a requirement for considerable memory space; hence memory becomes a vital part of a general-purpose digital computer.

As stated, a general-purpose digital computer works on step-by-step execution of simple operations. The information the computer works with is supplied in the form of a computer "word," which generally consists of two or more parts. One typical arrangement is that of a portion of the word comprising a code to define what operation is to be performed, and the remainder of the word providing the address of the data that is to be worked with. In normal operation, the computer goes to its memory at the address defined as the starting point of the program and obtains a computer word. This information is stored in temporary registers so that the memory may be released for other operations if necessary. The operation code portion of the computer word is examined, and the operation to be performed is determined. Affected registers, gates, adders, and so on, are enabled while the computer returns to memory to obtain the data and place it in proper temporary storage for action. While the data is being collected, the location of the next word is determined, and the memory is enabled to make this word available upon completion of the previous operation. Step-by-step the cycle is repeated according to the directions of the computer program, until all steps have been completed. With proper design and programming, many operations can be carried on concurrently, reducing the total time needed to obtain results.

Several different approaches may be taken to accomplish the functional requirements for a general-purpose digital computer. The memory-centered organization shown in Fig. 17—4, where all information into and out of the computer passes through the memory, is commonly encountered. Another somewhat more flexible organization of the general-purpose digital computer is seen in the "bus" organized concept of Fig. 17—5. The computer assemblies that require input (Control Logic, for example) are connected to a group of lines that have input

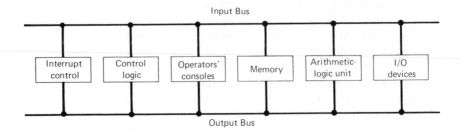

Figure 17—5   A "Bus" organized digital computer

signals on them. Similarly, those assemblies that furnish outputs will be connected to the output bus. When necessary, interrupt and control buses are provided. The "all-purpose bus" shown in Fig. 17—6 consists of many lines; the interconnection of these is a function of the physical wiring of the computer and its programming system. In a typical minicomputer, the "all-purpose bus" may have as many as 96 separate but interconnecting lines. Only those needed are connected via the interconnecting plug on each assembly.

   Another common concept in general-purpose digital computers is to use semiconductor registers to hold information prior to and while the information is being manipulated. In fact, whenever possible, it is desirable to have registers available, and under program control, for more than one purpose. Thus, if a particular register is not being used for its main purpose, it may be assigned as a temporary store until later needed.

   A more detailed diagram of a bus-organized computer is shown in Fig. 17—6. The purpose of each block is described in the following sections.

Analysis of a Typical Small Computer

   The computer design shown in Fig. 17—6 is that of a general-purpose digital computer using a 12-bit word length. Its basic 4096 (4K) word random-access memory cycle time is either 1.2 or 1.4 microseconds, depending on the type of cycle being performed. Five 12-bit registers (Ch. 12) are used to control computer operations, address memory, operate on data, and store data. The control console provides means for addressing and loading of memory, and indicators display the results. The computer may be programmed by means of a teletypewriter, a paper tape reader, or numerous other input/output (I/O) devices.

   The I/O, interrupt, and control busses interrelate the basic components (a *central processor, memory*, and *I/O equipment*) of the computer (Fig. 17—6). All arithmetic, logic, and system control operations are performed by the central processor. Information storage and retrieval operations are performed by the memory. The memory is continuously cycling and automatically performing a read-and-write operation during each computer cycle. Input and output address-

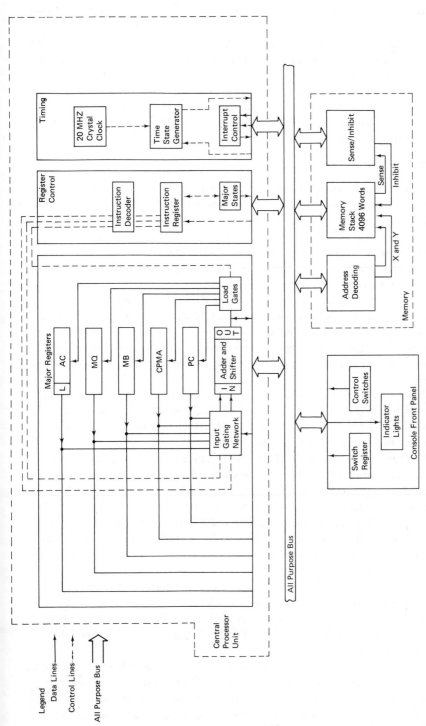

Figure 17-6 Small computer block diagram (courtesy Digital Equipment Corp., Maynard, Mass.)

485

ing and data buffering of the memory (Ch. 13) are performed by registers in the central processor. Operation of the digital computer memory is controlled by timing signals that are produced by the central processor (Ch. 11).

Connections to I/O equipment are provided by central processor interface circuits. Each I/O device detects its own selection code and provides all necessary gating. Data is transferred to I/O equipment via the accumulator register in the central processor.

THE CENTRAL PROCESSOR

The *central processor* contains the *major registers*, *register control*, and *timing* circuits. In order to store, retrieve, control, and modify information and to perform the required logical, arithmetic, and data processing operations, the memory and the central processor employ the *major registers* and *assemblies* that are described next and shown in diagram form in Fig. 17–6.

Accumulator (AC). The AC is a 12-bit register in which arithmetic and logic operations are performed. The Accumulator Register can be cleared, complemented, or rotated (shifted) left or right. Register operations include arithmetic addition (via the adder circuit), logical ANDing, and inclusive ORing. In these operations, the final results remain in the accumulator. The accumulator also functions as an I/O register, with all programmed information transferred between the memory and I/O devices passing through the AC to the data lines in the I/O bus.

Multiplier Quotient (MQ) Register. The MQ is a 12-bit bidirectional shift register that acts as an extension of the AC during multiplication and division operations. The Multiplier Quotient Register is also available for temporary storage upon request by the programmer.

Program Counter (PC). The PC is a 12-bit register that is used to control program sequence; that is, the order or sequence in which instructions are performed is determined by the Program Counter. The PC contains the address of the memory location from which the *next* instruction is to be taken.

Central Processor Memory Address (CPMA) Register. The CPMA is a 12-bit register that contains the address of the location in memory that is currently in use. It works in conjunction with the Program Counter to control the sequence of computer operation.

Memory Buffer (MB) Register. The Memory Buffer Register is a 12-bit register that is used for *all* information transfers between central processor registers and memory. Information can be transferred and temporarily held in the MB from the AC to the PC. In addition, information can be loaded into the MB from an I/O device.

Data Gates and Adders. The major registers also contain the gating needed to move data from one register to another. At the heart of the data gating (Chs. 3 and 4) is a 12-bit parallel adder (Ch. 13). Information from a register is gated to the adder inputs, then the output from the adder is applied to a set of shift gates. Next the output of the shift gates becomes data input to any or all of the major registers.

REGISTER CONTROL

The Register Control contains the *Link*, the *Major Register Control Circuits*, the *Major State Register,* the *Instruction Register,* and the necessary control circuits for the Major State Register and the Instruction Register.

Link. The Link is a 1-bit register that is used to extend the arithmetic facilities of the AC. It is used as the carry register for arithmetic operations.

Major Register Control Circuits. These circuits are used to enable the adder input and shift gates of the Major Register section. They also gate timing pulses to cause loading of the appropriate major registers.

Major State Register. The Major State Register is the control for the three major states (FETCH, DEFER, or EXECUTE) of the computer. Each of the major state signals, when asserted, is used to enable the corresponding register control circuitry.

Instruction Register (IR). The IR is a 3-bit register that contains the operation code of the instruction currently being performed by the machine. The three most-significant bits of the current instruction are loaded into the Instruction Register from the memory during a FETCH cycle. The contents of the IR are decoded to produce the eight basic instructions and effect the cycles and states entered during each step of the program.

TIMING GENERATOR

The Timing Generator section contains the *Time Pulse Generator*, *Interrupt Control Circuits*, the *Processor IOT Decoder*, and miscellaneous control circuits.

Time Pulse Generator. The Time Pulse Generator provides the timing pulses that determine the computer cycle time. The timing pulses are used to initiate sequential time-synchronized gating operations. Pulses that reset registers and control circuits during power off—on operations are produced by a power clear pulse generator. Several of the timing pulses are available for control of external I/0 devices.

Four time states are provided by the Time Pulse Generator. In addition, four time pulses are generated for use as gating pulses throughout the system. Memory timing is also provided by this circuit.

Interrupt Control Circuits. The Interrupt Control Circuits comprise the major portion of the Interrupt System. This circuitry responds whenever an INTERRUPT REQUEST signal is received from an external or internal device.

Processor IOT Decoder. This circuit decodes the last nine memory bits and determines the type of Input/Output Transfer (IOT) instruction to be performed.

CONTROL CONSOLE

The Control Console contains an array of switches, keys, and indicators that are used specifically for operation and maintenance. By means of the proper control, the operator can STOP, START, EXAMINE, MODIFY, or CONTINUE a program by actuation of switches and keys. The indicators, when properly selected, display the machine status and contents of the major registers.

MEMORY SYSTEM

The basic computer memory system is a 4096 (4K) word, 12-bit random-access memory that performs all normal functions of data storage and retrieval. The memory system contains circuits such as read/write switches, address decoders, inhibit drivers, and sense amplifiers. These circuits perform the electrical conversions needed to transfer information to or from the memory array. The memory system performs no arithmetic or logic operations upon the data.

XY Driver and Current Source. This circuit contains the components required to decode the address lines and drive the XY inputs of a 4K word memory (i.e., address, decoding, selection switches, XY current sources, and the like).

Memory Stack. The Memory Stack contains 4096 words of 12-bit memory and the X-axis and Y-axis selection matrix. There are no direct connections from the Memory Stack to the I/0 bus.

Sense/Inhibit Section. The Sense/Inhibit Section contains the sense amplifiers, memory register, and the inhibit drivers for a word length of 12 bits. Miscellaneous memory control circuits are also included in this unit.

MAJOR PROCESSOR STATES

FETCH (F) State. The computer enters the FETCH state to obtain a 12-bit instruction word. At the start of the FETCH cycle, the contents of the PC are loaded into the CPMA, giving the first memory address and starting the memory cycle. The computer obtains the contents of the addressed memory location and places the 12 bits on the memory data lines of the I/0 bus. The contents of these lines are decoded to determine the kind of instruction the processor must next perform. When the processor identifies the kind of instruction it must do, it begins performing the instruction, entering either a DEFER or an EXECUTE state. When the instruction has been executed, the computer returns to the FETCH state to obtain the next instruction.

Let us assume that the computer is in a fully automatic, running condition; that is, that the computer is continuously functioning to FETCH and/or EXECUTE instructions. The PC register's contents are transferred to the MA register at the end of the previous cycle, initiating the FETCH state. When this is accomplished, the MA is simultaneously incremented by one and the result transferred to the PC register. The memory is then interrogated, and the contents of the memory appear at the output of the sense amplifiers and are transferred to the memory data lines. The contents of the first three bits of the memory data are transferred to the IR. This completes the initial activity of the FETCH state.

If the instruction word has more than one instruction, the memory address is computed but no further action is taken until the DEFER or EXECUTE cycle. If a single-cycle instruction is given, the instruction is carried out immediately.

DEFER (D) State.   At the end of a FETCH cycle, the current instruction is directed to DEFER whenever more than one instruction is required to perform the indicated operation. The DEFER state allows sufficient time for the computer to obtain the additional instruction(s). It effectively delays the EXECUTE operation until the required instruction(s) are available for use.

EXECUTE (E) State.   The EXECUTE state is entered for practically all memory reference instructions. For example, during an AND or ADD instruction, the contents of the memory location specified by the address portion of the instruction are read into the MB and the operation specified by the IR is performed.

DIRECT MEMORY ACCESS (DMA) State.   A fourth state exists when any one of the other 3 major states is not enabled. This state, called DMA, is used to independently address memory and to store or read out information without the aid of processor instructions.

17–5 INDUSTRIAL CONTROLS

Concepts of Automation

Since the beginning of the Industrial Revolution, human skill, intelligence, and experience have been used to operate metal-working machines accurately and quickly. Now digital logic is being used to control lathes, drills, milling machines, and other industrial machinery that must repeat a series of work steps within close tolerances and in perfect sequence.

The operation of practically any metal-working machine may be viewed in the manner shown in the block diagram of Fig. 17–7. Even human operations of the machine fit into the diagram. The *input* is the blueprint of the finished product, showing required dimensions and specifications. Such information is examined by the human operator (the *control unit*), and decisions are made concerning the tools to use and the method of machine operation. Actual motive force for the tool, such as an electric motor, is the *drive*. As the material is being worked, measurements are taken to determine how much additional machining is required. This information is *feedback*, which the control unit uses to determine any changes required in the machining operation. For example, relatively large "cuts" may be made when a milling operation on a piece of metal is initially begun. As the finished part comes closer and closer to final dimensions and specifications, smaller and smaller "cuts" are taken, so that the finished product is as close as possible to exact requirements.

As mechanization replaced hand work, complex mechanical contrivances were developed to make the machine more self-sufficient, but accuracy was

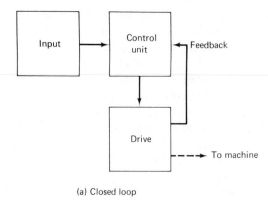

(a) Closed loop

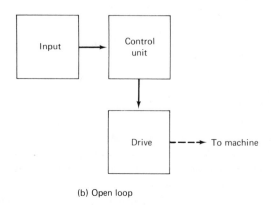

(b) Open loop

Figure 17—7   Functional block diagram — machine control

lacking, and precision metal work remained the domain of skilled machinists. Then electrical analog controls were added, with operator input that was set up by means of dials and with feedback supplied by special position-indicating devices mounted on the machine. Conventional motors powered these machine tools and the feedback information was compared in the control unit to determine if the machine had performed its "programmed" operations. If not, for example, the power to the driving motors could be modified to reduce the error between a desired position and an actual position for a tool. (Primitive digital controls, like switches and relays, were also used in analog control systems.) Accuracy and speed were improved, but due to the difficulty of realizing highly accurate measurements with these analog systems, much of the complex precision machining was still relegated to the human craftsman.

Numerical Control

The early 1950s saw the beginning of digital control of machines, when *numerical control* (the *control of tools and materials movement by numeric data*) began to be applied widely. Substitution of digital logic as the control unit, with input information furnished by punched paper tape, resulted in a much desired increase in production and machine accuracy. Feedback still came from the actual machine or the material being worked, but it was digitized so that maximum use could be made of the principles of digital control. A typical numerical control (NC) device is shown in Fig. 17−8, and is discussed in this section. The photograph of Fig. 1−7 shows a NC device in action.

An *open loop system*, as shown in Fig. 17−7, merely positions the tools and materials into specified locations, then actuates the necessary operations. With the capability to position to close tolerances in modern NC equipment, the requirements for the feedback used in closed loop systems is sometimes nullified. Both open and closed loop NC devices are in common use today.

Greater accuracy, speed, and coordination of control between tools and materials were the goals of digital control, and these goals have been met. Accuracies on the order of 0.0001″ are attained easily, whereas skilled craftsmen reach only 0.001″ under *ideal* conditions. Due to this increase in accuracy, losses of time and materials have dropped, and production rates of complex parts have soared astronomically. The machining of shapes that could not be attempted manually are the everyday tasks of numerically controlled equipment. These improvements can be attributed to the use of high-speed digital control programs in place of switch/relay controls or the analog control scheme.

Numerical control does not require a digital computer. Machine instructions are punched into paper tape, and the actual commands are read one at a time, performed, and the next command read, performed, and so on to the end of the tape. It isn't really that simple, because the *digital·controller* must perform many logic operations, compare results of operations with the required operation, store information, move data from one location to another, and so on. The digital controller itself uses most of the basic logic circuits discussed in earlier chapters of this text.

Analysis of a Computer-Operated Numerical Control Device

In grossly oversimplified terms, the computer shown in the block diagram of Fig. 17−8 serves as a storage and sequencing device for the *digital controller* (represented by the remainder of the block diagram).

INPUT / OUTPUT ORGANIZATION

The list of steps to be performed in the machining operation is punched into paper tape (Ch. 14), read by the tape reader, and stored within the memory (Ch. 13) of the computer. The input bus (consisting of eight lines, one each for

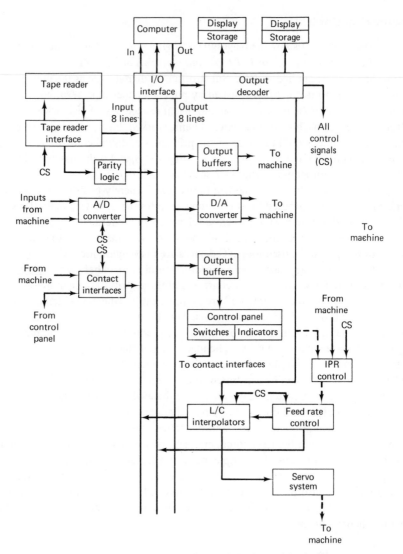

Figure 17—8   Numerical control device — block diagram
(courtesy Bendix Industrial Controls Division, Detroit, Mich.)

the 8 bits in a character) is used for the punching, loading and storing operations.
Note that parity is checked when tape reading operations are being performed.
Improper parity actuates an input to the interrupt bus, which stops the tape
reader.

Each machine operation to be performed is controlled by two groups of
8 bits. The first 8-bit group identifies the device to be operated; the second

8-bit group describes the operation. All of the devices in the network are connected to the output bus (also 8 lines). When a device recognizes its address, it connects itself to the output bus and awaits the data word. For example, if the device operates directly from digital outputs, such as relays or indicators, the output buffer that corresponds to the addressed device accepts the data word, then routes the word to the device. Following a fixed time period that corresponds to the time needed to transfer the data word, the device disconnects itself from the output bus.

ANALOG INPUT / OUTPUT

Many machine tools require analog control voltages for such operations as spindle speed control (the spindle carries the tool that performs the metal cutting operation). Information in the NC program is in digital form, and when it is made available on the output bus it must be converted into analog form in order to change the speed. Digital-to-analog converters, which use registers (Ch. 12) and gates (Chs. 3 and 4), are used to perform this function.

Some devices that provide feedback from the machine tools are analog in nature. These feedback inputs are applied to an analog-to-digital converter (comparators, gates, counters, and registers), and upon completion of the conversion, an interrupt is generated so that the now digitized feedback can be applied to the input bus. This digital feedback is then stored in, or acted upon, by the computer, whichever is required by the program(s) in use when the feedback is made available.

DISPLAY AND CONTROL

A control panel with switches and indicators is used to operate the NC machine. When instructed by the computer, output buffers supply information to the indicators for display, and to switches for selective re-routing of information. Contact interface circuits standardize logic levels for the input bus and provide the capability to supply switch position information both from the control panel and from the machine being controlled. Typical switches at the machine include limit switches, which signal discrete worktable or tool-holder positions.

Note that nearly all of these input and output devices have control signal inputs. These signals identify the device to be used and actuate the various processes going on within the machine. The information is contained in the address and command portion of the data on the output bus and is separated and routed by the output decoder (registers and gates). In addition, the output decoder supplies data to storage registers and decoders (Ch. 15) which actuate the cold-cathode, gas-filled, or segmented digital display devices (Ch. 15). Data that is commonly displayed includes the program being executed, step sequence numbers, and predefined tool offset positions.

CONTINUOUS PATH CONTROL

The full advantages of NC begin to be realized when a means of precisely controlling *each* working axis of the machine is integrated to furnish continuous-path operation of the tool and positioning of the material being worked. When effectively programmed, the tool and the material work together to follow a mathematically defined pattern of the finished product. In addition, the speed at which the tools work the material must be considered.

The movement of the axis of operation is integrated with tool/material feed in an *interpolator* (made up of registers, gates, adders, and the like). It is in the interpolator that the motion command and feed-rate command are combined to furnish a magnitude, direction, and rate of tool/material feed on a specific machine axis. Each interpolation results in movement of the axis; the movement that results from each command is returned to the computer memory to update its knowledge of tool/material positions. If movement is to be in a straight line, linear interpolation is used. *When a circular movement is required, operation of the interpolator is modified to provide for circular positioning.* One interpolator is used for each machine axis to be controlled.

The real advantage of the computer may be seen in these discrete interpolation operations. Machine tool/material movement can be controlled in increments of $0.0001''$. Such movements are usually described by the number of increments in one direction. Complex paths require great skill and care in defining the required movements accurately, but by using built-in computer programs, much labor and potential error are removed. Also, the computer receives feedback to use in a comparison mode, so that motion commands can be modified as required, according to predefined rules of operation. Effectively, each machining operation ends up with many more discrete steps, more accurately controlled, by use of the computer.

A special control is provided for lathe operations by use of the Inches per Revolution (IPR), or Threadcutting, control. Feed rate is provided in such cases by the IPR control rather than directly from a feedrate control. Feedback from the lathe is developed by a special pulse generator that is mounted directly on the lathe.

The actual motion control of each axis of the machine tool is developed by a servo system. This servo controls the power to the driving motor of each axis, which produces motion that is proportional to the command of the interpolator.

Advantages of Numerical Control (NC)

Conventional digital controllers are usually prewired to perform specific operations; modification of the operations is time-consuming and expensive. A computer-controlled NC machine uses the normal functions of the digital controller, but also requires that the correct procedural steps be a part of the computer's operating program. To change from one operation to another, a different program is inserted, or an existing program may be called from storage in the computer's memory. The stored program capability allows production operations to progress smoothly from one phase to another without a major interruption to load a new tape program.

Multiple operations can be handled by the computer-controlled NC device. In a modern general-purpose digital computer, it takes only a few microseconds to route information within the controller. Since the actions performed by the controller and machine tools may take milliseconds or longer, the computer is free to update its memory, interrogate sensors or switches in other parts of the machine being used (or other machines), or perform other important operations. In fact, there is usually so much free time that one computer can be used to control more than one operation (or machine) at the same time. Actually each operation (or machine) is controlled separately, but the rate of change makes it appear that simultaneous control exists.

Modern computer-controlled NC devices may have as much as 85 per cent of their capability in the form of software (computer programs). A simple change of software changes the entire production operation. This versatility makes NC machine tools and control systems almost universally applicable.

17–6 SUMMARY

Each system, subsystem, or assembly described in this chapter should bring to mind the basic circuit elements that could be used to implement the function under discussion. For example, storage functions should recall the registers of Ch. 12 and the memories of Ch. 13. The counters of Ch. 11 should be visualized when timing and control are discussed. Selective control and routing of logic signals are obviously the domain of the gates of Chs. 3 and 4.

When these types of mental responses occur, the reader is on his way toward becoming a knowledgeable member of the digital logic profession. He will find an almost endless number of opportunities, depending only on his desires and background. Circuitry design, logic design, field engineering and maintenance, systems analysis and troubleshooting, and computer programming are but a few. WELCOME TO THE NEW WORLD OF DIGITAL LOGIC.

Appendix **A**

Postulates, Theorems, and Laws of Boolean Algebra

POSTULATES

$0 \cdot 0 = 0$	(P–1)	$0 + 1 = 1$	(P–6)
$0 \cdot 1 = 0$	(P–2)	$1 + 0 = 1$	(P–7)
$1 \cdot 0 = 0$	(P–3)	$1 + 1 = 1$	(P–8)
$1 \cdot 1 = 1$	(P–4)	$0 = \bar{1}$	(P–9)
$0 + 0 = 0$	(P–5)	$1 = \bar{0}$	(P–10)

THEOREMS

$A \cdot 0 = 0$	(T–1)	$0 + A = A$	(T–8)
$0 \cdot A = 0$	(T–2)	$A + 1 = 1$	(T–9)
$A \cdot 1 = A$	(T–3)	$1 + A = 1$	(T–10)
$1 \cdot A = A$	(T–4)	$A + A = A$	(T–11)
$A \cdot A = A$	(T–5)	$A + \bar{A} = 1$	(T–12)
$A \cdot \bar{A} = 0$	(T–6)	$\bar{\bar{A}} = A$	(T–13)
$A + 0 = A$	(T–7)		

LAWS OF BOOLEAN ALGEBRA

The Laws of Identity	$A = A,$	$\bar{A} = \bar{A}$
The Commutative Laws	$AB = BA,$	$A + B = B + A$
The Associative Laws	$A(BC) = ABC,$	$A + (B + C) = A + B + C$
The Idempotent Laws	$AA = A$	$A + A = A$
The Distributive Laws	$A(B + C) = AB + AC,$	$A + BC = (A + B)(A + C)$
The Laws of Absorption	$A + AB = A,$	$A(A + B) = A$
The Laws of Expansion	$AB + A\bar{B} = A,$	$(A + B)(A + \bar{B}) = A$
DeMorgan's Laws	$\overline{AB} = \bar{A} + \bar{B},$	$\overline{A + B} = \bar{A}\bar{B}$

COMMON IDENTITIES OF BOOLEAN ALGEBRA

$A(\bar{A} + B) = AB$

$A + \bar{A}B = A + B$

$(AB)(A + B) = AB$

$(\overline{AB})(A + B) = A\bar{B} + \bar{A}B$

$\overline{A\bar{B} + \bar{A}B} = AB + \bar{A}\bar{B}$

$(A + B)(B + C)(A + C) = AB + BC + AC$

$(A + B)(\bar{A} + C) = AC + \bar{A}B$

$AC + AB + B\bar{C} = AC + B\bar{C}$

$(A + B)(B + C)(\bar{A} + C) = (A + B)(\bar{A} + C)$

Appendix **B**

MIL–STD–806B Extracts

AND. The symbol shown below represents the AND function.

The AND output is high (H) if and only if all the inputs are high.

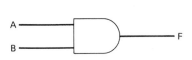

Input		Output
A	*B*	*F*
L	L	L
L	H	L
H	L	L
H	H	H

OR. The symbol shown below represents the INCLUSIVE OR function.

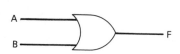

The OR output is high (*H*) if and only if any one or more of the inputs are high (*H*).

Input		Output
A	*B*	*F*
L	L	L
L	H	H
H	L	H
H	H	H

STATE INDICATOR (Active).   The presence of the small circle symbol at the input(s) or output(s) of a function indicates:

(a) *Input Condition.*   The electrical condition at the input terminal(s) that control the active state of the respective function.

(b) *Output Condition.*   The electrical condition existing at the output terminal(s) of an activated function.

A small circle(s) at the input(s) to any element (logical or nonlogical) indicates that the relatively low (L) input signal activates the function. Conversely, the absence of a small circle indicates that the relatively high (H) input signal activates the function.

A small circle at the symbol output indicates that the output terminal of the activated function is relatively low (L). This small circle should never be drawn by itself on a diagram.

EXCLUSIVE OR.   The symbol shown below represents the EXCLUSIVE OR function.

The EXCLUSIVE OR output is high if and only if any one input is high (H) and all other inputs are low (L).

Input		Output
A	*B*	$F = A(H)$ and $B(L)$ or $B(H)$ and $A(L)$
L	L	L
L	H	H
H	L	H
H	H	L

AND	OR	A	B	X
		H	H	H
		H	L	L
		L	H	L
		L	L	L
		H	H	L
		H	L	L
		L	H	H
		L	L	L
		H	H	L
		H	L	H
		L	H	L
		L	L	L
		H	H	L
		H	L	L
		L	H	L
		L	L	H
		H	H	H
		H	L	H
		L	H	H
		L	L	L
		H	H	H
		H	L	L
		L	H	H
		L	L	H
		H	H	H
		H	L	H
		L	H	L
		L	L	H
		H	H	L
		H	L	H
		L	H	H
		L	L	H

501

FLIP-FLOP.   The FLIP-FLOP (FF) is a device that stores a single bit of information. It has three possible inputs,* SET (*S*), CLEAR (RESET) (*C*), and TOGGLE (TRIGGER) (*T*), and two possible outputs, 1 and 0.** When not used, the TRIGGER input may be omitted.

Polarity.   The two outputs are normally of opposite polarity. A "1" is stored in the FF when the "1" output level is active and the "0" output level is inactive. A "0" is stored in the FF when the above condition is reversed.

States.   The FF assumes the "1" state when an active signal appears at the *S* input regardless of the original state. It assumes the "0" state when an active signal appears at the *C* input regardless of the original state. It reverses its state when an active signal appears at the *T* input. There are several possible variations to normal FF operations, depending on the response of the device when active inputs are applied simultaneously to more than one input.

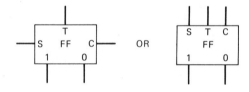

———
   * Other input designations such as *D* and *J–K* are commonly used. These inputs are defined in Ch. 8.
      ** The 1 output is commonly called *Q*, and the 0 output is commonly called $\bar{Q}$.

Appendix C

Equivalent Logic Symbols

Standard AND gate symbol and equivalents

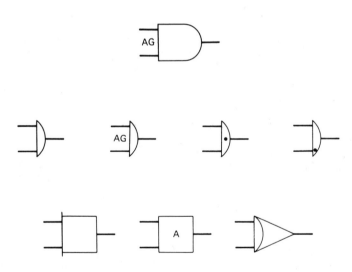

Standard OR gate symbol and equivalents

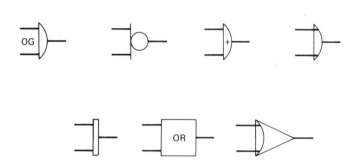

Standard NAND gate symbol and equivalents

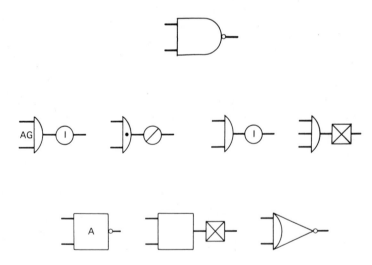

Standard NOR gate symbol and equivalents

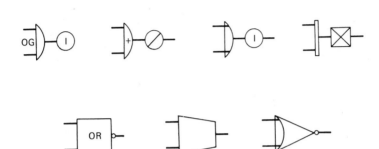

Standard INVERTER symbol and equivalents

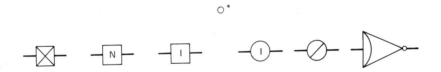

Standard AMPLIFIER symbol — common throughout symbology

* May not appear unless attached to another symbol.

Standard FLIP-FLOP symbol and equivalents

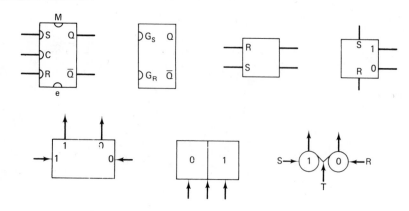

Appendix D

Typical Logic-element Circuit Diagrams

3-input diode AND gate using discrete components

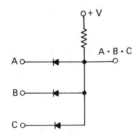

$A \cdot B \cdot C$

3-input Diode-Transistor Logic (DTL) NOR gate using discrete components

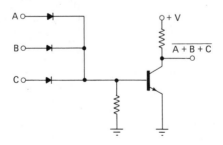

$\overline{A + B + C}$

3-input Resistor-Transistor Logic (RTL) NOR gate using discrete components

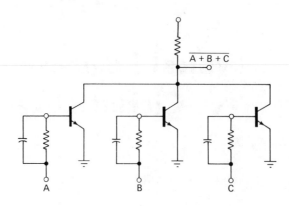

3-input Transistor-Transistor Logic (TTL) NAND gate using discrete components

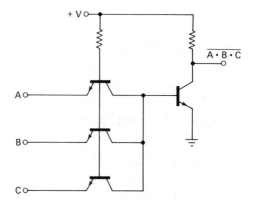

4-input Emitter Coupled Logic (ECL) OR/NOR gate

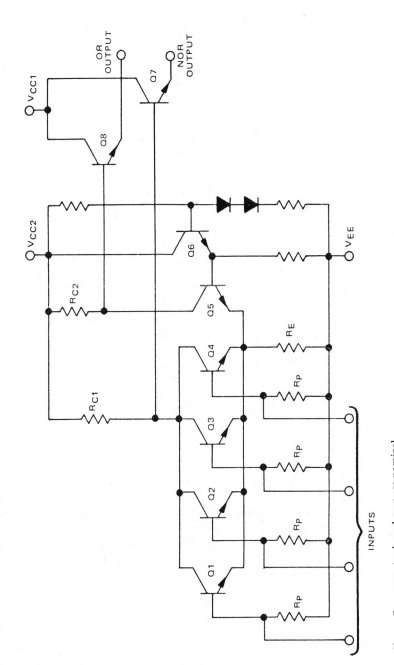

Note: Component values shown are nominal.

509

IC Implemented 3-input TTL AND gate

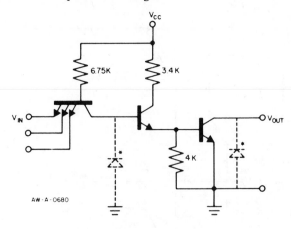

AW-A-0680

NOTE: 1/3 of unit shown. Component values are typical.

*Isolation diode

NOTE: Component values shown are nominal.

A typical $R-S$ FF implemented with discrete components

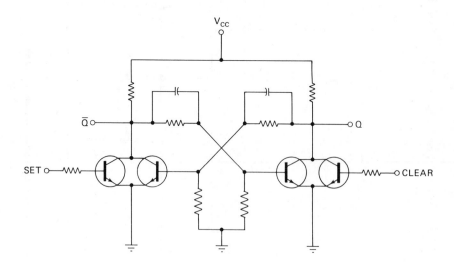

IC Implemented *D* Flip-Flop

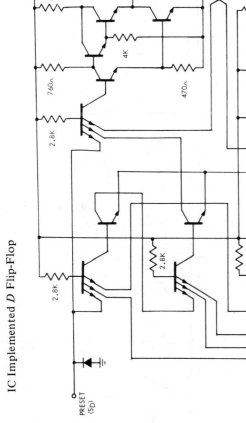

Component values shown are typical.

511

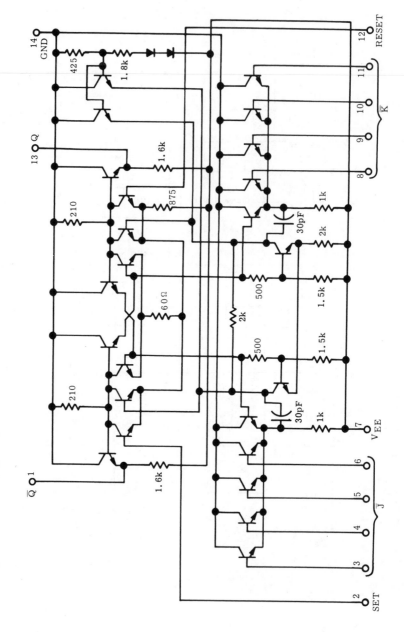

IC Implemented *J–K* Flip-Flop

Appendix **E**

Number Tables

Powers of Two

2^n	n	2^{-n}
1	0	1.0
2	1	0.5
4	2	0.25
8	3	0.125
16	4	0.062 5
32	5	0.031 25
64	6	0.015 625
128	7	0.007 812 5
256	8	0.003 906 25
512	9	0.001 953 125
1 024	10	0.000 976 562 5
2 048	11	0.000 488 281 25
4 096	12	0.000 244 140 625
8 192	13	0.000 122 070 312 5
16 384	14	0.000 061 035 156 25
32 768	15	0.000 030 517 578 125
65 536	16	0.000 015 258 789 062 5
131 072	17	0.000 007 629 394 531 25
262 144	18	0.000 003 814 697 265 625
524 288	19	0.000 001 907 348 632 812 5
1 048 576	20	0.000 000 953 674 316 406 25

Decimal-Binary and Binary-Decimal Conversion Tables

Decimal	Binary	Decimal	Binary	Decimal	Binary	Decimal	Binary
0	00000000	50	00110010	100	01100100	150	10010110
1	00000001	51	00110011	101	01100101	151	10010111
2	00000010	52	00110100	102	01100110	152	10011000
3	00000011	53	00110101	103	01100111	153	10011001
4	00000100	54	00110110	104	01101000	154	10011010
5	00000101	55	00110111	105	01101001	155	10011011
6	00000110	56	00111000	106	01101010	156	10011100
7	00000111	57	00111001	107	01101011	157	10011101
8	00001000	58	00111010	108	01101100	158	10011110
9	00001001	59	00111011	109	01101101	159	10011111
10	00001010	60	00111100	110	01101110	160	10100000
11	00001011	61	00111101	111	01101111	161	10100001
12	00001100	62	00111110	112	01110000	162	10100010
13	00001101	63	00111111	113	01110001	163	10100011
14	00001110	64	01000000	114	01110010	164	10100100
15	00001111	65	01000001	115	01110011	165	10100101
16	00010000	66	01000010	116	01110100	166	10100110
17	00010001	67	01000011	117	01110101	167	10100111
18	00010010	68	01000100	118	01110110	168	10101000
19	00010011	69	01000101	119	01110111	169	10101001
20	00010100	70	01000110	120	01111000	170	10101010
21	00010101	71	01000111	121	01111001	171	10101011
22	00010110	72	01001000	122	01111010	172	10101100
23	00010111	73	01001001	123	01111011	173	10101101
24	00011000	74	01001010	124	01111100	174	10101110
25	00011001	75	01001011	125	01111101	175	10101111
26	00011010	76	01001100	126	01111110	176	10110000
27	00011011	77	01001101	127	01111111	177	10110001
28	00011100	78	01001110	128	10000000	178	10110010
29	00011101	79	01001111	129	10000001	179	10110011
30	00011110	80	01010000	130	10000010	180	10110100
31	00011111	81	01010001	131	10000011	181	10110101
32	00100000	82	01010010	132	10000100	182	10110110
33	00100001	83	01010011	133	10000101	183	10110111
34	00100010	84	01010100	134	10000110	184	10111000
35	00100011	85	01010101	135	10000111	185	10111001
36	00100100	86	01010110	136	10001000	186	10111010
37	00100101	87	01010111	137	10001001	187	10111011
38	00100110	88	01011000	138	10001010	188	10111100
39	00100111	89	01011001	139	10001011	189	10111101
40	00101000	90	01011010	140	10001100	190	10111110
41	00101001	91	01011011	141	10001101	191	10111111
42	00101010	92	01011100	142	10001110	192	11000000
43	00101011	93	01011101	143	10001111	193	11000001
44	00101100	94	01011110	144	10010000	194	11000010
45	00101101	95	01011111	145	10010001	195	11000011
46	00101110	96	01100000	146	10010010	196	11000100
47	00101111	97	01100001	147	10010011	197	10000101
48	00110000	98	01100010	148	10010100	198	11000110
49	00110001	99	01100011	149	10010101	199	11000111

Bibliography

Baron, Robert C., and Albert T. Piccirilli. *Digital Logic and Computer Operations.*
New York: McGraw-Hill, 1967.

Bartee, Thomas C. *Digital Computer Fundamentals,* 3rd edition. New York:
McGraw-Hill, 1972.

Burroughs Corporation, Technical Training Department — Defense, Space, and
Special Systems Group. *Digital Computer Principles*, 2nd edition. New
York: McGraw-Hill, 1969.

Churchman, Lee W. *Survey of Electronics.* San Francisco: Rinehart Press, 1971.

Davenport, William P. *Modern Data Communication.* New York: Hayden,
1971.

Dietmeyer, Donald L. *Logic Design of Digital Systems.* Boston: Allyn and
Bacon, 1971.

Hoernes, Gerhard E., and Melvin F. Heilweil. *Introduction to Boolean Algebra
and Logic Design.* New York: McGraw-Hill, 1964.

Karnaugh, M. "The Map Method for Synthesis of Combinational Logic Circuits,"
AIEE Proceed., 593: November 1953.

Ketchum, Donald J. *Applications of Digital Logic.* Farmingdale, N.J.: Buck
Engineering Co., 1966.

Maley, Gerald A. *Manual of Logic Circuits.* Englewood Cliffs, N.J.: Prentice-
Hall, 1970.

Maley, Gerald A., and John Earle. *The Logic Design of Transistor Digital
Computers.* Englewood Cliffs, N.J.: Prentice-Hall, 1963.

Malmstadt, H. V., and C. G. Enke. *Digital Electronics for Scientists.* New York:
Benjamin, 1969.

Mandl, Matthew. *Fundamentals of Electronic Computers: Digital and Analog.*
Englewood Cliffs, N.J.: Prentice-Hall, 1967.

Marcus, Mitchell P. *Switching Circuits for Engineers,* 2nd edition. Englewood
Cliffs, N.J.: Prentice-Hall, 1967.

Phister, Montgomery, Jr. *Logical Design of Digital Computers*. New York: Wiley, 1958.

Sifferlen, Thomas P., and Vartan Vartanian. *Digital Electronics with Engineering Applications*. Englewood Cliffs, N.J.: Prentice-Hall, 1970.

Thomas, Harry E. *Handbook of Pulse-Digital Devices for Communications and Data Processing*. Englewood Cliffs, N.J.: Prentice-Hall, 1970.

Veitch, E. W. "A Chart Method for Simplifying Truth Functions," *Ass. Computing Machinery, Proceed.*, 127: May 1952.

——— , *Application Notes*. Plainfield, N.J.: Burroughs Corp., 1971.

——— , *Application Notes*. Phoenix, Az.: Motorola Semiconductor Products, Inc., 1972.

——— , *Application Notes*. Sunnyvale, Ca.: Signetics (a subsidiary of Corning Glass Works), 1972.

——— , *Application Notes*. Worcester, Mass.: Sprague Electric Company, 1971.

——— , *Digital Logic Handbook*. Maynard, Mass.: Digital Equipment Corp., 1971.

Index

Index